国外优秀数学著作原版系列

"十三五"重点出版物规划项目

他 山 之 石 系 列

动力系统——短期课程

Dynamical System—A Short Course

（英文）

[印] 南德奥·柯布拉加德（Namdeo Khobragade）著

哈尔滨工业大学出版社
HARBIN INSTITUTE OF TECHNOLOGY PRESS

黑版贸审字 08－2019－179 号

图书在版编目(CIP)数据

动力系统:短期课程＝Dynamical System:A
Short Course:英文/(印)南德奥·柯布拉加德
(Namdeo Khobragade)著. —哈尔滨:哈尔滨工业大学
出版社,2021.8
　ISBN 978－7－5603－4327－3

　Ⅰ.①动…　Ⅱ.①南…　Ⅲ.①动力系统(数学)－英文
Ⅳ.①O19

中国版本图书馆 CIP 数据核字(2021)第 156576 号

策划编辑　刘培杰　杜莹雪
责任编辑　刘立娟
封面设计　孙茵艾
出版发行　哈尔滨工业大学出版社
社　　址　哈尔滨市南岗区复华四道街 10 号　邮编 150006
传　　真　0451－86414749
网　　址　http://hitpress.hit.edu.cn
印　　刷　哈尔滨市工大节能印刷厂
开　　本　880 mm×1 230 mm　1/32　印张 15.75　字数 285 千字
版　　次　2021 年 8 月第 1 版　2021 年 8 月第 1 次印刷
书　　号　ISBN 978－7－5603－4327－3
定　　价　68.00 元

PREFACE

The book entitled '**Dynamical System – A Short Course**' contains eight chapters. This book is written for UG and PG students. **Chapter 1** contains matrices and operators, subspaces, bases and dimension. determinants, trace and rank. direct sum decomposition. real eigen values. differential equations with real distinct eigen values. complex eigen values.

Chapter 2 includes complex vector spaces. real operators with complex eigen values. application of complex linear algebra to differential equations. review of topology in R^n. new norms for old. exponential of operators. **Chapter 3** contains homogeneous linear systems. a non homogeneous equation. higher order systems. the primary decomposition. the S+N decomposition. nilpotent canonical forms.

Chapter 4 includes Jordan and real canonical forms. canonical forms and differential equations. higher order linear equations on function spaces. sinks and sources. hyperbolic flows. generic properties of operators. significance of Genericity.

Chapter 5 contains dynamical systems and vector fields. the fundamental theorem. existence and uniqueness. continuity of solutions in initial conditions. on extending solutions. global solutions. the flow of a differential equation. **Chapter 6** includes nonlinear sinks. stability. Liapunov function. gradient systems. gradients and inner products.

Chapter 7 contains limit sets, local sections and flow boxes, monotone sequences in planar dynamical systems. the Poincare-Bendixson theorem, applications of Poincare-Bendixson theorem; one species, predator and prey, competing species. **Chapter 8** includes asymptotic stability of closed orbits, discrete dynamical systems. stability and closed orbits. non autonomous equations and differentiability of flows. persistence of equilibria, persistence of closed orbits. structural stability.

CONTENTS

CHAPTER 1
VECTOR SPACE

Let F be a field and V an additive abelian group. We say that V is a vector space over the field F if there exists a map for $F \times V$ into V such that for r in F and v in V

$$r(v_1 + v_2) = xv_1 + rv_2 ; r \in F \quad and \quad v_1, v_2 \in V$$

$$(r+s)v = rv + sv; r, s \in F, \quad v \in V$$

$$(r.s)v = r(sv); \quad r, s \in F ; v \in V$$

$$1.v = v ; \quad v \in V, 1 \in F$$

If all this conditions satisfy, then we say V is **vector space** over field F.

When $F \equiv R$; the field of reals, we say that V is a real vector space.

When $F \equiv C$, the field of complex numbers, we say that V is complex vector space.

LINEAR MAP

Let V and W be two vector spaces over the same field F. A map $f : V \to W$ is called as Linear Map from V to W **iff**

$$f(\lambda v_1 + \mu v_2) = \lambda f(v_1) + \mu f(v_2)$$

where v_1 and v_2 are in V, λ and μ are in F.

If W is same as V then a linear Map $f : V \to W$ is known as (linear) operator.

Ex 1: Consider R^n, a collection of n - tuples where the components are all real.

Obviously, R^n with component wise addition is a real vector space.

If V is vector space over F, then $f : v \to V$ defined as

$$f(v) = 0 \quad \forall v \in V$$

DYNAMICAL SYSTEM – A SHORT COURSE

And $g:v \to V$ defined as $g(v)=v$, $\forall\ v \in V$

are linear operations on V.

Thus the collection of linear operators on a vector space is never empty.

As R^n is a real vector space, we denote by $L(R^n)$, the collection of all linear operation on R^n.

Let A be a $n \times n$ real matrix. We shall denote any member of R^n by a column matrix.

For any $\begin{pmatrix} r_1 \\ : \\ r_n \end{pmatrix} \in R^n$, $A\begin{pmatrix} r_1 \\ : \\ r_n \end{pmatrix} \in R^n$

Obviously,

$\lambda, \mu \in R$ and $\begin{pmatrix} r_1 \\ : \\ r_n \end{pmatrix} \in R^n$ $\begin{pmatrix} s_1 \\ : \\ s_n \end{pmatrix} \in R^n$ then

$$A\left[\lambda\begin{pmatrix} r_1 \\ : \\ r_n \end{pmatrix} + \mu\begin{pmatrix} s_1 \\ : \\ s_n \end{pmatrix} \right] = A\left[\begin{pmatrix} \lambda r_1 \\ : \\ \lambda r_n \end{pmatrix} + \begin{pmatrix} \mu s_1 \\ : \\ \mu s_n \end{pmatrix} \right] = A\begin{pmatrix} \lambda r_1 + \mu s_1 \\ : \\ \lambda r_n + \mu s_n \end{pmatrix}$$

From the properties of matrix multiplication, we have

$$= A\left[\lambda\begin{pmatrix} r_1 \\ : \\ r_n \end{pmatrix} + \mu\begin{pmatrix} s_1 \\ : \\ s_n \end{pmatrix} \right]$$

$$= \lambda\ A\begin{pmatrix} r_1 \\ : \\ r_n \end{pmatrix} + \mu\ A\begin{pmatrix} s_1 \\ : \\ s_n \end{pmatrix}$$

Thus if A is any $n \times n$ real matrix, then under multiplication on left, A maps an element of R^n to an element of R^n and is linear.

2

i.e. every $n \times n$ real matrix (under left multiplication) can be considered as a linear operator on R^n.

For $k = 1.....n$, we denote by e_k the n-tuples all of whose entries are zero excepting the kth entry which is 1.

i.e.,
$$e_1 = \begin{pmatrix} 1 \\ 0 \\ : \\ 0 \end{pmatrix} \quad e_1 = \begin{pmatrix} 0 \\ 1 \\ 0 \\ : \\ 0 \end{pmatrix} \quad and \quad e_i = \begin{pmatrix} 0 \\ 0 \\ : \\ 1 \\ : \end{pmatrix}$$

Let,

$$A = \begin{pmatrix} a_{11} & \cdots & a_{1n} \\ a_{21} & \cdots & a_{2n} \\ a_{n1} & \cdots & a_{nn} \end{pmatrix}$$

$$\therefore \quad A_{e_1} = \begin{pmatrix} a_{11} & \cdots & a_{1n} \\ a_{21} & \cdots & a_{2n} \\ a_{n1} & \cdots & a_{nn} \end{pmatrix} \begin{pmatrix} 1 \\ 0 \\ : \\ 0 \end{pmatrix} = \begin{pmatrix} a_{11} \\ a_{21} \\ : \\ a_{n1} \end{pmatrix} = \text{first column of A}$$

$$A.e_k = a_{1k} e_1 + a_{2k} e_2 + ... + a_{nk} e_n = \sum_{i=1}^{n} a_{ik} e_i = \begin{pmatrix} a_{1k} \\ a_{2k} \\ : \\ a_{nk} \end{pmatrix}$$

Note: $(x, y) \in R^2 \Rightarrow x(1,0) + y(0,1)$

$$\begin{pmatrix} r_1 \\ : \\ r_n \end{pmatrix} = r_1 e_1 + r_2 e_2 .... + r_n e_n$$

DYNAMICAL SYSTEM – A SHORT COURSE

Considering $R^n (n \geq 1)$ as real vector spaces, every $n \times n$ real matrix give rise to as a linear operator R^n.

The linear operator that we so obtain is given by $A : r \to A_r$ for any $r \in R^n$ and A the $n \times n$ matrix and $x \in R^n$ considered as a column matrix.

Further we know that if

$$e_k = \begin{pmatrix} 0 \\ \vdots \\ 1 \\ 0 \end{pmatrix}$$ a member of R^n with 1 at k^{th} positive and 0 elsewhere then

A_{ek} in the kth column of A and

$$A_{ek} = \sum_{i=1}^{n} a_{ik} e_k$$

Conversely: We show that every linear operator on R^n also gives rise to an $n \times n$ matrix.

Let T be a linear operator on R^n.

T_{ej} is known for $j = 1 \ldots n$.

As $T_{ej} \in R^n$ we can write T_{eij} as a linear combination of $e_1 \ldots e_n$.

Let a_{ij} denote the ith component of T_{ej}.

Since we can let $j = 1 \ldots n$.

All the n^2 entries a_{ij} $1 \leq i \leq n$, $1 \leq j \leq n$ are known.

Thus we can define matrix A as (a_{ij}).

We now show that the matrix A gives rise to the same operator T.

VECTOR SPACE

For this consider.

A_{e1} = First column of A.

$$= a_{11} e_1 + a_{21} e_2 + \ldots + a_{n1} e_n$$

$$T_{e2} = a_{12} e_1 + \ldots + a_{n2} e_n$$

In general,

$$A_{ej} = T_{ej}$$

As a matter of fact if $r \in R^n$

i.e. $r = \begin{pmatrix} r_1 \\ : \\ r_n \end{pmatrix}$

$$\text{LHS} = A_r = A\left(r_1 e_1 + r_2 e_2 + \ldots + r_n e_n\right)$$

$$= r_1 A e_1 + r_2 A e_2 + \ldots + r_n A e_n$$

$$= r_1 T e_1 + r_2 T e_2 + \ldots + r_n T e_n$$

$$= T\left(r_1 e_1 + r_2 e_2 + \ldots + r_n e_n\right)$$

$$= T_r$$

Thus with every $n \times n$ real matrix we can associate a linear operator on R^n and conversely with every linear operator on R^n we can associate an $n \times n$ matrix.

Let S and T be two linear operators on R^n.

Let A and B be the corresponding $n \times n$ matrix

i.e. $A_{e_j} = S_{e_j}$, $B_{e_j} = T_{e_j}$ $\quad i \leq j \leq n$

Consider the operator ST.

Consider $ST\left(e_j\right), \ 1 \leq j \leq n$

5

$$ST\,(e_j) = S(T(e_j)) = S(B(e_j))$$

$$= S\left(\sum_{k=1}^{n} b_{kj}\, e_k\right)$$

As $S =$ is an operator then,

$$= S\left(\sum_{k=1}^{n} b_{kj}\, e_k\right)$$

$$= \sum_{k=1}^{n} b_{kj}\,(S_{ek})$$

$$= \sum_{k=1}^{n} b_{kj}\,(A_{ek})$$

$$= \sum_{k=1}^{n} b_{kj}\left(\sum_{i=1}^{n} a_{ik}\, e_i\right)$$

$$= \sum_{i=1}^{n} \left(\sum_{k=1}^{n} a_{ik}\, b_{kj}\right) e_i$$

If $C = AB$ then $C_{ij} = \sum_{k=1}^{n} a_{ik}\, b_{kj}$

Thus $ST(e_j) = \sum_{i=1}^{n} C_{ij}\, e_i$

i.e., $ST(e_j) = C(e_j) = AB(e_j)$

The composition ST corresponds to the matrix multiplication AB .

If T is an operator on R^n , we say that T is **invertible** iff there exists an operator S on R^n such that

$ST = TS =$ identity map on R^n .

The matrices corresponding to the linear operators, zero operator and identity operator are obviously the zero matrix

6

VECTOR SPACE

(all entries is zero) and the identity matrix (unit) matrix.

If A correspond to T and B corresponds to S, then

$ST = TS$ = identity map ,

means $A.B = B.A$ = unit matrix

i.e. B is inverse of A.

i.e. A is invertible.

Thus a linear operator is invertible **iff** the corresponding matrix is invertible.

If $S,T \in L\ (R^n)$ and $\lambda \in R$.

We can define $S+T$ and λT as

$S+T\ (r) = S(r) + T(r)$

$\lambda (r) = \lambda.T(r) \qquad \forall\ r \in R^n$

Thus $L(R^n)$ itself is a real vector space.

NOTE:

A subset F of E is called Vector subspace of E **iff** F is an additive abelian.

Subgroup of E and it is closed for multiplication by scalars (real numbers)

Thus $F \subseteq E$ is a real subspace of E **iff**

$\forall\ v_1\ v_2 \in F$, $v_1 - v_2 \in F$ and

$\lambda \in R$, $v \in F$, $\lambda v \in F$

Let F_1 and F_2 be two subspaces of E.

We can define a linear map $f : F_1 \rightarrow F_2$.

Let $f : F_1 \rightarrow F_2$ be a linear map.

We denote by Ker f the subset of f_1 such that

$$f(v)=0$$

Thus ker $f = \{v \in F_1 \ \text{such that} \ f(v)=0\}$

Also,

$$\text{Im} f = \{w \in F_2 \quad \exists \, v \in F_1 \ \text{with} \ f(v)=w\}$$

$\{0\}$ is always a subspace of E.

It is known as a trivial subspace.

If F is a subspace of E and $F \subseteq E$ then we say that F is a proper subspace of E.

Result: If $f : F_1 \to F_2$ is a linear map then ker f and Im f are subspace of F_1 and F_2 respectively.

Ker f will be a subspace of F_1 **iff**

(1). whenever v_1 , $v_2 \in \ker f$ then

$\quad v_1 - v_2$ also is in ker f.

(2). and whenever $\lambda \in R$ and $v \in \ker f$

$\quad \lambda v \in \ker f$.

Proof: 1). Let v_1 , $v_2 \in \ker f$

By definition $f(v_1)=0=f(v_2)$

As $\quad f(v_1 - v_2) = f(v_1) - f(v_2) \qquad \{\because f \ \text{is linear}\}$

$\qquad\qquad\qquad = 0 - 0 = 0$

$\therefore v_1 - v_2 \in \ker f$

2). let $\lambda \in R$ and $v \in \ker f$

Consider $\quad f(\lambda v) = \lambda f(v) = \lambda . 0 = 0$, as $\ v \in \ker f$

VECTOR SPACE

$\therefore \quad \lambda \ v \in \ker f$

Ker f is subspace of F_1 .

Similarly Im f is subspace of F_2 .

Let $w_1 , w_2 \in \operatorname{Im} f$

By definition $\quad \exists \quad v_1 , v_2$ in F_1 such that

$w_1 = f(v_1)$ and $\quad w_2 = f(v_2)$

$\therefore \quad w_1 - w_2 = f(v_1) - f(v_2)$

$\qquad\qquad = f(v_1 - v_2) \qquad \{ \because \ f \ \text{is linear}\}$

But v_1 and v_2 are in F_1 and F_1 is a vector space (linear space)

$\therefore \quad v_1 - v_2 \in F_1$

Thus $w_1 - w_2$ is image of some element $v_1 - v_2$ of F_1 under f.....

$w_1 - w_2 \in im \ f$

Further,

Let $\lambda \in R$ and $w \in im \ f$

As $w \in im \ f \quad \exists \quad v \in F_1$ such that

$w = f(v)$

$\therefore \quad \lambda w = \lambda f(v)$

$\qquad\quad = f \lambda(v) \quad \{ \because \ f \ \text{is linear}\}$

and $\lambda v \in F_1 \qquad \{ \because \ \lambda \in R, v \in F_1 \ and \ F \ \text{is linear space} \}$

$\therefore \qquad \lambda w \in im \ f$

Let F is vector space (F is subspace of $E \equiv R^n$)

9

DYNAMICAL SYSTEM – A SHORT COURSE

Let $\{v_1....v_j\}$ be a collection of members in F.

We say that the set $\{v_1....v_j\}$ is linearly independent **iff** whenever $\alpha_1......\alpha_j \in R$ such that $\alpha_1 v_1 + \alpha_2 v_2 + ...\alpha_j v_j = 0$

Then $\alpha_1 = \alpha_2 = ... = \alpha_j = 0$.

Further let $\{w_1....w_r\}$ be a collection of member of F. We say that F is span by the set $\{w_1....w_r\}$ **iff** every member of F is expressible as a linear combination of $w_1....w_r$.

The collection $\{w_1....w_s\}$ of members of F is known as basics of F **iff** $\{w_1....w_s\}$ is linearly independent and $\{w_1....w_s\}$ spans F.

Let F be vector space (i.e. F is subspace of $E = R^n$)

If we have a system of simultaneous linear homogeneous equation. Then a zero tuple is always a solution of such a system.

A solution of a system of homogeneous linear equation which is not identical zero.

i.e., a zero tuple is unknown **no-trivial solution**.

Theorem 1. Any system of n simultaneous homogeneous linear equation with $n+1$ unknown always has a non-trivial solution. Here n is a +ve integer.

Proof: The statement holds for $n = 1$.

This is because for $n = 1$, we can consider equation $ax + by = 0$ in 2 unknown x and y. Obviously $a \neq 0$, $b \neq 0$

$$ax + by = 0 \Rightarrow y = -\tfrac{a}{b} x$$

Every non-zero values of x gives a non zero values of y.

VECTOR SPACE

Thus this equation has infinitely many non trivial solutions.

For some integer $|( )|$, let us assume that the statement is true for $n = k$.

Thus any homogeneous system of k lines equations in $k+1$ unknown has at least one non trivial solution.

For induction to hold true, we show that any system of $k+1$ homogenous linear equations in $k+2$ unknowns has at least one non-trivial solution.

We use one of this $k+1$ equations to express one of the unknowns in terms of the remaining unknowns. Make the substitution for the above unknown in the remaining equation; we thus get a system of R homogenous linear equations in $k+1$ unknowns.

By Induction hypothesis this has a non trivial solution. We use the expression for the $k+1$ unknown to get the expression for $k+2$ unknown. Thus we get a non-trivial solution to the system of $k+1$ equation

Thus the equation is true for every ' n '. By the similar argument we can show that any system of n simultaneous homogeneous linear equation in k unknowns where R is i.e. $k \geq n+1$ must have a non trivial solution.

Theorem 2: Let $\{e_1....e_m\}$ be a basis of vector space. If $\{v_1, v_2....v_r\}$ is a linearly independent set in F , then $r \leq m$

Proof: In order to show that $r \leq m$ we show that $r > m$

We show that $r \neq m+1$.

If possible let us assume that $r = m+1$.

As $\{e_1....e_m\}$ is the basis of F and $v_i \in F$, $1 \leq i \leq r = m+1$

We must have v_i expressed as linear combination of $e_1....e_m$.

Thus we must have,

$$v_1 = a_{11}e_1 + a_{12}e_2 + \ldots a_{1m}e_m$$
$$v_2 = a_{21}e_2 + a_{22}e_2 + \ldots a_{2m}e_m$$
$$\vdots$$
$$v_{m+1} = a_{m+1,1}e_1 + a_{m+1,2}e_2 + \ldots a_{m+1,m}e_m$$

where, $a_{ij} \in R$; $1 \le i \le m+1$, $1 \le j \le m$

We consider a system of homogeneous equation as,

$$x_1 a_{11} + x_2 a_{21} + x_3 a_{31} + \ldots + x_{m+1} a_{m+1,1} = 0$$
$$x_1 a_{12} + x_2 a_{22} + x_3 a_{32} + \ldots + x_{m+1} a_{m+1,2} = 0$$
$$\vdots$$
$$x_1 a_{1m} + x_2 a_{2m} + x_3 a_{3m} + \ldots + x_{m+1} a_{m+1,m} = 0$$

It is a system of m homogeneous linear equations in $m+1$ unknowns.
j^{th} equation of the system where $1 \le j \le m$ can be written as

$$\sum_{i=1}^{m+1} x_i a_{ij} = 0$$

By thm. 1, this system has at least one non trivial solution.

Thus we can find $m+1$ tuple $(x_1, x_2 \ldots x_{m+1})$ a non zero $m+1$ tuple which will satisfy the given system

i.e., for $1 \le j \le m$, $\displaystyle\sum_{i=1}^{m+1} x_i a_{ij} = 0$

Consider $\displaystyle\sum_{i=1}^{m+1} x_i v_i$

But we have

$$v_i = \sum_{j=1}^{m} a_{ij} e_j \qquad 1 \le i \le m+1$$

$$\therefore \sum_{i=1}^{m+1} x_i v_i = \sum_{i=1}^{m+1} x_i \left(\sum_{j=1}^{m} a_{ij} e_j \right)$$

$$= \sum_{j=1}^{m} \left(\sum_{i=1}^{m} x_i a_{ij} \right) e_j$$

(for every j $\sum_{i=1}^{m+1} x_i a_{ij} = 0$)

Thus we have, $\sum_{i=1}^{m+1} x_i v_i = 0$

But the set $\{ v_1 ... v_{m+1} \}$ is linearly independent.

$\therefore \ 0 = x_1 = x_2 = ... = x_{m+1}$

i.e. the $m+1$ tuple $(x_1, x_2 .... x_{m+1})$ is a 0-tuple.

This is a contradiction as $(x_1, x_2 .... x_{m+1})$ is a non zero tuple.

$\therefore \ r \ne m+1$

Similarly $r \ne m+1$, $m+2$, $m+3$ .....

$\therefore \ r \le m$.

Theorem 3. In a vector space F if every maximal collection $\{ v_1 .... v_r \}$ of linearly independent elements form a basis of F.

Proof: Since, the set $\{ v_1 .... v_r \}$ is maximal collection of linearly independent elements any proper superset of $\{ v_1 .... v_r \}$ is linearly dependent.

We show that $\{ v_1 .... v_r \}$ spans F.

Let $v_1 \in F$ such that $v \notin \{v_1 .... v_2\}$

Consider the set $\{v_1, v_1, v_2 .... v_r\}$

$\{u_1, u_1, u_2 ... u_r\}$ is a proper superset of $\{v_1, v_2 .... v_r\}$.

As such $\{v_1, v_1, v_2 .... v_n\}$ is linearly dependent.

Hence we can find scalars i.e., real numbers $\{\lambda_1, \lambda_1, \lambda_2, \lambda_3 .... \lambda_r\}$ are all not zero such that

$$\lambda v + \lambda_1 v_1 + ... + \lambda_r v_r = 0 \qquad (1.1)$$

λ must be non-zero.

If possible, let $\lambda = 0$ then (1.1) becomes,

$$\lambda_1 v_1 + \lambda_2 v_2 + ... + \lambda_r v_r = 0$$

$\Rightarrow$ $\lambda_1 = \lambda_2 = ... = \lambda_r = 0$ as the set $\{v_1 .... v_r\}$ is linearly independent.

Thus, in this case, we shall get

$$\lambda = \lambda_1 = \lambda_2 = ... = \lambda_r = 0$$

This is not possible.

$\therefore$ $\lambda \neq 0$

From equation (1.1) we get,

$$\lambda v = -\left(\lambda_1 v_1 + \lambda_2 v_2 + ... + \lambda_r v_r\right)$$

$$\therefore \quad v = -\frac{\lambda_1}{\lambda} v_1 - \frac{\lambda_2}{\lambda} v_2 .... \frac{\lambda_r}{\lambda} v_r$$

$\therefore$ $\{v_1 .... v_r\}$ span F.

Theorem 4:(*a*). Let F be a vector space (i.e. F is a subspace of $E \equiv R^n$). Then F has a basis.

(b) Any two basis of F contain same number of elements.

VECTOR SPACE

(c) If $\{v_1....v_r\}$ is a linearly independent set in F which is not a basis then it can be extended into a basis by adjoining elements from F.

Proof: If $F = (0)$ then the result holds as a zero space does not have a basis.

Thus, we exclude the case $F = (0)$

Let $F \neq (0)$

Let $v_1 \in F$ such that $v_1 \neq 0$

If $\{v_1\}$ is a basis and F then the result holds.

So, consider, the case where $\{v_1\}$ is not a basis of F.

Then, the space generated by v_1 is a proper subspace of F.

Let $v_2 \in F$ such that v_2 does not lie in the space generated by v_1.

Obviously, the set $\{v_1, v_2\}$ is linearly independent if $\{v_1, v_2\}$ is a basis of F, the we have got a basis.

If not then the space generated by $\{v_1, v_2\}$ would be a proper subspace of F.

We can find an element say $v_{r+1} \in F$ such that v_{r+1} does not lie in the span of $\{v_1....v_r\}$.

Hence set $\{v_1....v_r\}$ is linearly independent.

If $\{v_1....v_r, v_{r+1}\}$ is a basis then, we are through, otherwise we proceed ahead.

This process cannot go in indefinitely and must stop when we set a superset $\{v_1....v_r\}$ which spans F and in linear independent.

15

Thus we get a basis of which is superset of $\{v_1, v_2 \ldots v_r\}$.

Theorem 5: Let $f : v \to w$ be a linear map f is 1-1 **iff** ker f=$\{0\}$

Proof: We know that ker f is a subspace of v.

Obviously, $0 \in \ker f$.

Assume that f is 1-1

Here we have $(0) \le \ker f$

Let $v \in \ker f$

But $f(0) = 0$ $(\because f \ is \ lineaer)$

But f is 1-1

$\therefore \ f(u) = f(0)$

$\therefore \ v = 0$

$\ker f \le (0)$

$\ker f = (0)$.

Conversely. Let $\ker f = (0)$

To show that f is 1-1

Let us assume that $v_1, v_2 \in v$ if

$f(v_1) = f(v_2)$

$f(v_1 - v_2) = f(v_1) - f(v_2) = 0$

$\therefore \ v_1 - v_2 \in \ker f$

But $\ker f = 0$

$v_1 - v_2 = 0$

$v_1 = v_2$

16

VECTOR SPACE

$\therefore$ f is 1-1

Note: Let f be a vector space. Then dimension of F denoted by dim F is the number of elements in a basis of F.

Theorem 6: Let V and W be the two vector spaces over the same field F. Let f be a linear map from v to w. If f is 1-1 and onto then we say that f is an isomorphism from v onto w.

Proof. Let $f: v \to w$ be an isomorphism.

Since f is 1-1 and onto.

Let $g: w \to v$ be its inverse

i.e. $\forall$ $v \in v$, $gf(v) = v$ and $\forall$ $w \in w$, $fg(w) = w$

g is also a linear map.

Let $w_1, w_2 \in w$

To show that,

$$g(w_1 + w_2) = g(w_1) + g(w_2)$$

$w_1 + w_2 = w_1 + w_2$

$$= fg(w_1) + fg(w_2) \qquad \because fg(w) = w$$

$$= f(g(w_1) + g(w_2)) \qquad \because f \text{ is linear}$$

Thus, $w_1 + w_2 = f(g(w_1) + g(w_2))$

$\therefore g(w_1 + w_2) = g(f(g(w_1) + g(w_2)))$

$$= gf(g(w_1) + g(w_2))$$

$$= g(w_1) + g(w_2)$$

i.e. $g(w_1 + w_2) = g(w_1) + g(w_2)$

Similarly

let $\lambda \in R$ and $w \in W$

We show $g(\lambda w) = \lambda g(w)$

let $\lambda w = \lambda w$

$$= \lambda f(g(w))$$

$\lambda w = f \lambda g(w)$ ($\because f$ is linear)

$g(\lambda w) = g(f(\lambda g(w))) = g f(\lambda g(w))$

$\therefore g(\lambda w) = \lambda g(w)$

Let $f : R^n \to R^m$ be a linear map with f we can associate a matrix of real numbers.

The matrix associated with f is an $m \times n$ matrix.

Let this matrix be denoted by $A = (a_{ij})$

An n-tuple $\begin{pmatrix} x_1 \\ x_2 \\ \vdots \\ x_n \end{pmatrix}$ will be in kernel of f iff

$$A \begin{pmatrix} x_1 \\ x_2 \\ \vdots \\ x_n \end{pmatrix} = \begin{pmatrix} 0 \\ 0 \\ \vdots \\ 0 \end{pmatrix} \qquad m-\text{tuple}$$

$$\begin{pmatrix} a_{11} & a_{12} & \cdots & a_{1n} \\ a_{21} & a_{22} & \cdots & a_{2n} \\ a_{m1} & \cdots & \cdots & a_{nn} \end{pmatrix} \begin{pmatrix} x_1 \\ x_2 \\ \vdots \\ x_n \end{pmatrix} = \begin{pmatrix} 0 \\ 0 \\ \vdots \\ 0 \end{pmatrix}$$

VECTOR SPACE

$$a_{11} x_1 + a_{12} x_2 + \ldots + a_{1n} x_n = 0$$
$$a_{21} x_1 + a_{22} x_2 + \ldots + a_{2n} x_n = 0$$
i.e., :
$$a_{m1} x_1 + a_{m2} x_2 + \ldots + a_{mn} x_n = 0$$

Thus an n-tuple $\begin{pmatrix} x_1 \\ x_2 \\ : \\ x_n \end{pmatrix}$ (i.e. an element in R^n) is in the ker f **iff**

$(x_1 \ldots x_n)$ is a solution of the system of m homogeneous linear equation in n unknowns.

NOTE: If the domain is greater than image it cannot be 1-1 for the elements to be distinct they must be non zero i.e. linear independent.

Thus, if $n > m$, the system will have a non-trivial solution i.e. Ker f will contain at least one non zero element.

Thus f cannot be 1-1.

By convention we define dimension of the zero space as 0.

Theorem 7: (a). Two vector spaces are isomorphism iff they have the same dimension.

(b). In particular every n-dimensional vector space is isomorphic to R^n.

Proof: Let V and W be two vector spaces which are isomorphic to each other.

Let be the isomorphism from V onto W.

Let dim $V = m$.

We show that dim $W = m$

As dim $v = m$. Let $\{v_1, v_2 \ldots v_m\}$ be a basis of V.

19

Consider the set $\{f(v_1), f(v_2)...f(v_m)\}$ in W .

We show that $\{f(v_1), f(v_2)...f(v_m)\}$ is a basis of W .

Consider

$$\lambda_1 f(v_1) + \lambda_2 f(v_2) + ... + \lambda_m f(v_m) = 0$$

where $\lambda_1, \lambda_2....\lambda_m \in R$

But   LHS$= f (\lambda_1 v_1 + \lambda_2 v_2 + ...\lambda_m v_n)$ ($\because f$ is linear)

Thus we have,

$$f (\lambda_1 v_1 + \lambda_2 v_2 + ...\lambda_m v_m) = 0$$

$$\Rightarrow \lambda_1 v_1 + \lambda_2 v_2 + ...\lambda_m v_m = 0$$

$\{ \because f$ is isomorphism i.e. ker f$= (0) \}$

$$\Rightarrow \lambda_1 v_1 = \lambda_2 v_2 + \lambda_2 v_m = 0$$

$\{ \because$ As $v_1, v_2....v_m$ is a basis of $V \}$

Thus $\sum_{i=1}^{m} \lambda_i f(v_i) = 0 \Rightarrow \lambda_i = 0 \qquad 1 \le i \le m$

i.e., the set $\{f(v_1), f(v_2)...f(v_m)\}$ is linearly independent.

Now, we show that,

$\{f(v_1), f(v_2)...f(v_m)\}$ spans W.

Let $w \in W$.

But f is onto. So $\exists$ $v \in V$ such that $w = f(v)$

But $v \in V$ $\therefore$ $v = \alpha_1 v_1 + \alpha_2 v_2 + ... + \alpha_m v_m$

for some reals $\alpha_1, \alpha_2...\alpha_m$.

$$w = f(v) = f(\alpha_1 v_1 + \alpha_2 v_2 + ... + \alpha_m v_m)$$

$$= \alpha_1 f(v_1) + \alpha_2 f(v_2) + ... + \alpha_m f(v_m)$$

$\therefore \{f(v_1), f(v_2)....f(v_m)\}$ is a basis of W.

$\therefore$ dim $W = m =$ dim V

Conversely:

Let $m = \dim V = \dim W$

Let $\{e_1...e_m\}$ be a basis of V and say $\{f_1...f_m\}$ basis of W.

We define a map $T : V \to W$ as follows:

Any $v \in V$ can be written as

$v = \lambda_1 e_1 + \lambda_2 e_2 + ...\lambda_m e_m$

We define $T(v) = \lambda_1 f_1 + \lambda_2 f_2 + ...\lambda_n f_n$

T is well defined.

Let us show that T is linear.

Let $v, w \in V$

Let $v = \lambda_1 e_1 + ... + \lambda_m e_m$

$w = \mu_1 e_1 + ... + \mu_m e_m$

To show that, $T(v + w) = T(v) + T(w)$

Now consider,

$$T(v + w) = T((\lambda_1 \mu_1)e_1 + (\lambda_2 \mu_2)e_2 + ... + (\lambda_m \mu_m)e_m)$$
$$= (\lambda_1 + \mu_1)f_1 + (\lambda_2 + \mu_2)f_2 + ... + (\lambda_m + \mu_m)f_m$$
$$= (\lambda_1 f_1 + \lambda_2 f_2 + ...\lambda_m f_m) + (\mu_1 f_1 + \mu_2 f_2 + ...\mu_m f_m)$$
$$= T(v) + T(w).$$

Further consider,

$T(\lambda v) = \lambda T(v)$

$= T(\lambda(\lambda_1 e_1 + \lambda_2 e_2 + ...\lambda_m e_m))$

Consider,

21

$$T(\lambda v) = T(\lambda \lambda_1 e_1 + \lambda \lambda_2 e_2 + ... \lambda \lambda_m e_m)$$
$$= \lambda \lambda_1 f_1 + \lambda \lambda_2 f_2 + ... \lambda \lambda_m f_m$$
$$= \lambda (\lambda_1 f_1 + \lambda_2 f_2 + ... \lambda_m f_m)$$
$$= \lambda\, T(v)$$

Thus $T : V \to W$ is linear map.

Now show that T is one-one.

$v \in \ker T$ iff $T(v) = 0$

iff $\quad T(\lambda_1 e_1 + \lambda_2 e_2 + ... \lambda_m e_m) = 0$

iff $\quad \lambda_1 f_1 + ... + \lambda_m f_m = 0$

iff $\quad \lambda_1 = \lambda_2 = ... = \lambda_m = 0$

iff $\quad v = 0$

$\therefore\quad$ Ker $T = (0)$

$\therefore\quad T$ is one-one.

We show that T is onto.

Let $\quad w = W$

As $f_1 ... f_m$ is basis of W we have

$$w = \alpha_1 f_1 + ... + \alpha_m f_m$$

Obviously

$$w = T(\alpha_1 e_1 + ... + \alpha_m e_m) \qquad \because\ e_1 ... e_m \in v$$
$$w = T(v)$$

$\therefore T$ is onto

$\therefore T$ is isomorphism.

Theorem 8: Let $T : E \to F$ be a linear map.

VECTOR SPACE

Then dim $(\operatorname{Im} T) + \dim \ker (T) = \dim E$

In particular, suppose $\dim E = \dim F$.

Then the following are equivalent statement:

(a) $\ker T = (0)$

(b) $\operatorname{Im} T = F$

(c) T is an isomorphism.

Proof. Let $\{f_1 \cdots f_m\}$ be a basis of $\operatorname{Im} T = F$.

Let $\{g_1 \cdots g_r\}$ be a basis of $\ker (T)$

As $f_1, f_2 \cdots f_m \in \operatorname{Im} T$, we can find $e_1, e_2 \cdots e_m \in E$ such that

$T(e_1) = f_1$, $T(e_2) = f_2 \cdots T(e_m) = f_m$.

Here we have,

$\dim (\operatorname{Im} T) = m$ and $\dim (\ker T) = r$

We show that,

$\dim E = m + r$

For this we show that the set $\{e_1 \cdots e_m, g_1 \cdots g_r\}$ is a basis of E.

Now we show that it is linearly independent

i.e. the set $\{e_1 \cdots e_m, g_1 \cdots g_r\}$ is Linearly Independent.

Consider,

$\alpha_1 e_1 + \ldots + \alpha_m e_m + \beta_1 g_1 + \ldots + \beta_r g_r = 0$

where, $\alpha_1, \alpha_2 \cdots \alpha_m$, $\beta_1, \beta_2, \ldots \beta_r \in R$.

Operating by T we get

$T(\alpha_1 e_1 + \ldots + \alpha_m e_m + \beta_1 g_1 + \ldots + \beta_r g_r) = 0$

i.e., $\alpha_1 T(e_1) + \ldots + \alpha_m T(e_m) + \beta_1 T(g_1) + \ldots \beta_r T(g_r) = 0$

23

$$\therefore \quad \alpha_1 f_1 + \ldots + \alpha_m f_m + \beta_1 T(g_1) + \ldots + \beta_r T(g_r) = 0$$

$$\{ T \text{ is linear and } T(e_1) = f_1 \ldots T(e_m) = f_m \}$$

But $\quad T(g_1) = T(g_2) = \ldots T(g_r) = 0$

$$\{ w_z = g_1, g_2 \ldots g_r \in \ker T \}$$

$$\alpha_1 f_1 + \ldots + \alpha_m f_m = 0$$

$$\Rightarrow \quad \alpha_1 = \alpha_2 = \ldots = \alpha_m = 0 \quad \{ \because f_1 \ldots f_m \text{ is a basis of } \operatorname{Im} T \}$$

Thus,

$$\alpha_1 e_1 + \alpha_2 e_2 + \ldots + \alpha_m e_m + \beta_1 g_1 + \ldots + \beta_r g_r = 0 \text{ becomes,}$$

$$\beta_1 g_1 + \beta_2 g_2 + \ldots + \beta_r g_r = 0$$

But $\quad \{ g_1, g_2 \ldots g_r \}$ is a basis for ker T.

$$\therefore \quad \beta_1 = \beta_2 = \ldots = \beta = 0$$

Thus whenever we have

$$\alpha_1 e_1 + \alpha_2 e_2 + \ldots + \alpha_m e_m + \beta_1 g_1 + \ldots + \beta_r g_r = 0$$

then we must have

$$\alpha_1 = \alpha_2 = \ldots = \alpha_m = \beta_1 = \beta_2 = \ldots = \beta_r = 0$$

We next show that the set $\{ e_1, e_2 \ldots e_m, g_1 \ldots g_r \}$ spans E.

Let $\quad v \in E$

Consider $\quad T(v)$

$T(v) \in \operatorname{Im} T$.

But $\quad \{ f_1 \ldots f_m \}$ is a basis of $\operatorname{Im} T$.

$$\therefore \quad T(v) = \alpha_1 f_1 + \ldots + \alpha_m f_m \text{ for some reals } \alpha_1, \alpha_2 \ldots \alpha_m.$$

We define $w = \alpha_1 e_1 + \ldots + \alpha_m e_m$

Obviously , $T(w) = T(v)$

$$T(v-w)=0$$

$$v-w \in \ker T$$

But $g_1, g_2 \cdots g_r$ is a basis of ker T.

$\therefore$ $v-w = \beta_1 g_1 + \ldots + \beta_r g_r$ for some scalars $\beta_1, \beta_2 \ldots \beta_r$

$$v = w + \beta_1 g_1 + \ldots + \beta_r g_r$$

i.e. $v = \alpha_1 e_1 + \alpha_2 e_2 + \ldots + \alpha_m e_m + \beta_1 g_1 + \ldots + \beta_r g_r$

Thus $\{e_1, e_2 \ldots e_m, g_1 \ldots g_r\}$ is linearly independent and spans E.

$\therefore$ $\{e_1, e_2 \ldots e_m, g_1 \ldots g_r\}$ is a basis of E.

$\therefore$ dim $E = m + r$

$\therefore$ dim $E =$ dim ImT + dim ker T.

Next dim $E =$ dim F

Show that $a \Rightarrow b$

As $\ker T = (0)$

$\quad$ dim $(\ker T) = 0$

From the equation

dim $(\text{Im}T)$ + dim $(\ker T) =$ dim E $\hfill$ (1.2)

This gives,

dim $(\text{Im}T) =$ dim E

But we have

dim $E =$ dim F

We have,

dim $(\text{Im}T) =$ dim $E =$ dim F

But $\text{Im}T$ is a subspace of F and

$\dim\left(\mathrm{Im}\,T\right)=\dim F$

$\therefore\ \mathrm{Im}\,T=F$

$$(b)\Rightarrow(c)$$

We have $\ \mathrm{Im}\,T=F$

$\therefore\ T:E\to F\ \ \ $ is onto.

As $\ \ \mathrm{Im}\,T=F$,

We have $\ \dim\left(\mathrm{Im}\,T\right)=\dim F$

But $\ \ \dim F=\dim E$

$\therefore\ \dim\left(\mathrm{Im}\,T\right)=\dim E$

From equation (1)

$\dim\left(\ker T\right)=0$

$\therefore\ \ker T=\left(0\right)$

T is one-one.

T is isomorphism.

$$(c)\Rightarrow(a)$$

Obvious $\ \ \ T$ is isomorphism

i.e. T must be one-one

if it is one-one then,

$\ker T=\left(0\right)$.

COORDINATE TRANSFORMATION

Relation between matrices of a linear transformation under a coordinate transformation.

Let E be vector space of $\dim n$.

Let $\left\{e_1...e_n\right\}$ be a basis (not necessarily the standard basis) of E .

VECTOR SPACE

Let z be an element of E

z will be expressible as a linear combination of $e_1 ... e_n$ i.e. we can write

$$z = x_1 e_1 + x_2 e_2 + ... + \alpha_n e_n$$

where, $x_1, x_2 ... x_n$ are real numbers uniquely determined by g.

Thus with z we associate a unique n-tuple $(x_1, x_2 ... x_n)$.

Conversely. With every n-tuple $(\alpha_1, \alpha_2 ... \alpha_n)$ in R^n, we associate a unique point

$$\alpha_1 e_1 + \alpha_2 e_2 + ... + \alpha_n e_n \text{ of } E$$

Thus if a basis system $(e_1 ... e_n)$ is fixed then for every point we can associate its first coordinate x_1, second coordinate $x_2 .... n^{th}$ coordinate x_n.

Since different point in E will have different expression as a linear combination of $e_1, e_2 ... e_n$ the $1^{st}, 2^{nd}$ and $3^{rd} ..... n^{th}$ coordinate will in general vary with point in E.

Thus we write for every $z \in E$,

$$z = x_1(z)e_1 + x_2(z)e_2 + ... + x_n(z)e_n$$

$x_1(z), x_2(z) ... x_n(z)$ are known as **coordinate function**.

$x_i(z)$ gives i^{th} coordinate (i.e. coefficient of e_i) of the point z with respect to the basis $(e_1 ... e_n)$.

x_i, $1 \le i \le n$ is a linear map from E to R.

Since $e_1, e_2 ... e_n \in E$, we can find $x_i(e_j)$ $1 \le i \le n$, $1 \le j \le n$

as $\quad e_1 = 1.e_1 + 0.e_2 + ... + 0.e_n$

27

$$e_2 = 0.e_1 + 1.e_2 + \ldots + 0.e_n$$

and so on. It follows that

$$x_i(e_j) = 1 \quad \text{If } i = j \quad \text{and}$$

$$\qquad\qquad 0 \quad \text{If } i \neq j$$

i.e. $\quad x_i(e_j) = \delta_{ij}$

Let $\{f_1, f_2 \ldots f_n\}$ be another basis of E.

The point $z \in E$ will again be expressible as a linear combination of $f_1, f_2 \ldots f_n$.

say $z = y_1 f_1 + y_2 f_2 + \ldots + y_n f_n$.

We can call $y_1, y_2 \ldots y_n$ $\quad 1^{\text{st}}$, 2nd and n^{th} coordinate of z in the basis system $\{f_1 \ldots f_n\}$.

As $y_1, y_2 \ldots y_n$ will very with z .

we can write,

$$z = y_1(z)f_1 + y_2(z)f_2 + \ldots + y_n(z)f_n$$

y_i is the ith coordinate function for the basis system $\{f_1, f_2 \ldots f_n\}$.

Obviously

$$y_i(f_i) = \delta_{ij} \quad 1 \leq i \leq n,\ 1 \leq j \leq n$$

As $f_1, f_2 \ldots f_n \in E$ and $(e_1, e_2 \ldots e_n)$ is basis of E ,

we can write f_1 as a linear combination of $e_1, e_2 \ldots e_n$.

f_2 as L.C. of $e_1, e_2 \ldots e_n$ and so on.

For ex we can write

$$f_1 = p_{11} e_1 + p_{12} e_2 + \ldots + p_{1n} e_n$$

$$f_2 = p_{21}e_1 + p_{22}e_2 + \ldots + p_{2n}e_n$$

$$\vdots$$

$$f_n = p_{n1}e_1 + p_{n2}e_2 + \ldots + p_{nn}e_n$$

We this get a matrix $p = (p_{ij})$ of reals such that,

$$f_i = p_{ij}e_j \tag{1.3}$$

Similarly, we can express the coordinate functions $y_1, y_2 \ldots y_n$ in terms of the coordinate functions $x_1, x_2 \ldots x_n$.

thus we may write

$$y_i = q_{ij}x_j \quad 1 \le i \le n, \ 1 \le j \le n \tag{1.4}$$

We thus get the matrix $Q = (q_{ij})$ of reals.

We can find relation between the matrices Q and P.

For this we have,

$$y_i(f_j) = \delta_{ij}$$

i.e., $\quad \delta_{ij} = y_i(f_j)$

But we have,

$$f_j = p_{jk}e_k$$

$$\therefore \quad \delta_{ij} = y_i(p_{jk}e_k)$$

From (1.4) we have,

$$y_i = q_{il}x_i$$

$$\Rightarrow \quad \delta_{ij} = \sum_{l=1}^{n} q_{il} x_l \left(\sum_{k=1}^{n} p_{jk} e_k \right)$$

$$\Rightarrow \quad \sum_{l=1}^{n} q_{il} \sum_{k=1}^{n} p_{jk} \ x_l(e_k)$$

29

DYNAMICAL SYSTEM – A SHORT COURSE

$$\Rightarrow \quad \sum_{l=1}^{n} \sum_{k=1}^{n} q_{il}\, p_{jk}\, x_l(e_k)$$

$$\Rightarrow \quad \sum_{l=1}^{n} \sum_{k=1}^{n} q_{il}\cdot p_{jk}\cdot \delta_{lk} \quad \because\ x_l(e_k)=\delta_{lk}$$

$$\Rightarrow \quad \sum_{k=1}^{n} q_{ik}\, p_{jk}$$

We now define a new matrix $R=(r_{ij})$ such that

$$R = p^T$$

i.e. $\quad p_{kj} = r_{kj}$

thus we have,

$$\delta_{ij} = \sum_{k=1}^{n} q_{ik}\cdot r_{kj}$$

i.e., $\quad I = Q.R = Q.p^T$

i.e., $\quad Q = \left(p^T\right)^{-1} = \left(p^{-1}\right)^{T}$

thus the two matrices (i.e. the matrix of transform of coordinates and of basis are related by the relation

$$Q = \left(p^{-1}\right)^{T})$$

Thus if $\{e_1 ... e_n\}$ and $\{f_1 ... f_n\}$ are two basis in R^n and if x_i, $\quad 1 < i < n, y_i = 1 \le i \le n$ are the two respective coordinate system then we get

$$y_i = \sum_{j=1}^{n} q_{ij}\, x_j\ , \quad f_i = \sum_{j=1}^{n} p_{ij}\, e_j$$

$$Q = \left(p^T\right)^{-1} \quad \text{where} \quad Q = (q_{ij}),\ p = (p_{ij})$$

VECTOR SPACE

If we denote the n-tuple $\begin{pmatrix} y_1 \\ : \\ y_n \end{pmatrix}$ by y and the n-tuple $\begin{pmatrix} x_1 \\ : \\ x_n \end{pmatrix}$ by x

then we have

$y = Qx$ and $x = Q^{-1}y$.

Let T be a linear operator on R^n.

Consider a basis $\{e_1 ... e_n\}$

We know that, T is associated with an $n \times n$ matrix $A = (a_{ij})$ where

a_{ij} is the ith component of $A_e j$ i.e. T_{ej}.

Thus the matrix A associated with the linear operator T depends on the basis $\{e_1 ... e_n\}$.

So if $\{f_1 ... f_n\}$ is another basis of R^n.

Then T may have a different matrix say B with respect to the basis $\{f_1 ... f_n\}$.

We denote B by (b_{ij}).

Thus while a_{ij} , the coefficient of e_i of the element of A_{ej} i.e. T_{ej}

b_{ij} is the coefficient of f_i of the element $B f_j$ i.e. $T f_j$.

Let z be a point of $E \equiv R^n$.

We consider $T(z)$ a point of E.

If z is written as $z = \begin{pmatrix} x_1 \\ : \\ x_n \end{pmatrix}$ in the basis $\{e_1 ... e_n\}$ then

DYNAMICAL SYSTEM – A SHORT COURSE

$A\begin{pmatrix} x_1 \\ : \\ x_n \end{pmatrix}$ i.e. Ax is the element of $T(z)$ expressed in the basis

system $\{e_1 \dots e_n\}$.

Similarly, if we represent the point z by $\begin{pmatrix} y_1 \\ : \\ y_n \end{pmatrix}$ in the basis $\{f_1 \dots f_n\}$ then

$b\begin{pmatrix} y_1 \\ : \\ y_n \end{pmatrix}$ i.e. by is the representation of point $T(z)$ in the basis $\{f_1 \dots f_n\}$.

Thus the point $T(z)$ is given by Ax and by in the basis

systems $\{e_1 \dots e_n\}$ and $\{f_1 \dots f_n\}$ respectively.

Thus we must have,

$By = Q\,Ax$

But $\quad y = Q\,x \Rightarrow x = Q^{-1}y$

$\therefore \quad By = Q\,Ax = Q\,AQ^{-1}y$

Since this holds for each $y \in E$,

we must have, $\quad B = Q\,AQ^{-1}$

i.e., $\quad B = \left(p^{-1}\right)^{-1} A\left(p^{-1}\right)$

Thus, the two matrices associated with the same linear operator with respect to the basis $\{e_1 \dots e_n\}$, $\{f_1 \dots f_n\}$ are related to each other by relation $B = Q\,AQ^{-1}$,

where Q is the matrix transforming x coordinate into y-coordinate.

Two **matrices** S and T $(n \times n)$ are said to be **similar** to each other **iff** ∃

32

exist an invertible matrix say L such that $T = L S L^{-1}$

Similarly, two operator say A and B on R^n are similar to each other **iff** there is an invertible operator C such that $B = C A C^{-1}$.

Thus the matrices associated with the linear operator with respect to different basis are similar to each other.

DETERMINANT , TRACE AND RANK OF AN $n \times n$ REAL MATRIX:

With every $n \times n$ real matrix A we associate a unique real number denoted by det A which satisfies the following properties.

1) det $I = 1$

2) det $(A \cdot B) = $ det $A \cdot$ det B

3) A is invertible iff det $A \neq 0$.

Thus if A and B are similar matrices

i.e. $B = p A p^{-1}$ then

det $B = $ det $\left(p A p^{-1} \right) = $ det p det A det p^{-1}

$$= \text{det } p \text{ det } A \left(\text{det } p \right)^{-1}$$

$$= \text{det } A .$$

$\Rightarrow$ det $B = $ det A

Thus if T is a linear operator on R^n, the matrices associated with T with respect to different basis system are similar.

We can define,

det T as the determinant of any matrix.

Theorem 9: Let A be an operator. Then the following statements are equivalent.

a) $\det A \neq 0$

b) $\ker A = (0)$

c) A is 1-1

d) A is onto

e) A is invertible

TRACE:

If $A = \left(a_{ij}\right)$ is an $n \times n$ real matrix , then we define $T_r\, A$ (trace of A)

as

$$\sum_{i=1}^{n} a_{ii} = a_{11} + a_{22} + \ldots + a_{nn}$$

If A and B are two $n \times n$ real matrices then

$T_r\,(AB) = T_r\,(BA)$

$T_r\,(A+B) = T_r\,(A) + T_r\,(B)$

If B and A are similar i.e. $B = p\,A\,p^{-1}$

Then $T_r\, B = T_r\!\left(p\,A\,p^{-1}\right)$

$\qquad\qquad = T_r\, A \qquad \left\{ \because T_r\,(cD) = T_r\,(cD) \right\}$

RANK: If T is a linear operator on R^n .

We define Rank T as the dimension of $\mathrm{Im}\,T$.

Thus if A is an $n \times n$ matrix then we define rank of A as the rank of the

linear operator corresponding to matrix A .Hence

Rank of similar matrices is same.

DIRECT SUM OF SUBSPACES:

Let v be the vector space and $v_1 \ldots v_r$ be subspaces of v . we say that V

VECTOR SPACE

is the direct sum of $v_1, v_2...v_r$ **iff** every elements of $v_1, v_2...v_r$ for every $v \in V$, there exist unique $v_1 \in V_1$, $v_2 \in V_2$, ...$v_r \in V_r$ such that $v = v_1 + v_2 + ... + v_r$

When V is the direct sum of $v_1, v_2...v_r$, we write it as

$v = v_1 + v_2 + ... + v_r$

Since $0 \in V_i$ $\qquad$ $1 \le i \le r$

and $\quad 0 + 0 + 0 ... + 0 = 0$

1) When $V = V_1 \oplus V_2 \oplus V_3 \oplus ... \oplus V_r$ then

$\quad 0 = 0 + 0 + 0 ... + 0$ is the only representation of zero.

2) Further if $V = \oplus \sum_{i=1}^{r} V_i$

Then $\quad V_j \cap \sum_{\substack{k=1 \\ k \ne j}}^{r} V_k = (0); 1 \le j \le r$

i.e., $\quad V_1 \cap (V_2 + V_3 + ... + V_r) = (0)$

$\qquad V_2 \cap (V_1 + V_3 + ... + V_r) = (0)$

$\qquad V_3 \cap (V_1 + V_2 + + V_4 + ... + V_r) = (0)$

$\qquad V_n \cap (V_2 + V_3 + ... + V_{r-1}) = (0)$

Proof. To show $\quad V_j \cap \sum_{\substack{k=1 \\ k \ne j}}^{r} V_k = (0)$

We know that, $\quad 0 \in V_k$, $\quad k = 1...r$

$\therefore \ 0 \in V_j \cap \sum_{\substack{k=1 \\ k \ne j}}^{r} V_k$

If possible let $x \ne 0$ be in the intersection

35

$$V_j \cap \sum_{\substack{k=1 \\ k \neq j}}^{r} V_k$$

$$\therefore \quad x \in V_j \quad \text{and also} \quad x \in \sum_{\substack{k=1 \\ k \neq j}}^{r} V_k$$

We can this write

$$x = V_j \text{ for some } V_j \in V_j$$

As $\quad x \in \sum_{\substack{k=1 \\ k \neq j}}^{r} V_k \quad$ we have,

$$x \in \sum_{\substack{k=1 \\ k \neq j}}^{r} v_k$$

Thus we have

$$x = v_j = \sum_{\substack{k=1 \\ k \neq j}}^{r} v_k$$

i.e. $\quad 0 = \sum_{\substack{k=1 \\ k \neq j}}^{r} v_k - v_j$

$$= v_1 + ... v_{j-1} + v_{j+1} ... + v_r - v_j$$

$$= v_1 + v_2 + ... + v_{j-1} - v_j + v_{j+1} + ... v_r$$

Thus $\quad 0 = v_1 + v_2 + ... + v_{j-1} - v_j + v_{j+1} + ... + v_r$

is a representation of 0 as a sum of element of $V_1, V_2 ... V_r$.

$\therefore$ We must have

$$0 = v_1 = v_2 = ... = v_j = v_{j+1} = ... = v_r$$

i.e., $v_j = 0$

i.e., $x = 0$

It's a contradiction as x was assumed to be non zero.

Thus $V_j \cap \sum_{\substack{k=1 \\ k \neq j}}^{r} V_k = (0)$

Thus whenever,

$V = \oplus \sum_{k=1}^{r} V_k$ then,

$V_i \cap V_j = (0)$, $i \neq j$

Proof. Let $V = \oplus \sum_{i=1}^{r} V_i$

We can show that the union of the basis $V_1...V_r$ is a basis of V.

Let us assume that $V = V_1 \oplus V_2$

Let $\{e_1...e_r\}$ be a basis of V_1 and $\{f_1, f_2...f_s\}$ be a basis of V_2.

We show that the set $\{e_1...e_r, f_1...f_s\}$ is a basis of V.

We show that $\{e_1...e_r, f_1...f_s\}$ is linearly independent.

So let us assume that

$\lambda_1 e_1 + \lambda_2 e_2 + ... + \lambda_r e_r + \mu_1 f_1 + ...\mu_s f_s = 0$

for reals $\begin{matrix} \lambda_1, \lambda_1, ...\lambda_r \\ \mu_1, \mu_2, ...\mu_s \end{matrix}$

But as $V = V_1 \oplus V_2$

We must have

$0 = 0 + 0$ as the only representation of 0 .

37

DYNAMICAL SYSTEM – A SHORT COURSE

Since $\quad \lambda_1 e_1 + \lambda_2 e_2 + ... + \lambda_r e_r \in V_1$

$\qquad \mu_1 f_1 + \mu_2 f_2 + ... + \mu_s f_s \in V_2$

$\therefore \quad \underbrace{\dfrac{\lambda_1 e_1 + \lambda_2 e_2 + ... + \lambda_r e_r}{(V_1)}} + \underbrace{\dfrac{\mu_1 f_1 + \mu_2 f_2 + ... + \mu_s f_s}{(V_2)}} = 0$

$\Rightarrow \quad \lambda_1 e_1 + \lambda_2 e_2 + ... + \lambda_r e_r = 0$

$\qquad \mu_1 f_1 + \mu_2 f_2 + ... + \mu_s f_s = 0$

But $\{e_1 ... e_r\}$ and $\{f_1 ... f_s\}$ are basis of V_1 and V_2.

$\therefore \quad \lambda_1 = \lambda_2 = ... \lambda_r = 0 = \mu_1 = \mu_2 = ... \mu_s$

Next to show that $\{e_1 ... e_r, f_1 ... f_s\}$ spans V.

Let $v \in V$ then

$v = v_1 + v_2 \quad$ for some $\ v_1 \in V_1, \ v_2 \in V_2$

But $\ v_1 = \lambda_1 e_1 + \lambda_2 e_2 + ... + \lambda_r e_r$

$\qquad\qquad\qquad \because \{e_1 ... e_r\}$ is a basis of V_1.

and $\quad v_2 = \mu_1 f_1 + \mu_2 f_2 + ... + \mu_s f_s$

$\qquad\qquad\qquad \because \{f_1 ... f_s\}$ is basis of V_2.

$\therefore \ v = \lambda_1 e_1 + \lambda_2 e_2 + ... + \lambda_r e_r + \mu_1 f_1 + ... + \mu_s f_s$

$\therefore \ \{e_1 ... e_r, f_1 ... f_s\}$ is a basis of V.

$\therefore \ \dim V = r + s$

$\therefore \ \dim V = \dim V_2 + \dim V_2$

Thus in general, if $\quad V = \oplus \sum_{k=1}^{r} V_k$

Then $\ \dim V = \dim V_1 + \dim V_2 + ... + \dim V_r$.

Let T be a linear operation on V.

VECTOR SPACE

Let $\dim V = n$

Then a matrix associated with T is an $n \times n$ matrix.

Let us denote it by say A.

Further let us assume that,

$$V = \oplus \sum_{k=1}^{r} V_k$$

Further, let V_i be invariant under T for $1 \le i \le r$.

i.e. $T(V_i) \subseteq V_i$ $\qquad 1 \le i \le r$

and for $V \in v_i$, $T(v) = T_i(v)$ $\qquad 1 \le i \le r$

we then say that T is a direct sum of the operators $T_1, T_2 ... T_r$ and write it as

Let $V = \oplus \sum_{i=1}^{l} V_i$ and $V = \oplus \sum_{i=1}^{r} T_i$

be a direct sum decomposition of T.

A basis of V can be obtained by adjoining the basis of $V_1, V_2 ... V_r$ in that order.

We can get matrix representations of $T_1, T_2 ... T_r$ with respect to the basis of $V_1, V_2 ... V_r$.

Similarly we can get matrix representation of T with respect to the basis of V obtained by adjoining basis of $V_1, V_2 ... V_r$.

The matrix of T is given by planning matrices of $T_1, T_2 ... T_r$ along the diagonal.

For example: Let $V = V_1 \oplus V_2$ and $T = T_1 \oplus T_2$

where T_1 is linear operator on V_1 and T_2 is linear operator on V_2.

Let $\dim V_1 = r$

$\dim V_2 = s$

$\dim V = n$

$\therefore\ n = r + s$

Let $\{e_1...e_r\}$ and $\{f_1...f_s\}$ be a basis of V_1 and V_2 respectively.

Then $\{e_1...e_r, f_1...f_s\}$ is a basis of V.

Let A be the matrix of T with respect to this basis

and B be the matrix of T_1 with respect to the basis

$\{e_1...e_r\}$ and C be the matrix of T_2 with respect to the basis $\{f_1...f_s\}$.

First column A is obtained from $T(e_1)$.

But, $e_1 \in V_1$ $\qquad \therefore\ T(e_1) = T_1(e_1)$

$\therefore\ T(e_1) \in V_1$

As $V = V_1 \oplus V_2$, we must have

$T(e_1) = \alpha_1 e_1 + \alpha_2 e_2 + ... + \alpha_r e_r + 0\cdot f_1 + 0\cdot f_2 + ... + 0\cdot f_s$

Thus the first column of A is α_1

$$\alpha_2$$
$$\vdots$$
$$\alpha_r$$
$$0$$
$$\vdots$$
$$0$$

But $T(e_1) = T_1(e_1)$ and $T_1(e_1)$ is the first column of B.

Thus the column $\begin{pmatrix} \alpha_1 \\ \vdots \\ \alpha_r \end{pmatrix}$ must be the first column of B.

Thus, first column of A is the first column of B followed by S number of zeros.

By the similar argument; second column of A is given by 2nd column of B followed by S number of zeros. This will continue upto the nth column of A. To get $r+1^{th}$ column of A, we consider $r+1^{th}$ number of the basis i.e., f_1 and consider $T(f_1)$.

But $T(f_1) \in V_2$ and $T(f_1) \equiv T(f_2)$

As $T(f_i) \in V_2$, we have

$$T(f_1) = 0e_1 + 0e_2 + ... + 0e_r + B_1 f_1 + ... + B_s f_s$$

and the $r+1^{th}$ column of A is

$$\begin{pmatrix} 0 \\ \vdots \\ 0 \\ B_1 \\ \vdots \\ B_s \end{pmatrix}$$

But $T(f_1) = T_2(f_1)$ and $T_2(f_1)$ is the first column of the matrix C.

Thus $r+1^{th}$ column of A is r zeros followed by first column of C. In this way $r+2...r+s^{th}$ columns of A can be obtained.

Thus

$$A = \begin{pmatrix} B & 0 \\ 0 & C \end{pmatrix}.$$

EIGEN VALUES AND EIGEN VECTORS

Let T be a linear operator on a vector space E. A non zero element $x \in E$ is called **eigen vector** of the operator T if there exists a real numbers α such that

$$Tx = \alpha x$$

α is known as the eigen values of T and we say that eigen vector x belongs to eigen value α.

If x and y are eigen values belongs to eigen value α then $x + y$ as well as λx, $\lambda \in R$ are eigen vectors belongs to the same eigen value α.

We thus have an eigen space belonging to an eigen values.

By definition, α is an **eigen value** of operator T, provided there exists non-zero $x \in E$ such that

$$Tx = \alpha x = \alpha \cdot Ix$$

i.e., $(T - \alpha I)x = 0$

i.e. ker $(T - \alpha I)$ contains at least one non-zero element.

If A is a matrix of T, then

$(T - \alpha I) = 0$ becomes,

$$(T - \alpha I)\begin{pmatrix} x_1 \\ \vdots \\ \vdots \\ 0 \end{pmatrix}$$

which is a homogeneous system of n simultaneous linear Equations.

This will have a non-trivial solutions **iff** $\det (T - \alpha I) = 0$

The LHS is a polynomial in α of degree n.

This is known as **Characteristic polynomials**.

If A in a matrix representation of the linear operator T then the corresponding expression characteristic polynomial becomes

$\det (A - \alpha I) = 0$

where I in the $n \times n$ unit matrix.

VECTOR SPACE

If B in a matrix representation of T with respect to a new basis then the corresponding characteristic polynomial becomes

$$\det (B - \alpha I) = 0$$

But $B = q A q^{-1}$ for some:-

$$
\begin{aligned}
\therefore \ \det (B - \alpha I) &= \det \left(q A q^{-1} - \alpha I \right) \\
&= \det \left(q A q^{-1} - \alpha q I q^{-1} \right) \\
&= \det \left(q A q^{-1} - q \alpha I q^{-1} \right) \\
&= \det \left(q (A - \alpha I) q^{-1} \right) \\
&= \det (A - \alpha I)
\end{aligned}
$$

Thus there is a unique characteristic polynomial associated with a linear operator T. It is given by

$$\det (A - \alpha I) = 0$$

where A is any matrix representation of T and α is an indeterminate.

We say that $\det (A - \alpha I) = 0$ in the characteristic polynomial of linear operator T.

If in a polynomial of degree n (dimension of E domain of T) in the variable α. Real roots of this polynomial are known as real eigen values of T.

An operator T on a vector space E is said to be **diagonalizable iff** there exists a basis of E such that the matrix representation with respect to this basis in diagonal matrix.

Theorem 10: Let T be an operator for an n-dimension vector space E. If the characteristic polynomial of T has n distinct real roots. Then T can be diagonalized.

Proof: Let $\alpha_1, \alpha_2 ... \alpha_n$ be the distinct eigen values the operator T.

43

Let $e_1, e_2 ... e_n$ be respective eigen vector, thus we have

$$T e_1 = \alpha_1 e_1, T e_2 = \alpha_2 e_2, ... T e_n = \alpha_n e_n$$

We show that $\{e_1, e_2 ... e_n\}$ in a basis of E.

If possible set $\{e_1, e_2 ... e_n\}$ be not a basis of E.

Since $n = \dim E$. $\{e_1, e_2 ... e_n\}$ is not a basis means the set $\{e_1, e_2 ... e_n\}$ is not linearly independent.

Since each of the single sets $\{e_1\}, \{e_2\} ... \{e_n\}$ is linearly independent collection of linearly independent subsets of $\{e_1 ... e_n\}$ is non-empty.

From this non-empty collection we find a maximal one.

If required we consider the members $e_1, e_2 ... e_n$.

Then let us assume that $\{e_1, e_2 ... e_m\}$ is a maximal linearly independent subset of $\{e_1, e_2 ... e_n\}$.

Obviously $m < n$. Hence $\{e_1, e_2 ... e_m, e_n\}$ is not linearly independent. Hence we must have

$$e_n = \sum_{i=1}^{m} t_i e_i \tag{1.5}$$

As $e_1, e_2 ... e_n$ are eigen vectors belongs to eigen values $\alpha_1, \alpha_2 ... \alpha_n$ respectively we have

$$T e_i = \alpha_i e_i \qquad I \le i \le n.$$

In parameter,

$$\alpha_n e_n = T e_n$$

i.e. $\quad 0 = T e_n - \alpha_n e_n \tag{1.6}$

use expression (1.5) in (1.6) we get

$$0 = T\left(\sum_{i=1}^{m} t_i\, e_i\right) - \alpha_n \sum_{i=1}^{m} t_i\, e_i$$

$$= \sum_{i=1}^{m} t_i\, T\, e_i - \sum_{i=1}^{m} \alpha_n\, t_i\, e_i$$

$$= \sum_{i=1}^{m} t_i\, \alpha_i\, e_i - \sum_{i=1}^{m} \alpha_n\, t_i\, e_i$$

$$= \sum_{i=1}^{m} t_i\, (\alpha_i - \alpha_n) e_i$$

i.e., $\displaystyle\sum_{i=1}^{m} t_i\, (\alpha_i - \alpha_n) e_i = 0$

But $\{e_1, e_2 ... e_m\}$ is linearly independent

$\therefore\ t_i\, (\alpha_i - \alpha_n) = 0$ $\qquad\qquad 1 \le i \le m$

As $\quad T e_i = \alpha_i\, e_i$ $\qquad\qquad 1 \le i \le n$

It follows that the matrix of T with respect to the basis $\{e_1, e_2 ... e_n\}$ is

$$\begin{bmatrix} \alpha_1 & 0 & 0 & 0 \\ 0 & \alpha_2 & 0 & 0 \\ 0 & 0 & \alpha_3 & 0 \\ 0 & 0 & 0 & 0 \end{bmatrix} \quad \text{diagonal matrix}$$

$\therefore\ T$ is diagonalizable.

Theorem 11: Let A be an $n \times n$ matrix having n distinct real eigen values $\lambda_1, \lambda_2, ... \lambda_n$. Then there exists an invertible $n \times n$ matrix q such that $q\, A\, q^{-1} = diag\ \{\lambda_1 ... \lambda_n\}$

Proof: Let $\{e_1, e_2 ... e_n\}$ be the standard basis of R^n.

We know that the $n \times n$ matrix A leads to (or represents) a linear operator T on R^n

i.e., T is a linear operator on R^n whose matrix with respect to the basis $\{e_1, e_2 \dots e_n\}$ is A.

Since A has n real distinct eigen values, it follows that T has n distinct eigen values namely $\alpha_1, \alpha_2 \dots \alpha_n$.

Let $f_1, f_2 \dots f_n$ be corresponding eigen vectors of R^n.

We know that $(f_1, \dots f_n)$ is a basis of R^n.

Let B be the matrix of T with respect to basis $f_1, f_2 \dots f_n$.

We know that

$$B = \begin{bmatrix} \lambda_1 & 0 & 0 & 0 \\ 0 & \lambda_2 & 0 & 0 \\ 0 & 0 & \lambda_3 & 0 \\ 0 & 0 & 0 & \lambda_1 \end{bmatrix} = diag\ (\lambda_1, \lambda_2, \dots \lambda_n)$$

And $B = q\,A\,q^{-1} = \left(P^T\right)^{-1} A\,P^T$

where the matrix $P = (P_{ij})$ is such that $f_i = \sum_{j=1}^{n} P_{ij}\,e_j$.

Theorem 12: Let A be an operator on R^n having n distinct real eigen values. Then for all $x_0 \in R^n$, the differential equation $x' = A x,\ x(0) = x_0$ has unique solution.

Proof: Here it is assumed that t is independent variable and dependent variable in $x(t)$ i.e.

$$x(t) = \begin{bmatrix} x_1(t) \\ x_2(t) \\ \vdots \\ x_n(t) \end{bmatrix}$$ so that the DE

VECTOR SPACE

$x' = A x$, $x(0) = x_0$

stands for a system of n differential equation i.e.,

$$x_1'(t) = a_{11} x_1(t) + a_{12} x_2(t) + ... + a_{1n} x_n(t)$$
$$x_2'(t) = a_{21} x_1(t) + a_{22} x_2(t) + ... + a_{2n} x_n(t)$$
$$\vdots$$
$$x_n'(t) = a_{n1} x_1(t) + a_{n2} x_2(t) + ... + a_{nn} x_n(t)$$

$$\begin{bmatrix} x_1(0) \\ x_2(0) \\ \vdots \\ x_n(0) \end{bmatrix} = \begin{bmatrix} x_{01} \\ x_{02} \\ \vdots \\ x_{0n} \end{bmatrix}$$

where $\left(a_{i_j} = A \right)$ and $\begin{bmatrix} x_{01} \\ x_{02} \\ \vdots \\ x_{0n} \end{bmatrix}$

We are given that matrix A [representation] has n distinct real eigen values say .

We know that we can find an invertible matrix such that $\lambda_1, \lambda_2, ... \lambda_n$

We consider a change of variable $x(t)$ to new variable $y(t)$ given by

$$y(t) = q x(t) \tag{1.7}$$

i.e. $\quad y_i(t) = \sum_{j=1}^{n} q_{ij} x_j(t) \tag{1.8}$

where $\left(q_{ij} = Q \right)$

But from equation (1.8) we get

$$y_i'(t) = \sum_{j=1}^{n} q_{ij} x_j'(t) \qquad 1 \le i < n$$

i.e., $\quad y'(t) = Q x'(t)$ $\hfill$ (1.9)

Then $x(t)$ is a solution of the differential equation

$x'(t) = A x(t)$

iff $\quad y'(t) = Q A x(t)$

$\qquad\qquad = Q A Q^{-1} y(t)$ $\qquad\qquad \{\because\ y(t) = Q x(t)\}$

i.e., $\quad y'(t) = B y(t)$ $\qquad\qquad (\because B = Q A Q^{-1})$

then $x(t)$ is a solution of the DE

$x'(t) = A x(t)$

iff $y(t)$ is a solution of the diff. equation

$y'(t) = B y(t)$ $\hfill$ (1.10)

But $\quad B = diag\ \{\lambda_1, \lambda_2, \dots \lambda_n\}$

So equation (1.10) is equivalent to

$y_1'(t) = \lambda_1 y_1(t)$
$y_2'(t) = \lambda_2 y_2(t)$
$\vdots$
$y_n'(t) = \lambda_n y_n(t)$

$$B = \begin{bmatrix} \lambda_1 & 0 \dots 0 \\ 0 & \lambda_2 \dots \\ 0 & \quad \vdots \\ \vdots & \dots \lambda_1 \end{bmatrix}$$

i.e., $\quad y_i'(t) = \lambda_i y_i(t)$ $\qquad\qquad 1 \leq i \leq n$

i.e., $\quad y_i'(t) = y_i(0) \exp(\lambda_i t)$ $\qquad 1 \leq j \leq n$

then $x(t)$ is a solution of the diff. equation

iff $\quad y(t) = \begin{bmatrix} y_1(t) \\ : \\ y_n(t) \end{bmatrix}$

where $y_i(t) = y_i(0)\exp\left(\lambda_i t\right)$

Since $y(t) = Qx(t)$, we have $\quad x(t) = Q^{-1}y(t)$

i.e., $\quad \begin{bmatrix} x_1(t) \\ : \\ x_n(t) \end{bmatrix} = Q^{-1} \begin{bmatrix} y_1(t) \\ : \\ y_n(t) \end{bmatrix}$ $\hspace{3cm}$ (1.11)

where $y_i(t) = y_i(0)\ \exp\ \left(\lambda_i(t)\right)$

we can indeed show that $x(t)$ given by (1.11) is a solution of the

$DE \quad x'(t) = Ax(t)$

Equation (1.11) can be written as

$x(t) = Q^{-1}y(t)$

$\begin{aligned} x'(t) = Q^{-1}y'(t) &= Q^{-1}B\,y(t) \\ &= Q^{-1}QA\ Q^{-1}y(t) \\ &= AQ^{-1}y(t) \\ &= Ax(t) \end{aligned}$

Thus (1.11) gives us a solution of the DE

$x'(t) = Ax(t)$

Since we want $x(0) = x_0$

i.e., $\quad x_i(0) = x_{0i} \hspace{2cm} 1 \le i \le n$

and as from (1.11)

$x_i(0) = \sum_{j=1}^{n} (Q)_{ij}^{-1} y_j(0)$

49

$$x_{0i} = \sum_{j=1}^{n} \left(Q^{-1}\right)_{ij} \, y_j(0)$$

it follows that we must take

$$y(0) = Q(x_0)$$

We thus get a solution to the initial value problem

$$x'(t) = A x(t) \quad , \quad x(0) = x_0$$

If $\bar{x}$ and $\bar{\bar{x}}$ were two solutions of the diff. equation $x' = Ax$, $x(0) = x_0$

Then $Q\bar{x}$ and $Q\bar{\bar{x}}$ would be solution of the diff. equation

$$y' = B y \quad , \quad y(0) = y_0, \quad \text{where } y_0 = Q x_0$$

But as $B = \begin{bmatrix} \lambda_1 & 0 & \cdots & 0 \\ : & \lambda_2 & \cdots & : \\ : & & & : \\ 0 & & \cdots & \lambda_1 \end{bmatrix}$

The initial value problem

$$\therefore \; y' = B y \; ,$$

$$y(0) = y_0$$

has a unique solution

$$Q\bar{x} = Q\bar{\bar{x}}$$

As Q is invertible, we must have

$\bar{x} = \bar{\bar{x}}$. Thus the solution must be **unique**.

Theorem 13: Let the $n \times n$ matrix A have n distinct real eigen values $\lambda_1, \lambda_2, \ldots \lambda_n$. Then every solution to the DE.

$x' = A x, x(0) = \mu$ is of the form

VECTOR SPACE

$$x_i(t) = c_{i1} \exp(t\, \lambda_1) + c_{i2} \exp(t\, \lambda_2) + \ldots\ldots + c_{in} \exp(t\, \lambda_n)$$

for unique constants $c_{i1}, c_{i2}, \ldots c_{in}$ depending on μ.

Proof: We know that if Q is an invertible matrix such that

$$Q A Q^{-1} = B = diag\,\{\lambda_1 \ldots \lambda_n\}$$

Then the solution of the initial value problem

$$x'(t) = A x(t), \qquad x_0 = \mu$$

is given by,

$x(t) = Q^{-1} y(t)$, where,

$$y(t) = \begin{bmatrix} y_1(t) \\ y_2(t) \\ : \\ y_n(t) \end{bmatrix}$$

And $y_i(t) = y_i(0) \exp(t\, \lambda_i)$

with $\quad U = x(0) = Q^{-1}(y(0))$

i.e. $\quad y(0) = Q(\mu)$

Let us denote ρ^{-1} by $c = (c_{ij})$

Then $x(t) = Q^{-1}(y(t)) = c \cdot y(t)$

$$x_i(t) = \sum_{j=1}^{n} c_{ij}\, y_j(t)$$

$$= c_{i1}\, y_1(t) + c_{i2}\, y_2(t) + \ldots + c_{in}\, y_n(t)$$

$$= c_{i1}\, y_1(0) \exp(\lambda_1 t) + c_{i2}\, y_2(0) \exp(\lambda_2 t) + \ldots + c_{in}\, y_n(0) \exp(\lambda_n t)$$

with the coefficient $c_{i1}\, y_1(0), c_{i2}\, y_2(0), \ldots c_{in}\, y_n(0)$ depending on μ.

51

DYNAMICAL SYSTEM – A SHORT COURSE

COMPLEX EIGEN VALUES

Let a and b be two real numbers such that $b \neq 0$. Consider an operator T on R^n whose matrix is given $\begin{pmatrix} a & -b \\ b & a \end{pmatrix}$ characteristic polynomial of T is given by

$$\begin{vmatrix} a-\lambda & -b \\ b & a-\lambda \end{vmatrix} = 0$$

i.e. $(a-\lambda)^2 + b^2 = 0$

i.e. $a^2 - 2a\lambda + \lambda^2 + b^2 = 0$

i.e. $\lambda = a \pm ib$, Complex roots as $b \neq 0$.

Let $r = \sqrt{a^2 + b^2}$ and $\theta = \cos^{-1} \dfrac{a}{r}$

We show that the operator T is equal to rotation through θ about origin (anticlockwise if $b > 0$) followed by scaling by factor r.

Scaling by r means $(x, y) \to (rx, ry)$.

The matrix of scaling by factor r

i.e., $\begin{pmatrix} x \\ y \end{pmatrix} \to \begin{pmatrix} rx \\ ry \end{pmatrix}$

is given by $\begin{pmatrix} r & 0 \\ 0 & r \end{pmatrix}$ and the matrix of rotation through θ about origin is

(anticlockwise if $b > 0$) is

$$\begin{pmatrix} \cos\theta & -\sin\theta \\ \sin\theta & \cos\theta \end{pmatrix}$$

$\therefore$ The product

52

VECTOR SPACE

$$\begin{pmatrix} r & 0 \\ 0 & r \end{pmatrix} \begin{pmatrix} \cos\theta & -\sin\theta \\ \sin\theta & \cos\theta \end{pmatrix} = \begin{pmatrix} r\cos\theta & -r\sin\theta \\ r\sin\theta & r\cos\theta \end{pmatrix}$$

Thus if T_r and T_θ denote scaling by r and rotation through θ then we have,

$$T = T_r \cdot T_\theta = T_\theta \cdot T_r$$

Since $T_r \cdot T_r = T_r^2$

Similarly, $T_\theta \cdot T_\theta = T_{2\theta}$ we have,

$$T^2 = T_{2\theta} \cdot T_{r^2} = T_{r^2} \cdot T_{2\theta}$$

In general,

$$T^n = T_{r^n} \cdot T_{n\theta}$$

i.e. $\begin{pmatrix} a & -b \\ b & a \end{pmatrix} = \begin{pmatrix} r^n & 0 \\ 0 & r^n \end{pmatrix} \cdot \begin{pmatrix} \cos n\theta & -\sin n\theta \\ \sin n\theta & \cos n\theta \end{pmatrix}$

we can also have another interpretation of the operator T .

We identify the point $\begin{pmatrix} x \\ y \end{pmatrix}$ of R^2 with complex number $x + iy$.

Then $\begin{pmatrix} x \\ y \end{pmatrix} \rightarrow (x + iy)$

$\quad\quad T \uparrow \quad\quad \downarrow$ multiply by $a + ib$

$$\begin{pmatrix} a & -b \\ b & a \end{pmatrix} \begin{pmatrix} x \\ y \end{pmatrix} = \begin{pmatrix} ax - by \\ ay + bx \end{pmatrix} \rightarrow (ax - by) + i(ay + bx)$$

Thus if we identify points of R^2 by complex numbers then the operation T represents multiplication by complex number $a + ib$.

DYNAMICAL SYSTEM – A SHORT COURSE

Ex 2: Consider the system,

$$x_1' = 5x_1 + 3x_2$$

$$x_2' = -6x_1 - 4x_2 \quad , \quad x_1(0) = 1 \ and \ x_2(0) = 2$$

Solution. The matrix A is

$$\begin{pmatrix} 5 & 3 \\ -6 & -4 \end{pmatrix}$$

Eigen values are given by

$$\begin{vmatrix} 5-\lambda & 3 \\ -6 & -4-\lambda \end{vmatrix} = 0$$

$$(5-\lambda)(-4-\lambda) - (-6)(3) = 0$$

$$-20 - 5\lambda + 4\lambda + \lambda^2 + 18 = 0$$

$$-2 - \lambda + \lambda^2 = 0$$

$$\lambda^2 - \lambda - 2 = 0$$

$$\lambda = 2 \ or \quad \lambda = -1$$

Roots of the characteristic polynomial are 2,

Thus the matrix A has 2 distinct real eigen values. Thus A can be diagonalised a $\begin{pmatrix} 2 & 0 \\ 0 & -1 \end{pmatrix}$ and the diagonalised matrix $B = QAQ^{-1}$ is

$$\{2 \ -1\} = \begin{pmatrix} 2 & 0 \\ 0 & -1 \end{pmatrix}$$

Thus the equivalent system is,

$$y' = B y$$

i.e. $\quad y_1' = 2y_1 \qquad y_2' = -1y_2$

$$\therefore \ y_1(t) = y_1(0) \ \exp \ (2t)$$

$$y_2(t) = y_2(0) \ \exp \ (-t)$$

54

VECTOR SPACE

where, $y = Qx$ i.e. $x = Q^{-1}y$

We know that $Q = \left(P^T\right)^{-1}$

i.e. $Q^{-1} = P^T$

where $P = \left(P_{ij}\right)$ such that,

$f_i = \sum_{j=1}^{2} P_{ij} e_j$ and $\{f_1 \ f_2\}$ is a basis of eigen values, $\{e_1 ... e_n\}$ is the standard basis.

Thus to find the basis $\{f_1 \ f_2\}$ we find the eigen spaces of the eigen values 2, -1.

An eigen vector $\begin{pmatrix} \alpha \\ \beta \end{pmatrix}$ belong to eigen values of the matrix A iff

$$\begin{pmatrix} a_{11}-2 & a_{12} \\ a_{21} & a_{22}-2 \end{pmatrix}\begin{pmatrix} \alpha \\ \beta \end{pmatrix} = \begin{pmatrix} 0 \\ 0 \end{pmatrix}$$

i.e. $\begin{pmatrix} 5-2 & 3 \\ -6 & -4-2 \end{pmatrix}\begin{pmatrix} \alpha \\ \beta \end{pmatrix} = \begin{pmatrix} 0 \\ 0 \end{pmatrix}$

$\begin{pmatrix} 3 & 3 \\ -6 & -6 \end{pmatrix}\begin{pmatrix} \alpha \\ \beta \end{pmatrix} = \begin{pmatrix} 0 \\ 0 \end{pmatrix}$

i.e. $\begin{array}{l} 3\alpha + 3\beta = 0 \\ -6\alpha - 6\beta = 0 \end{array} \Rightarrow \alpha = -\beta$

i.e. $(t, -t)$ is an eigen vector belongs to eigen value 2 for all $t \in R$.

Thus the eigen space belonging to eigen value 2 is the space

$\{(t, -t), t \in R\}$.

We choose f_1 to be any non-zero vector of this space.

Thus we may choose $f_1 = (1, -1)$.

If (α, β) is an eigen vector belonging to eigen value -1, then we must have

$$\begin{pmatrix} 5+1 & 3 \\ -6 & -4+1 \end{pmatrix} \begin{pmatrix} \alpha \\ \beta \end{pmatrix} = \begin{pmatrix} 0 \\ 0 \end{pmatrix}$$

$\begin{aligned} 6\alpha + 3\beta &= 0 \\ -6\alpha - 3\beta &= 0 \end{aligned} \Rightarrow 3(2\alpha + \beta) = 0$

$$2\alpha + \beta = 0$$

$$2\alpha = -\beta$$

$\Rightarrow \quad 2\alpha = -\beta$

Thus the eigen space belonging to the eigen values -1 is $\{(t, -2t), t \in R\}$.

We can choose $f_2 = (1, -2)$

Since, $f_1 = e_1 - e_2$ and $f_2 = e_1 - 2e_2$

we have,

$P_{11} = 1 , \quad P_{12} = -1$

$P_{21} = 1 , \quad P_{22} = -2$

$$\therefore \quad P = \begin{pmatrix} 1 & -1 \\ 1 & -2 \end{pmatrix}$$

$$\therefore \quad Q^{-1} = P^{-1}$$

$$\therefore \quad Q^{-1} = \begin{pmatrix} 1 & 1 \\ -1 & -2 \end{pmatrix}$$

$$\therefore \quad \begin{pmatrix} x_1(t) \\ x_2(t) \end{pmatrix} = \rho^{-1} \begin{pmatrix} y_1(t) \\ y_2(t) \end{pmatrix}$$

$$= \begin{pmatrix} 1 & 1 \\ -1 & -2 \end{pmatrix} \begin{pmatrix} y_1 \cos e^{2t} \\ y_2 \cos e^{-t} \end{pmatrix}$$

$$\therefore \quad x_1(t) = y_1(0)e^{2t} + y_2(0)e^{-t}$$

VECTOR SPACE

$$x_2(t) = -y_1(0)e^{2t} - 2y_2(0)e^{-t}$$

To find $y_1(0)$ and $y_2(0)$

We note that,

$$\begin{pmatrix} y_1(t) \\ y_2(t) \end{pmatrix} = Q \begin{pmatrix} x_1(t) \\ x_2(t) \end{pmatrix} \qquad \text{for all } t.$$

Hence at $t = 0$ we have

$$\begin{pmatrix} y_1(0) \\ y_2(0) \end{pmatrix} = Q \begin{pmatrix} x_1(0) \\ x_2(0) \end{pmatrix} = Q \begin{pmatrix} 1 \\ 2 \end{pmatrix}$$

We have,

$$Q^{-1} = P^{-1} = \begin{pmatrix} 1 & 1 \\ -1 & -2 \end{pmatrix}$$

$$\therefore Q = \begin{pmatrix} 2 & 1 \\ -1 & -1 \end{pmatrix}$$

$$\begin{pmatrix} y_1(0) \\ y_2(0) \end{pmatrix} = \begin{pmatrix} 2 & 1 \\ -1 & -1 \end{pmatrix} \begin{pmatrix} 1 \\ 2 \end{pmatrix} = \begin{pmatrix} 0 \\ 1 \end{pmatrix}$$

$$\therefore y_1(0) = 0$$
$$y_2(0) = 1$$

$$\therefore x_1(t) = e^{-t}$$
$$x_2(t) = -2e^{-t}$$

Ex (3). Consider

$$x_1' = x_1$$
$$x_2' = x_1 + 2x_2$$
$$x_3' = x_1 - x_3$$
$$x(0) = (U_1 \ U_2 \ U_3)$$

DYNAMICAL SYSTEM – A SHORT COURSE

Solution. Let the matrix

$$A = \begin{pmatrix} 1 & 0 & 0 \\ 1 & 2 & 0 \\ 1 & 0 & -1 \end{pmatrix} = \begin{pmatrix} 1-\lambda & 0 & 0 \\ 1 & 2-\lambda & 0 \\ 1 & 0 & -1-\lambda \end{pmatrix}$$

Eigen values are given by

$$(1-\lambda)(2-\lambda)(-1-\lambda) = 0$$
$$(2-\lambda)(-2\lambda+\lambda^2)....$$
$$\lambda_1 = 1 , \quad \lambda_2 = 2 , \quad \lambda_3 = -1$$

A can be diagonalized.

Let B be diagonalized matrix

$$B = \begin{pmatrix} 1 & 0 & 0 \\ 0 & 2 & 0 \\ 0 & 0 & -1 \end{pmatrix} \text{ where } B = QAQ^{-1}.$$

So let $\quad y = Qx$

then we know that

$$y' = By$$

$$y_1'(t) = y_1(t)$$
$$y_2'(t) = 2y_2(t)$$
i.e. $\quad y_3'(t) = -y_3(t)$
$$y_1(t) = y_1(0)e^t$$
$$y_2(t) = y_2(0)e^{2t}$$
$$y_3(t) = y_3(0)e^{-t}$$

where $\quad y = Qx$

$$\therefore \quad x = Q^{-1}y$$

We next find matrix $\quad Q = \left(P^{-1}\right)^{-1}$

VECTOR SPACE

To find P we find basis of eigen vectors for the eigen value $\lambda = 1$ eigen vector V satisfies.

$$(A - 1 \cdot I)v = \begin{pmatrix} 0 \\ 0 \\ 0 \end{pmatrix}$$

i.e. $(a_{11} - 1)v_1 + a_{12}v_2 + a_{13}v_3 = 0$

$a_{21}v_1 + (a_{22} - 1)v_2 + a_{23}v_3 = 0$

$a_{31}v_1 + a_{32}v_2 + (a_{32} - 1)v_3 = 0$

i.e.   0   i.e. $v_1 = t$

$v_1 + v_2 = 0$ $v_2 = -t$

$v_1 - 2v_3 = 0$ $v_3 = t/2$

Thus the eigen space corresponding to eigen value 1 is given by $\{(t, -t, t/2), t \in R\}$

So we can choose, $f_1 = \left(1, -1, \frac{1}{2}\right)$

Similarly, Eigen vector V corresponding to $\lambda = 2$

$\therefore$ $(a_{11} -)v_1 + a_{12}v_2 + a_{13}v_3 = 0$

$a_{21}v_1 + (a_{22} - 2)v_2 + a_{23}v_3 = 0$

$a_{31}v_1 + a_{32}v_2 + (a_{32} - 2)v_3 = 0$

i.e. $-v_1 = 0$

i.e. $v_1 = 0$, $v_3 = 0$

$v_1 = 0$

$v_1 - 3v_3 = 0$

Thus the eigen space corresponding to eigen value 2 is

$$\{(0, t, 0)\} \quad t \in R.$$

we choose $f_2 = (0, 1, 0)$

Again,

$$2v_1 = 0$$
$$v_1 + 3v_2 = 0 \qquad v_1 = 0$$
$$v_1 + 0v_2 + 0v_3 = 0 \qquad v_2 = 0$$

i.e., $(0, 0, t) \quad t \in R$

we take $f_3 = (0, 0, 1)$

Thus we consider eigen vector

$$f_1 = (1, -1, \tfrac{1}{2})$$
$$f_2 = (0, 1, 0)$$
$$f_3 = (0, 0, 1)$$

Obviously,

$$f_1 = e_1 - e_2 + {}^{e_3}\!/_2$$
$$f_2 = 0e_1 + e_2 + 0e_3$$
$$f_3 = 0e_1 + 0e_2 + e_3$$

$$P = \begin{pmatrix} 1 & -1 & \tfrac{1}{2} \\ 0 & 1 & 0 \\ 0 & 0 & 1 \end{pmatrix},$$

$$P^{-1} = \begin{pmatrix} 1 & 0 & 0 \\ -1 & 1 & 0 \\ \tfrac{1}{2} & 0 & 1 \end{pmatrix}$$

VECTOR SPACE

Solve

a) $x' = -x$

$y' = x + 2y$, $x(0) = 0$, $y(0) = 3$

b) $x_1' = 2x_1 + x_2$

$x_2' = x_1 + x_2$, $x_1(1) = 1$, $x_2(1) = 1$

3) $x' = Ax$ $A = \begin{bmatrix} 0 & 3 \\ 1 & -2 \end{bmatrix}$, $x(0) = (3, 0)$

4) $x' = Ax$

$x(0) = (0, -b, b)$

$A = \begin{bmatrix} 2 & 0 & 0 \\ 0 & -1 & 0 \\ 0 & 2 & -3 \end{bmatrix}$

2) Find a 2×2 matrix A such that one solution to $x' = Ax$ is

$$x(t) = \left(e^{2t} - e^{-t} , e^{et} + 2e^{-t} \right)$$

3) Show that the only solution to

$x_1' = x_1$

$x_2' = x_1 + x_2$, $x_1(0) = a$, $x_2(0) = b$

is $x_1(t) = ae^t$, $x_2(t) = e^t (b + at)$

4) Consider the second order differential equation

$x'' + bx' + cx = 0$, b and C are constants

Solve it such that $x(0) = \mu$, $x'(0) = v$

5) $x' = -y$

a) $y' = x$, $x(0) = 1$, $y(0) = 1$

61

b) $x_1' = -2x_2$

$x_2' = 2x_1$, $x_1(0) = 0$, $x_2(0) = 2$

CHAPTER 2

COMPLEX VECTOR SPACES

Let T be an operator on real vector space E. We can obtain the characteristic polynomial T as

$\det |T - \lambda I| = 0$.

If a root say λ of this polynomial is non real then the equation

$Tx = \lambda Ix$ for $x \in E$

is meaningless as λ is non real and $x \in E$ with E as a real vector space.

Thus we need to extend the real vector space E into a complex vector space.

Similarly the operator T also needs to be extended into an operator on a complex vector space.

A complex vector space is a vector space the field of complex numbers.

For $n \geq 1$, C^n is a complex vector space.

Although R^n can be considered as a subset of C^n, R^n is not a complex subspace of C^n. C^n is real as well as complex vector space (considered as a subspace of C^n) and T an operator on F.

We can then consider the matrix represented of T and also the characteristic polynomial of T.

Roots of this polynomial (real as well as non are known as eigen values of T and it is an eigen value of T then a non trivial (non zero) element say $v \in F$ will be known eigen vector belonging to the eigen value λ provided,

$Tx = \lambda Ix$

Theorem 1: Let $T : F \to F$ be an operator on an n-dimensional complex vector space F. If the characteristic polynomial has distinct roots then T

can be diagonalized. This implies that when these roots are distinct, then one may find a basis $\{e_1 ... e_n\}$ of eigen vectors for F so that if

$z = \displaystyle\sum_{j=1}^{n} z_j e_j$ is in F, then $T(z) = \displaystyle\sum_{j=1}^{n} \lambda_j z_j e_j$, e_j is the eigen vector belong

to the eigen value λ_j.

Proof: Let the n distinct roots of the characteristic polynomial of T be

$\lambda_1, \lambda_2 ... \lambda_n$.

Further let $e_1, e_2 ... e_n$ be the corresponding eigen vectors respective.

We show that the set $(e_1, e_2 ... e_n)$ is a basis of F.

For this we show that the set $(e_1 ... e_n)$ is linearly independent.

If possible let us assume that $(e_1 ... e_n)$ set is not linearly independent. If required set, we re-index and then let $(e_1 ... e_m)$ be a maximal linearly independent subset of $(e_1, e_2 ... e_m)$.

We must have $m < n$.

Hence the set $(e_1, e_2 ... e_m, e_n)$ is linearly dependent as such,

$e_n = \displaystyle\sum_{j=1}^{m} \alpha_j e_j \qquad \alpha_j \in C$.

We have,

$T_{en} = \lambda_n e_n$

$\therefore \; 0 = T_{en} - \lambda_n e_n$

$= T \left(\displaystyle\sum_{j=1}^{m} \alpha_j e_j - \lambda_n \sum_{j=1}^{m} \alpha_j e_j \right)$

$$= \sum_{j=1}^{m} \alpha_j \, T\!\left(e_j\right) - \lambda_n \sum_{j=1}^{m} \alpha_j e_j$$

$$= \sum_{j=1}^{m} \alpha_j \, e_j - \sum_{j=1}^{m} \lambda_n \alpha_j e_j$$

Then we have,

$$0 = \sum_{j=1}^{m} \left(\alpha_j \lambda_j - \lambda_n \alpha_j\right) e_j \quad = \sum_{j=1}^{m} \alpha_j \left(\lambda_j - \lambda_n\right) e_j$$

But $\{e_1, e_2 \ldots e_m\}$ is linearly independent.

$\therefore$ for $j = 1 \ldots m$, $\alpha_j \left(\lambda_j - \lambda_n\right) = 0$

But $\lambda_j \neq \lambda_n$ $1 \leq j \leq m$

$\therefore$ $\lambda_j = 0$ $1 \leq j \leq m$

But then $e_n = \sum_{j=1}^{m} \alpha_j e_j \quad \Rightarrow \quad\quad e_n = 0$

This is a contradiction, as ' e_n ' being an vector is non-zero.

Therefore $\left(e_1, e_2 \ldots e_n\right)$ is linearly independent.

Since dim $F = n$, we must have $\left(e_1, e_2 \ldots e_n\right)$ a basis of F.

Obviously, matrix of T with respect to the basis $\left(e_1 \ldots e_n\right)$ is diagonal $\{\lambda_1, \lambda_2 \ldots \lambda_n\}$.

$\therefore$ T can be diagonalized.

Definition: An operator T on a complex vector space F is Semi Simple if it is diagonalizable.

COMPLEXIFICATION AND DE-COMPLEXIFICATION

Let F be a complex vector space (a subspace of C^n).

Consider $F \cap R^n$. It is a real vector space. We denote it by F_R. F_R is known as a vector space of real vectors in F.

Let E be a real vector space (subspace of R^n).

Let E_c denote the collection of all terms of the type, $\sum \lambda_j v_j$ (sum over finite suffixes) where $\lambda_j \in C$ and $v_j \in E$.

It can be shown that E_C is a complex vector space (If λ_j is real then $\lambda_j v_j$ usual product of the real λ_j and element $v_j \in E$ as E is a real vector space.)

Also we have $F_R \subseteq F$ and $E \subseteq E_C$.

As E_C is a complex vector space, we can obtain E_{CR} from E_C.

Now we have $E = E_{CR}$.

i)   As $E \subseteq E_c$ and $E \subseteq R^n$ we have $E \subseteq E_c \cap R^n$

$\therefore \ E \subseteq R^n$

ii) Since $E_{CR} = E_c \cap R^n$ and a typical element of E_c is of the type

$\sum \lambda_i v_i$, where $v_i \in E$ (and as such i.e. a real $n-$tuple)

$\sum \lambda_i v_i \in E_{CR}$ iff $\lambda_i \in R$

but with $\lambda_i \in R$, $\sum \lambda_i v_i \in E$

$\therefore E_{CR} \subseteq E \subseteq E_{CR}$

$\therefore \ E = E_{CR}$

Let F be a complex vector space. If $z = (z_1, z_2 \ldots z_n)$ is a typical element of F. We define σ on F as $\sigma z = (\overline{z_1}, \overline{z_2} \ldots \overline{z_n})$

σ is thus a well defined map from $F \to C^n$

$\therefore \sigma$ is not a linear map.

Since,

$$\sigma(\lambda z) = \sigma(\lambda z_1, \lambda z_2 \ldots \lambda z_n)$$
$$= (\overline{\lambda z_1}, \overline{\lambda z_2} \ldots \overline{\lambda z_n})$$
$$= \overline{\lambda} \, \sigma(z)$$

$\therefore \quad \sigma$ is not linear.

Further we can show that,

$$F_R = \{ z \in F \cdot z \quad \sigma(z) = z \}$$

For a complex vector space F, $F = F_{RC}$ If and only if $\sigma(F) \subseteq F$

As by definition , $F_R \subseteq F$ and F is a complex vector space, it follows that

F_{RC} is a complex subspace of F.

Thus, $F = F_{RC}$ iff $F \subseteq F_{RC}$.

Hence we show that

$F \subseteq F_{RC}$ iff $\sigma(F) \subseteq F$

Let us assume that $\sigma(F) \subseteq F$

Let $\phi \in F$ hence we can write

$$\phi = (z_1, z_2, \ldots z_n) \quad , \quad i \in c \quad z_i \in C$$

and further we can write

$$z_j = x_j + i\, y_j$$

As $\quad \sigma(F) \subseteq F, \quad \sigma(\phi) \in F$

But $\quad \sigma(\phi) = (\overline{z_1}, \overline{z_2} \ldots \overline{z_n})$
$$= (x_1 - i\, y_1, x_2 - i\, y_2, \ldots \ldots x_n - i\, y_n)$$

$$\therefore \quad \frac{\phi + \sigma(\phi)}{2} \in F$$

i.e. $\quad (x_1, x_2 ... x_n) \in F$

Similarly, $\quad \frac{\phi - \sigma(\phi)}{2i} \in F$

i.e. $\quad (y_1, y_2 ... y_n) \in F$

$\therefore \ (x_1, x_2 ... x_n) \in F \cap R^n, \ (y_1, y_2 ... y_n) \in F \cap R^n$

$\therefore \ (x_1, x_2 ... x_n) \in F_R, \quad (y_1, y_2 ... y_n) \in F_R$

$(x_1, x_2 ... x_n) + i(y_1, y_2 ... y_n) \in F_{RC}$

$\big((x_1 + i y_1), (x_2 + i y_2), ... (x_n + i y_n) \in F_{RC}\big)$

i.e. $\quad \phi \in F_{RC}$

$\therefore \phi \in F \Rightarrow \phi \in F_{RC}$

$\therefore \ F \subseteq F_{RC} \subseteq F$

$\therefore \ F = F_{RC}$

Conversely. We are given that

$F = F_{RC} \quad$ i.e. $\quad F \subseteq F_{RC}$

We show that $\quad \sigma(F) \subseteq F$

Let ϕ be an element of F

i.e. $\phi \in F$ and $\phi \in F_{RC}$

$\therefore \ \phi = \sum \lambda_i v_i$ a finite sum, where $\quad \lambda_i \in C$, $v_i \in F_R$

As $\quad v_i \in F_R$, $v_i \in F \cap R^n$

$\therefore \ \sigma(\phi) = \sum \overline{\lambda_i} v_i \qquad (\because \ v_i \in R^n)$

$\qquad \in F$

COMPLEX VECTOR SPACES

$\{\because \ v_i \in F$ and F is complex vector space$\}$

Thus $\phi \in F \Rightarrow \sigma(\phi) \in F$

$\therefore \ \sigma(F) \subseteq F$

Consider a real vector space E.

Let E_C be its complexification.

Let $(e_1, e_2 ... e_n)$ be a basis of E as a real vector space.

We show that $(e_1, e_2 ... e_n)$ is also a basis of E_C as a complex vector space.

Any element of E_C is of the type $\sum \lambda_i v_i$ with $\lambda_i \in C$, $v_i \in E$.

As $(e_1, e_2 ... e_n)$ is a basis of E, each v_i is a linear combination of $e_1, e_2 ... e_n$ with real coefficients using this expressions for $v_i's$.

$\sum \lambda_i v_i$ becomes a linear combination of $e_1 ... e_n$ with complex coefficients.

Thus the set $(e_1, e_2 ... e_n)$ spans E_C as a complex vector space.

To show that $(e_1, e_2 ... e_n)$ is linearly independent in E_C , We consider

$\alpha_1 e_1 + \alpha_2 e_2 + ... + \alpha_n e_n = 0$ in E_C.

Further, let $\alpha_j = i a_j + i b_j$ $\qquad 1 \leq j \leq n$

As $e_1, e_2 ... e_n \in E$, each of them is a real n-tuple.

$\therefore \ \sum_{j=1}^{n} \alpha_j e_j = \sum_{j=1}^{n} a_j e_j + i \sum_{j=1}^{n} b_j e_j$

Thus, $\sum_{j=1}^{n} \alpha_j e_j = 0$ in E_C

$$\Rightarrow \sum_{j=1}^{n} a_j \, e_j + i \sum_{j=1}^{n} b_j \, e_j = 0$$

Hence we must have,

$$\therefore \quad \sum_{j=1}^{n} a_j \, e_j = 0 \text{ in } R^n \text{ and } \sum_{j=1}^{n} b_j \, e_j = 0 \text{ in } R^n$$

i.e. $\quad \sum_{j=1}^{n} a_j \, e_j = 0$ in E and $\quad \sum_{j=1}^{n} b_j \, e_j = 0$ in E

But as the set $(e_1, e_2 \dots e_n)$ is linearly independent in F we have,

$$0 = a_j = b_j \quad , \quad 1 \le j \le n$$

$$\therefore \quad a_j = 0 \quad , \quad 1 \le j \le n$$

{ a_j and b_j are real and img part they are zero}

Thus, if $(e_1, e_2 \dots e_n)$ is a basis of E as a real vector space then $\{e_1 \dots e_n\}$ is a basis of E_C as a complex vector space.

If T is a linear operator on real vector space E then we can extend T to an linear operator T_C on the complex vector space E_C as follows.

As any element of E_C is a finite sum of the type

$$\sum \lambda_j v_j \quad , \quad \lambda_j \in C \quad , \quad v_j \in E,$$

We define $T_C \left(\sum \lambda_j v_j \right) = \sum \lambda_j T(v_j)$

Obviously T_C is a linear operator on E_C.

As if we know that,

If $\{e_1 \dots e_n\}$ is a basis of E and $\{e_1 \dots e_n\}$ is a basis of E_C.

Also as $T_C \left(e_j \right) = T \left(e_j \right)$

COMPLEX VECTOR SPACES

It follows that the matrix of T_c and matrix of T are same in the basis $\{e_1 \ldots e_n\}$. Hence characteristic polynomial of T is same as characteristic polynomial of T_C.

Thus if a root of characteristic polynomial of T is non-real, we can still consider it as an eigen value of the complexification operator T_C and get corresponding eigen vector in E_C.

Definition: A real operator T is said to be semi simple if its complexified operator T_C can be diagonalised.

Theorem 2: Let $E \subseteq R^n$ be a real vector space and $E_C \subseteq C^n$ its complexification. If $Q \in L(E_C)$ (i.e. linear operator on E_C then $Q = T_C$ i.e. Q is a complexification of a real operator) for some $T \in L(E)$ iff $Q\sigma = \sigma Q$, where $\sigma : E_C \to E_C$ is complex conjugation.

Proof: Let us assume that $Q \in L(E_C)$ and is such that $Q\sigma = \sigma Q$

We show that $Q(E) \subseteq E$.

For this, let $v \in E$ as E is a real vector V is a real n-tuple and hence
$$\sigma(v) = v$$

$$\therefore \ Q\sigma(v) = \sigma Q(v) \text{ gives us}$$

$$Qv = \sigma Q(v) \qquad \{ \text{ i.e. } \sigma y = y \text{ for } y = Qv \}$$

Thus, writing $Q(E)$ as $\{Qv, v \in E\}$ we get

$$Q(E) \subseteq \{y \in E_C \text{ such that } \sigma y = y\} = E_{CR} = E$$

$$\therefore \ Q(E) \subseteq E$$

Let T denote the restriction of Q to E. Then T is a linear operator on E.

We show that $\qquad Q = T_C$

As any element of E_C is of the type $\sum \lambda_i v_j$ with $v_i \in E$, $\lambda_i \in C$

$$Q\left(\sum \lambda_i v_i\right) = \sum \lambda_i Q(v_i) = \sum \lambda_i T(v_i) \quad \{\because on\, E\, Q \equiv T\}$$

$$= T_c \left(\sum \lambda_i v_i\right)$$

$$\therefore \quad Q = T_C$$

Conversely: Let $Q = T_C$ for some linear operator T on E.

We show that, $Q\sigma = \sigma Q$ on E_C

For any element $\sum \lambda_i v_i$ of E_C.

$$Q\sigma\left(\sum \lambda_i v_i\right) = Q\left(\sigma\left(\sum \lambda_i v_i\right)\right)$$

$$= Q\left(\sum \overline{\lambda_i} v_i\right)$$

$$= \sum \overline{\lambda_i}\, Q(v_i)$$

$$= \sum \overline{\lambda_i}\, T(v_i)$$

Similarly,

$$\sigma Q\left(\sum \lambda_i v_i\right) = \sigma\left(Q\left(\sum \lambda_i v_i\right)\right)$$

$$= \sigma\left(T_c\left(\sum \lambda_i v_i\right)\right)$$

$$= \sigma\left(\sum \lambda_i T(v_i)\right)$$

$$= \left(\sum \overline{\lambda_i}\, T(v_i)\right)$$

Thus, $Q\sigma = \sigma Q$ on E_C.

Definition: If E is a real vector space then E_C is known as complexification of E.

COMPLEX VECTOR SPACES

If F is a complex vector space then F_R is known as the space of real vector in F.

If F is a complex vector space and if there is a real vector space E such that $E_C = F$. Then we say that E is a decomplexification of F.

An operator T on F a real vector space E semi simple if its complexification T_C is a diagonalizable operator on E_C.

Ex 1: Let $F \subseteq C^2$ be the subspace spanned by the vector $(1,i)$.

a) Prove that F is not invariant under conjugation and hence is not the complexification of subspace of R^2.

b) Find F_R and $(F_R)_C$.

Solution: a). Since F is generated by a single element $(z_1 z_2)$ of C^2 where,

$$z_1 = 1 \quad , \quad z_2 = i$$

Any element of F is of the type

$\lambda(z_1, z_2) = (\lambda z_1, \lambda z_2)$, where λ is a complex number.

Thus any element of F is of the type (λ, λ_i) ,where $\lambda \in C$.

i.e. $F = \{(\lambda, \lambda_i) \, \lambda \in C\}$

Hence, $\sigma(F) = \{(\bar{\lambda}, -\bar{\lambda}_i) \, , \, \lambda \in C\}$

As for a non zero complex number we cannot find a complex number 'μ' such that,

$$(\bar{\lambda}, -\bar{\lambda}_i) = \mu(1,i) = (\mu, \mu_i)$$

It follows that $\sigma(F) \not\subset F$

Thus F cannot be decomplexified.

b) As a typical element of F is of type (λ, λ_i) where $\lambda \in C$,

$F_R = F \cap R^2$ will contain those elements of the type (λ, λ_i)

where λ is real, λ_i is also real.

So, this can happen only when $\lambda = 0$

$\therefore$ $F_R = \{(0,0)\}$

$\therefore$ $F_{RC} = \{(0,0)\}$

(2) Let $E \subseteq R^n$ and let $F \subseteq C^n$ be subspace what relations if any exist

between dim E and dim E_C? between dim F and dim E

$\dim E \le \dim E_C$, $\dim F \ge \dim F_R$

(3) If $F \subseteq C^n$ is any subspace what relation is there between F and F_{RC}

F_{RC} is subspace of F

$\therefore$ $\dim F_{RC} \le \dim F$

(4) Let E be a real vector space and $T \in L(E)$

Show that $(\ker T)_c = \ker(T_c)$

$(\operatorname{Im} T)_C = (\operatorname{Im} T_C)$

and $(T^{-1})_c = (T_c)^{-1}$ if T is invertible.

Theorem 3: If T is an operator on real vector space E, then the set of its

eigen values is preserved under complex conjugation. Thus if λ is an

eigen value so is $\bar{\lambda}$.

Consequently we may write the eigen values of T as $\lambda_1, \lambda_2 ... \lambda_r$ all real

and $\mu_1, \overline{\mu_1}, \mu_2, \overline{\mu_2} ... \mu_s, \overline{\mu_s}$ all non real.

Proof. If λ is a real eigen value then as $\bar{\lambda} = \lambda$ then the theorem obviously holds.

Then let λ be a non-real root of the characteristic polynomial of T. Thus we consider λ as an eigen value of the complexification operator T_C on E_C.

By definition, there exist a non zero element $\phi \in E_C$ such that

$$T_c \phi = \lambda I \phi = \lambda \phi$$

$\therefore$ If σ is the complex conjugation then we have,

$$\sigma(T_c \phi) = \sigma(\lambda \phi) = \overline{\lambda} \overline{\phi}$$

i.e. $\quad (\sigma T_C)\phi = \overline{\lambda} \, \overline{\phi}$

But the operator T_C on E_C is by definition, complexification of the real operator T. We must have

$$\sigma T_C = T_C \sigma \text{ on } E_C.$$

$\therefore \quad \overline{\lambda} \, \overline{\phi} = (\sigma T_C)\phi$

$$\qquad = T_c(\sigma \phi)$$

$$\qquad = T_c \overline{\phi}$$

i.e. $\quad T_C \overline{\phi} = \overline{\lambda} \overline{\phi}$

i.e., $\overline{\lambda}$ is a non real eigen value and $\overline{\phi}$ is corresponding eigen vector.

Theorem 4: Let $T : E \to E$ be a real operator with distinct eigen values $\lambda_1, \lambda_2 \ldots \lambda_r$ (all real) $\mu_1, \overline{\mu_1}, \mu_2, \overline{\mu_2} \ldots \mu_s, \overline{\mu_s}$ (all non real) then E and T have a direct sum decomposition.

$$E = E_a \oplus E_b, \qquad T = T_a \oplus T_b$$

$$T_a : E_a \to E_a$$

$$T_b : E_b \to E_b$$

where T_a has real eigen values and T_b has non real eigen values.

Proof: By the definition, $\lambda_1 , \lambda_2 \ldots \lambda_r , \mu_1 , \overline{\mu_1} , \mu_2 , \overline{\mu_2} \ldots \mu_s , \overline{\mu_s}$ are eigen values of the complexification operator T_C on the space E_C.

Let $e_1 , e_2 \ldots e_r , f_1 , \overline{f_1} \ldots f_s , \overline{f_s}$ be the eigen vectors belonging to eigen values $\lambda_1 \ldots \overline{\mu_s}$.

As all the eigen values are distinct, the set $\left\{ e_1 \ldots e_r , f_1 , \overline{f_1} \ldots \overline{f_s} \right\}$ is a basis of E_C as a complex vector space.

Let F_a be the subspace of E_C (complex subspace) spanned by $\{ e_1 , e_2 \ldots e_r \}$.

Similarly, let F_b be the subspace of E_c spanned by $\{ f_1 , \overline{f_1} \ldots f_s , \overline{f_s} \}$.
Then we have

$$E_C = F_a \oplus F_b .$$

Also F_a and F_b are invariant under the operator T_C i.e.,

$$T_C (F_a) \subseteq F_a , \qquad T_C (F_b) \subseteq F_b$$

Also, if 'σ' is the complex conjugation operator then,

$$\sigma (F_a) \subseteq F_a \text{ and } \sigma (F_b) \subseteq F_b$$

Hence, the spaces F_a and F_b can be decomplexification and

$$F_a = ((F_a)_R)_C$$

Similarly, $F_b = ((F_b)_R)_C$

Let, $E_a = F_a \cap E$, $E_b = F_b \cap E$

COMPLEX VECTOR SPACES

Obviously, E_a and E_b are real subspaces.

$$E = E_a \oplus E_b$$

Further, we then define T_a to be the restriction of T to t_a and t_b to the restriction of T to E_b.

Then it is clear that

$$E = E_a \oplus E_b, \quad T = T_a \oplus T_b$$

Further eigen values of T_a are $\lambda_1, \lambda_2 \dots \lambda_r$ and eigen values of T_b are $\mu_1, \overline{\mu_1} \dots \mu_s, \overline{\mu_s}$.

Theorem 5: Let $T : E \to E$ be an operator on a real vector space with distinct non real eigen values $\mu_1, \overline{\mu_1}, \mu_2, \overline{\mu_2} \dots \mu_s, \overline{\mu_s}$.

Then there is an invariant direct sum decomposition for E and corresponding direct sum decomposition for T.

$E = E_1 \oplus E_2 + \dots \oplus E_r$, $T = T_1 \oplus T_2 + \dots \oplus T_s$ such that, each E_i is two dimensional and $T_i \in L(E_i)$ has eigen values μ_i and $\overline{\mu_i}$.

Proof: We consider the operator T_C on vector space E_C. The operator T_C has eigen values $\mu_1, \overline{\mu_1} \dots \mu_s, \overline{\mu_s}$ all distinct and on real.

Let $f_1, \overline{f_1} \dots f_s, \overline{f_s}$ be the eigen vectors belonging to $\mu_1, \overline{\mu_1} \dots \mu_s, \overline{\mu_s}$ respectively.

Let F_i be the subspace of E_c generated by $f_i, \overline{f_i}$, $1 \le i \le s$.

It is clear that f_i is a two dimensional complex subspace of E_C .

Also, $T_C \left(F_i \right) \subseteq F_i$, $\sigma \left(F_i \right) \subseteq F_i$

Let $E_i = F_i \cap E$

DYNAMICAL SYSTEM – A SHORT COURSE

It is clear that, E_i is a real subspace of E and

$$E = \oplus \sum_{i=1}^{s} E_i$$

We let T_i to be the restriction of T to E.

(As F_i is invariant under T_c , E_i is invariant under T), we then get, the direct sum decomposition

$$E = \oplus \sum_{i=1}^{s} E_i , \quad T = \oplus \sum_{i=1}^{s} T_i$$

where $T_i : E_i \rightarrow E_i$ with $\mu_1 , \overline{\mu_1}$ as eigen values of T_i.

Theorem 6: Let T be an operator on a two-dimensional vector space $E \subseteq R^n$ with non real eigen values $\mu_1 , \overline{\mu_1}$, $\mu = a + ib$ then there is a matrix representation A for T with

$$A = \begin{bmatrix} a & -b \\ b & a \end{bmatrix}$$

Proof: Considering the complexification T_C and E_C of T and E respectively.

We know that μ and $\overline{\mu}$ are the eigen values of T_C.

Hence by definition, we can find an eigen vector say $\phi \in E_C$ such that

$T_C \phi = \mu \phi$ (and $T_C \overline{\phi} = \overline{\mu} \overline{\phi}$)

As $\phi \in E_C \subseteq C^n$ $\left(\because E \subseteq R^n \right)$

We can write $\phi = (z_1 , z_2 ... z_n)$ with $z_i \in C$, $1 \leq i \leq n$

Let $z_j = x_j + i y_j$, $1 \leq j \leq n$

Let, $\mu = (x_1 , x_2 ... x_n), \quad v = (y_1 , y_2 ... y_n)$

78

COMPLEX VECTOR SPACES

If $\phi, \bar{\phi} \in E_C$, $\quad \dfrac{\phi + \bar{\phi}}{2} \in E_C \quad$ i.e. $U \in E_C$

and $\quad \dfrac{\phi - \bar{\phi}}{2i} \in E_C \quad$ i.e. $v \in E_C \quad$ ($\left\{\begin{array}{l}\phi = x_1 + i\, y_1 ... \\ \bar{\phi} = x_1 - i\, y_1 ...\end{array}\right\} \Rightarrow \dfrac{2x_1...}{2} = x_1 \in U$)

But $\quad \mu \in R^n \quad , \quad v \in R^n$

So, $\quad \mu \in E_C \cap R^n \quad , \quad U \in E_{CR} = E$

Similarly, $v \in E_C \cap R^n \quad , \quad v \in E_{CR} = E$

We next show that $\{U, v\}$ forms a basis of E

Let $\quad \alpha \mu + \beta v = 0 \quad$ for real α and β.

i.e. $\quad \alpha \left(\dfrac{\phi + \bar{\phi}}{2} \right) + \beta \left(\dfrac{\phi - \bar{\phi}}{2i} \right) = 0$

$\therefore \quad \left(\dfrac{\alpha}{2} + \dfrac{\beta}{2i} \right) \phi + \left(\dfrac{\alpha}{2} - \dfrac{\beta}{2i} \right) \bar{\phi} = 0$

But ϕ and $\bar{\phi}$ being eigen vector corresponding to distinct eigen values $\mu_1, \overline{\mu_1}$ are linearly independent so we must have

$\dfrac{\alpha}{2} + \dfrac{\beta}{2i} = 0$

Similarly $\quad \dfrac{\alpha}{2} - \dfrac{\beta}{2i} = 0$

i.e. $\quad \alpha - \beta i = 0 \quad , \quad \alpha + \beta i = 0$

$\alpha = \beta = 0$

As $\dim E = 2$ and $u, v \in E$ and $\{u, v\}$ is linearly independent , then the set $\{u, v\}$ is the basis of E.

We show that with respect to the ordered basis $\{v,u\}$ the operator T has the required matrix.

Since the operator T_C has eigen values $a \pm ib$

We known that, $T_C \phi$ is also given by the complex multiplication

$$(a+ib)(u+iv) \quad = au - bv + i(bu + av)$$

$$T_c(u+iv) = (a+ib)(u+iv) \qquad \{\because a+ib \text{ is an eigen value and}$$

$$u+iv \text{ is an eigen vector.}\}$$

Thus, $\quad T_C(u+iv) = (au - bv) + i(bu + av)$ \hfill (2.1)

But by the definition of T_C, we have

$$T_C(u+iv) = T(u) + i\, T(v) \hspace{3cm} (2.2)$$

Then (2.1) and (2.2) $\Rightarrow T(u) + i(Tv) = (au - bv) + i(bu + av)$

$$\therefore T(u) = au - bv$$

$$T(v) = av + bu$$

It is clear that the matrix of T with respect to ordered basis

$\{v,u\}$ is $\begin{bmatrix} a & -b \\ b & a \end{bmatrix}$.

Theorem 7: Let $\phi \in E$ be an eigen vector of T belonging to $a+ib$, $b \neq 0$ if $\phi = u + iv \in C^n$, then $\{v,u\}$ is an ordered basis for E giving ϕ the matrix $\begin{bmatrix} a & -b \\ b & a \end{bmatrix}$.

Proof.

1) For an operator T on R^3 find an invariant two dimension subspace and a basis of it giving restriction of T to it a matrix of the form $\begin{bmatrix} a & -b \\ b & a \end{bmatrix}$.

COMPLEX VECTOR SPACES

$$T = \begin{bmatrix} 1 & 0 & 1 \\ 0 & 0 & -2 \\ 0 & 1 & 0 \end{bmatrix}$$

We first find the eigen values of the operator given by

$$\begin{bmatrix} 1 & 0 & 1 \\ 0 & 0 & -2 \\ 0 & 1 & 0 \end{bmatrix}.$$

They are given by

$$\begin{vmatrix} 1-\lambda & 0 & 1 \\ 0 & -\lambda & -2 \\ 0 & 1 & -\lambda \end{vmatrix} = 0$$

$$\left(\lambda^2 + 2\right)\left(1 - \lambda\right) = 0$$

Thus the eigen values are 1 and $\pm\sqrt{2}i$.

We find the eigen space corresponding to real eigen values 1.

The eigen space belonging to the real eigen value will be a subspace of R^3 and is given by (α, β, γ) where

$$\left(a_{11} - 1\right)\alpha + a_{12}\beta + a_{13}\gamma = 0$$

$$a_{21}\alpha + \left(a_{22} - 1\right)\beta + a_{23}\gamma = 0$$

$$a_{32}\alpha + a_{23}\beta + \left(a_{33} - 1\right)\gamma = 0$$

i.e. $\gamma = 0$

$$-\beta - 2\gamma = 0$$

$$\beta - \gamma = 0$$

i.e. $\beta = \gamma = 0$

Thus the eigen space belonging to real eigen value 1 is given by $\{(t, 0, 0),\ t \in R\}$

81

Thus we can take $(1,0,0)$ R^3 as an eigen vector belonging to real eigen value 1.

Now to find eigen vector corresponding to non real eigen value $+\sqrt{2}\,i$; we have to consider E_C with $E = R^3$. Thus eigen vector belonging to eigen value $\sqrt{2}\,i$ will be of the type (z_1, z_2, z_3) such that

$$\left(a_{11} - \sqrt{2}\,i\right)z_1 + a_{12}z_2 + a_{13}z_3 = 0$$

$$a_{21}z_1 + \left(a_{22} - \sqrt{2}\,i\right)z_2 + a_{23}z_3 = 0$$

$$a_{31}z_1 + a_{32}z_2 + \left(a_{33} - \sqrt{2}\,i\right)z_3 = 0$$

i.e. $\left(1 - \sqrt{2}\,i\right)z_1 + z_3 = 0$

$$-\sqrt{2}\,iz_2 - 2z = 0$$

$$z_2 - \sqrt{2}\,iz = 0$$

i.e. $z_1 = \dfrac{-z_3}{\left(1 - \sqrt{2}\,i\right)} = \dfrac{-z_3}{-i^2 - \sqrt{2}\,i} = \dfrac{z_3}{i\left(\sqrt{2} + i\right)} = \dfrac{-iz_3}{\sqrt{2} + i}$

$\therefore\quad z_2 = \dfrac{2z_3}{-\sqrt{2}\,i} = i\sqrt{2}\,z_3$

Thus an element of the type

$$\left(\frac{-iz_3}{\sqrt{2} + i}, \ i\sqrt{2}\,z_3, \ z_3\right)$$

where $z_3 \in C$ is an eigen vector belong to non real eigen value $\sqrt{2}\,i$.

We take $z_3 = 1$ then eigen vector belonging to eigen value $\sqrt{2}\,i$ is

$$\left(\frac{-i}{\sqrt{2} + i}, \ i\sqrt{2}, \ 1\right).$$

COMPLEX VECTOR SPACES

As $\quad \dfrac{-i}{\sqrt{2}+i} = \dfrac{-i}{\sqrt{2}+i}\left(\dfrac{\sqrt{2}-i}{\sqrt{2}-i}\right) = \dfrac{-1-\sqrt{2}\,i}{3}$

the eigen vector belonging to eigen value $+\sqrt{2}\,i$ is

$\phi = \left(\dfrac{-1}{3}\;\dfrac{-\sqrt{2}\,i}{3},\, 0+\sqrt{2}\,i,\, 1+0i\right)$

Then we let $\; u = \left(\dfrac{-1}{3},\,0,\,1\right), \quad v = \left(\dfrac{-\sqrt{2}}{3},\,\sqrt{2},\,0\right)$

Thus the required invariant two dimensional subspace of R^3 would be generated space by the two elements

$\left(\dfrac{-1}{3},\,0,\,1\right)$ and $\left(\dfrac{-\sqrt{2}}{3},\,\sqrt{2},\,0\right)$

We find $\qquad T\begin{bmatrix} -\dfrac{1}{3} \\ 0 \\ 1 \end{bmatrix}$ and $T\begin{bmatrix} -\dfrac{\sqrt{2}}{3} \\ \sqrt{2} \\ 0 \end{bmatrix}$

$T(u) \Rightarrow \begin{bmatrix} 1 & 0 & 1 \\ 0 & 0 & -2 \\ 0 & 1 & 0 \end{bmatrix}\begin{bmatrix} -\frac{1}{3} \\ 0 \\ 1 \end{bmatrix} = \begin{bmatrix} 0 \\ 0 \\ 0 \end{bmatrix}$

$\Rightarrow \begin{bmatrix} -\frac{1}{3} & 0 & 1 \\ 0 & 0 & -2 \\ 0 & 1 & 0 \end{bmatrix} = \begin{bmatrix} 0 \\ 0 \\ 0 \end{bmatrix} \Rightarrow \begin{bmatrix} +\frac{2}{3} \\ -2 \\ 0 \end{bmatrix}$

$T(v) \Rightarrow \begin{bmatrix} 1 & 0 & 1 \\ 0 & 0 & -2 \\ 0 & 1 & 0 \end{bmatrix}\begin{bmatrix} -\frac{2}{3} \\ \sqrt{2} \\ 0 \end{bmatrix} = \begin{bmatrix} -2\sqrt{3} \\ 0 \\ \sqrt{2} \end{bmatrix}$

$T(v) = \alpha u + \beta v = \alpha\left(-\frac{1}{3},0,1\right) + \beta\left(-\frac{2}{3},\sqrt{2},0\right)$

DYNAMICAL SYSTEM – A SHORT COURSE

$$= \sqrt{2}\,u + 0v = \left(\tfrac{-\sqrt{2}}{3}, 0, \sqrt{2}\right)$$

$$T(u) = \gamma u + \delta v = \gamma\left(-\tfrac{1}{3}, 0, 1\right) + \delta\left(-\tfrac{\sqrt{2}}{3}, \sqrt{2}, 0\right)$$

The matrix T with respect to basis $\{v, u\}$ is $\begin{bmatrix} 0 & -\sqrt{2} \\ \sqrt{2} & 0 \end{bmatrix}$

Ex 5: $T = \begin{bmatrix} 0 & 0 & +15 \\ 1 & 0 & -17 \\ 0 & 1 & 7 \end{bmatrix}$

$\begin{bmatrix} -\lambda & 0 & 15 \\ 1 & -\lambda & -17 \\ 0 & 1 & 7 \end{bmatrix} = 0$

$$((7-\lambda)\,(-\lambda)+17)(-\lambda)+15 = 0$$

$$(-7\lambda + \lambda^2 + 17)(-\lambda) + 15 = 0$$

$$7\lambda^2 - \lambda^3 - 17\lambda + 15 = 0$$

$$-\lambda^3 + 7\lambda^2 - 17\lambda + 15 = 0$$

$$\lambda^3 - 7\lambda^2 + 17\lambda - 15 = 0$$

$$(\lambda - 3)(\lambda^2 - 4\lambda + 5) = 0$$

$$\lambda = 3, \quad 2 \pm i$$

We consider the real eigen value 3.

So the eigen space belonging to eigen value 3 will be $\{(\alpha, \beta, \gamma) \in R^3\} - 3$ such that

$$\begin{bmatrix} a_{11} - 3 & a_{12} & a_{13} \\ a_{21} & a_{22} - 3 & a_{23} \\ a_{31} & a_{32} & a_{33} - 3 \end{bmatrix} \begin{bmatrix} \alpha \\ \beta \\ \gamma \end{bmatrix} = \begin{bmatrix} 0 \\ 0 \\ 0 \end{bmatrix}$$

(i) $\quad -3\alpha + 15\gamma = 0$

84

(ii) $\quad \alpha - 3\beta - 17\gamma = 0 \quad\Rightarrow\quad \alpha = 5\gamma$

(iii) $\quad \beta + 4\gamma = 0 \quad\Rightarrow\quad \beta = -4\gamma$

(ii) $\quad\Rightarrow\quad 5\gamma + 12\gamma - 17\gamma = 0$

i.e. γ can be chosen arbitrarily.

So setting $\quad \gamma = t, \quad t \in R$

The eigen space belonging to eigen value 3 is given by

$\{5t, -4t, t\} \quad t \in R$

We now consider then non real eigen values $2 + i$ to find eigen vector belonging to eigen value $2 + i$, we note that it will be subspace of C^3.

Thus (z_1, z_2, z_3) will belong to eigen value $2 + i$ provided

$$\begin{bmatrix} a_{11} - 2 + i & a_{12} & a_{13} \\ a_{21} & a_{22} - 2 + i & a_{23} \\ a_{31} & a_{32} & a_{33} - 2 + i \end{bmatrix} \begin{bmatrix} z_1 \\ z_2 \\ z_3 \end{bmatrix} = \begin{bmatrix} 0 \\ 0 \\ 0 \end{bmatrix}$$

$(-2 + i)z_1 + 15z_3 = 0 \quad\Rightarrow$

$z_1 - (2 + i)z_2 - 17z_3 = 0 \quad\Rightarrow$

$z_2 + (7 - (2 + i))z_3 = 0 \quad\Rightarrow$

$z_1 = \dfrac{15z_3}{(2 + i)}, \quad z_2 = -(7 - 2 - i)z_3 = -(5 - i)z_3$

$\left(\dfrac{15z_3}{(2 + i)}, -(5 + i)z_3, z_3 \right)$

$\left(\dfrac{15}{(2 + i)}, (-5 + i), 1 \right)$

$\dfrac{15}{2 + i} \times \dfrac{(2 - i)}{(2 - i)} = \dfrac{15(2 - i)}{5} = 3(2 - i) = 6 - 3i$

$\phi = ((6 - 3i), (-5 + i), (1 + 0i))$

$$u = (6, -5, 1) \qquad v = (-3, 1, 0)$$

$$T(u) = \begin{bmatrix} 0 & 0 & 15 \\ 1 & 0 & -17 \\ 0 & 1 & 7 \end{bmatrix} \begin{bmatrix} 6 \\ -5 \\ 1 \end{bmatrix} = \begin{bmatrix} 15 \\ -11 \\ 2 \end{bmatrix}$$

$$T(v) = \begin{bmatrix} 0 & 0 & 15 \\ 1 & 0 & -17 \\ 0 & 1 & 7 \end{bmatrix} \begin{bmatrix} -3 \\ 1 \\ 0 \end{bmatrix} = \begin{bmatrix} 0 \\ -3 \\ 1 \end{bmatrix}$$

$$T(u) = \alpha u + \beta v$$

$$(15, -11, 2) = \alpha(6, -5, 1) + \beta(-3, 1, 0)$$
$$= (6\alpha - 3\beta, \ -5\alpha + \beta, \ \alpha)$$

$$\Rightarrow \quad 6\alpha - 3\beta = 15$$

$$-5\alpha + \beta = -11 \qquad \Rightarrow \quad \beta = -1$$

$$\alpha = 2$$

Thus $T(u) = 2u - v$

$$T(v) = \gamma u + \delta v$$

$$(0, -3, 1) = \gamma(6, -5, 1) + \delta(-3, 1, 0)$$
$$= (6\gamma - 3\delta, \ -5\gamma + \delta, \ \gamma)$$

$$\Rightarrow \quad 6\gamma - 3\delta = 0$$

$$-5\gamma + \delta = -3 \qquad \Rightarrow \quad \delta = 2$$

$$\gamma = 1$$

$$T(v) = u + 2v$$

$\therefore$ The matrix will be $\begin{bmatrix} a & -b \\ b & a \end{bmatrix} = \begin{bmatrix} 2 & -1 \\ 1 & 2 \end{bmatrix}$.

Consider the D.E.

COMPLEX VECTOR SPACES

$$x_1' = -2x_2$$
$$x_2' = x_1 + 2x_2$$

Here we assume that x_1, x_2 are coordinates in the standard basis system. Hence the matrix representation will be and the operator is given by matrix in standard basis system is $\begin{bmatrix} 0 & -2 \\ 1 & 2 \end{bmatrix}$

$\therefore$ The eigen values corresponding to this matrix are given by the solution of the equation

$$\begin{vmatrix} -\lambda & -2 \\ 1 & 2-\lambda \end{vmatrix} = 0$$

$$-\lambda(2-\lambda) + 2 = 0$$

$$-2\lambda + \lambda^2 + 2 = 0$$

$$\lambda = \frac{2 \pm \sqrt{4-8}}{2} = 1 \pm i$$

We next find the eigen space belonging to the eigen value $1+i$, it is given by $\{z_1, z_2\}$ in C such that

$$\begin{bmatrix} a_{11}-1+i & a_{12} \\ a_{21} & a_{22}-1+i \end{bmatrix} \begin{bmatrix} z_1 \\ z_2 \end{bmatrix} = \begin{bmatrix} 0 \\ 0 \end{bmatrix}$$

$$\Rightarrow \quad (-1+i)z_1 - 2z_2 = 0$$

$$\Rightarrow z_1 + (2-(1+i))z_2 = 0$$

$$\Rightarrow z_1 + (1-i)z_2 = 0$$

$$\Rightarrow \quad z_1 = \frac{-2z_2}{(1+i)} = \left(\frac{-2}{1+i}\right)z_2$$

Thus the eigen space belonging to the non-real eigen value $1+i$ is the subspace of C^2 given by

$$\left\{ \left(\frac{-2}{1+i} z_2 \, , \, z_2 \right) \quad z_2 \in C \right\}$$

We choose $z_2 = 1$. Thus an eigen vector belonging to eigen value $1+i$ is

$$\left(\frac{-2}{1+i}, 1 \right)$$

$$\Rightarrow \quad \frac{-2}{1+i} = \frac{-2}{1+i} \times \frac{(1-i)}{(1-i)} = \frac{-2(1-i)}{2} = -1+i$$

Thus we consider an eigen vector $(-1+i \, , \, 1)$ belonging eigen value $1+i$.

$\therefore$ Hence we have, $u = (-1,1)$ and $v = (1,0)$

$$T(u) = \alpha u + \beta v$$

$$(-2,1) = \alpha(-1,1) + \beta(1,0)$$

$\therefore \quad -\alpha + \beta = -2 \quad$ and $\quad \alpha = 1$

$\therefore \quad \beta = -1$

$$T(v) = \begin{bmatrix} 0 & -2 \\ 1 & 2 \end{bmatrix} \begin{bmatrix} 1 \\ 0 \end{bmatrix} = \begin{bmatrix} 0 \\ 1 \end{bmatrix}$$

$$T(v) = \gamma u + \delta v$$

$$(0,1) = \gamma(-1,1) + \beta(1,0)$$

$\therefore \quad -\gamma + \beta = 0 \quad \Rightarrow \quad \beta = 1$

It is clear that, $T(v) = v + u \quad$ and $\quad T(u) = -v + u$

$\therefore$ Thus the matrix of T with respect to ordered basis $\{v, u\}$ is $\begin{bmatrix} 1 & -1 \\ 1 & 1 \end{bmatrix}$.

COMPLEX VECTOR SPACES

Thus the system of D.E. in the basis system $(1,0)\,(0,1)$ given by,

$$x_1' = -2x_2$$
$$x_2' = x_1 + 2x_2$$

will get transform to a new system

$$y' = B \cdot y \quad \text{where } B \text{ is}$$

$$B = \begin{bmatrix} 1 & -1 \\ 1 & 1 \end{bmatrix}$$

To get $(y_1\,,\,y_2)$ in terms of $(x_1\,,\,x_2)$ we note if a point P is given by (x_1,x_2) in the standard basis then it is given by $(y_1\,,\,y_2)$ in the basis (v,u).

Thus $x_1(1,0) + x_2(0,1) = y_1\,v + y_2\,u$

i.e. $(x_1,x_2) = y_1(1,0) + y_2(-1,1)$

$$x_1 = y_1 - y_2$$
$$x_2 = y_2$$

We know that,

$$y' = \begin{bmatrix} 1 & -1 \\ 1 & 1 \end{bmatrix} y$$

$$\therefore \ y_1' = y_1 - y_2$$
$$y_2' = y_1 + y_2$$

We know that if a point of R^2 i.e. $(y_1\,,\,y_2)$ expressed as a complex number $y_1 + i y_2$ then

$\begin{bmatrix} 1 & -1 \\ 1 & 1 \end{bmatrix} \begin{bmatrix} y_1 \\ y_2 \end{bmatrix}$ is given by complex multiplication $(1+i)\,(y_1 + i\,y_2)$.

Thus if we write $y_1 + i\,y_2$ as a complex number z then

$$\begin{bmatrix} y_1' \\ y_2' \end{bmatrix} = \begin{bmatrix} 1 & -1 \\ 1 & 1 \end{bmatrix} \begin{bmatrix} y_1 \\ y_2 \end{bmatrix}$$

becomes $\quad z' = (1+i)z$

$z(t) = z_0 \exp\,(1+i)t$

Writing $z_0 = u + i\,v$ for some real u and v we get

$$z(t) = (u+iv)\,e^t\,(\cos t + i\sin t)$$

i.e., $\quad (y_1(t) + i\,y_2(t)) = e^t[(u\cos t - v\sin t) + i(u\sin t + v\cos t)]$

$\therefore \;\; y_1(t) = e^t(u\cos t - v\sin t)$

$\quad\;\; y_2(t) = e^t(u\sin t - v\cos t)$

$x_1(t) = e^t\,(u\cos t - v\sin t) - e^t(u\sin t + v\cos t)$

$\qquad = e^t\,((u-v)\cos t - (u+v)\sin t)$

$x_2(t) = e^t\,(u\sin t + v\cos t)$

Ex 6: Consider on R^3 the D.E.

$$\frac{dx}{dt} = Ax\,, \quad A = \begin{bmatrix} 1 & 0 & 0 \\ 0 & 2 & -3 \\ 1 & 3 & 2 \end{bmatrix}$$

Solution. The given matrix $\quad A = \begin{bmatrix} 1 & 0 & 0 \\ 0 & 2 & -3 \\ 1 & 3 & 2 \end{bmatrix}$

The eigen values are given by polynomial equation

$$\begin{vmatrix} 1-\lambda & 0 & 0 \\ 0 & 2-\lambda & -3 \\ 1 & 3 & 2-\lambda \end{vmatrix} \Rightarrow$$

i.e. $\left((2-\lambda)^2+a\right)(1-\lambda)=0$ $\lambda=1$

i.e. $\lambda=1$, $(2-\lambda)^2+9=0$

i.e., $\lambda=1$, $\lambda^2-4\lambda+4+9=0$

$\lambda^2-4\lambda+13=0$

$$\frac{4\pm\sqrt{16-52}}{2}=\frac{4\pm6i}{2}=2\pm3i$$

Thus the eigen values are $1,\ 2\pm3i$

So now we find eigen space belonging to eigen value 1.

$\therefore (a_{11}-1)\alpha+a_{12}\beta+a_{13}\gamma=0$

$a_{21}\alpha+(a_{22}-1)\beta+a_{23}\gamma=0$

$a_{31}\alpha+a_{32}\beta+(a_{33}-1)\gamma=0$

$\Rightarrow \qquad \alpha=10\gamma \Rightarrow \alpha=-10\gamma$

$\beta-3\gamma=0 \Rightarrow \beta=3\gamma$

$\alpha+3\beta+\gamma=0 \Rightarrow \gamma=0$

Thus the eigen space belonging to eigen value 1 is given by $\{-10\gamma,3\gamma,\gamma \qquad \gamma\in R\}$

Choosing $\gamma=1$ we get eigen vector, $(-10,3,1)$ belonging to eigen value 1.

We denote it by e_1 .

We next find an eigen space (subspace of C^s) belonging to $2+3i$.

It is given by set $\left\{\{z_1, z_2, z_3\}, \; z_1, \; z_2, \; z_3 \in C\right\}$

such that

$$\left(a_{11}-(2+3i)\right)z_1 + a_{12}z_2 + a_{13}z_3 = 0$$

$$a_{21}z_1 + \left(a_{22}-(2+3i)\right)z_2 + a_{23}z_3 = 0$$

$$a_{31}z_1 + a_{32}z_2 + \left(a_{33}-(2+3i)\right)z_3 = 0$$

$$(-1+3i)z_1 = 0 \quad \Rightarrow z_1 = 0$$

$$-3iz_2 - 3z_3 = 0$$

$$3z_2 - 3iz_3 = 0$$

i.e. $z_2 = iz_3$

Thus the eigen space belonging to eigen value $2+3i$ is $\left\{0, iz_3, z_3, z_3 \in C\right\}$

Choose $z_3 = 1$, we get eigen vector belonging to $e.v.$ $2+3i$ as $\{0, i, 1\}$.

Writing ϕ as $u+iv$, we get

$$u = (0,0,1)$$

$$v = (0,1,0)$$

Further we note that, we can write space R^s

i.e. E as $E = E_a \oplus E_b$

where E_a will be generated by $(-10,3,1)$ and E_b will be generated by $(0,1,0)$ and $(0,0,1)$.

We can also write the operator T as $T = T_a \oplus T_b$.

where T_a is the restriction of T to the subspace $(1\,\text{dim})$ generated by $(-10,3,1)$ and T_b is the restriction of T to the subspace $(2\,\text{dim})$ generated by $\{u,v\}$.

Matrix of T_a with respect to basis $\{(-10,3,1)\}$ is $[1]$

Further, the matrix of T_b with respect to basis $\{v,u\}$ is $\begin{bmatrix} 2 & -3 \\ 3 & 2 \end{bmatrix}$ as the eigen value is $2+3i$.

Thus if we consider the basis system (e_1,v,u) the matrix of the operator T will be given by

$$B = \begin{bmatrix} 1 & 0 & 0 \\ 0 & 2 & -3 \\ 0 & 3 & 2 \end{bmatrix}$$

If y_1,y_2,y_3 is the coordinate system in this basis then the corresponding system is

$$\begin{bmatrix} y_1' \\ y_2' \\ y_3' \end{bmatrix} = \begin{bmatrix} 1 & 0 & 0 \\ 0 & 2 & -3 \\ 0 & 3 & 2 \end{bmatrix} \begin{bmatrix} y_1 \\ y_2 \\ y_3 \end{bmatrix}$$

where since, (y_1,y_2,y_3) are the coordinates of a point in basis system (e_1,v,u) we have

$y_1 e_1 + y_2 v + y_3 u = x_1(1,0,0) + x_2(0,1,0) + x_3(0,0,1)$

$\therefore\ y_1(-10,3,1) + y_2(0,1,0) + y_3(0,0,1)$

$\qquad\qquad = x_1(1,0,0) + x_2(0,1,0) + x_{3_1}(1,0,0)$

$\therefore \quad x_1 = -10\, y_1$

$x_2 = 3\, y_1 + y_2$

$x_3 = y_1 + y_2$

$x_3 = y_1 + y_3$

Thus after solving the system

$\quad y' = B\ y$

We can get $x_1(t),\ x_2(t), x_3(t)$ by using the above relations:

$y_1^{'} = B\ y \quad$ gives us

$y_1^{'} = y_1$

$y_2^{'} = 2\, y_2 - 3\, y_3$

$y_3^{'} = 3\, y_2 + 2\, y_3$

i.e. $\quad y_1(t) = a\,e^t$ for some constant ' a '.

To solve the equation

$y_2^{'} = 2\, y_2 - 3\, y_3$

$y_3^{'} = 3\, y_2 + 2\, y_3$

We use the fact that if we represent the point $\begin{bmatrix} y_2 \\ y_3 \end{bmatrix}$, yes by the complex

number $\left(y_2 + i\, y_3 \right)$

Then the operator $\begin{bmatrix} 2 & -3 \\ 3 & 2 \end{bmatrix}$ is given by complex multiplication by complex

number $2 + 3i$.

These two equation are equivalent to

COMPLEX VECTOR SPACES

$$z' = (2+3i)z \qquad \text{where} \quad z = y_2 + i\, y_3$$

i.e. $\quad z(t) = z_0 \exp(2+3i)t$

For some constant complex number z_0.

We write $z_0 = u + i v$. Thus we get

$$y_2(t) + i\, y_3(t) = (u + iv)e^{2t}(\cos 3t + i\sin 3t)$$

$$y_2(t) + i\, y_3 = e^{2t}(u\cos 3t + i\sin 3t)$$

$$y_3(t) = e^{2t}(u\sin 3t + v\cos 3t)$$

Ex 7. Solve the system,

$$\frac{dx}{dt} = A x \ , \ \text{where}$$

$$(1) \qquad A = \begin{bmatrix} 1 & 0 & 1 \\ 0 & 0 & -2 \\ 0 & 1 & 0 \end{bmatrix} \quad \text{and} \quad (2) \qquad A = \begin{bmatrix} 0 & 0 & 15 \\ 1 & 0 & -17 \\ 0 & 1 & 7 \end{bmatrix}$$

Solution 1. The given matrix is

$$A = \begin{bmatrix} 1 & 0 & 1 \\ 0 & 0 & -2 \\ 0 & 1 & 0 \end{bmatrix}$$

The eigen values are given by characteristic polynomial

$$\begin{bmatrix} 1-\lambda & 0 & 1 \\ 0 & -\lambda & -2 \\ 0 & 1 & -\lambda \end{bmatrix} \ \Rightarrow \ (1-\lambda)\lambda^2 + 2 = 0$$

$$\Rightarrow (1-\lambda)(\lambda^2 - 2) = 0$$

$$\Rightarrow \lambda = 1 \ , \ \lambda = \pm\sqrt{2i}$$

Now we find the eigen space belonging to eigen value 1.

The eigen space belonging to real eigen value 1 will be the subspace of R^3 and is given by (α, β, γ)

where

$$(a_{11} - 1)\alpha \quad a_{12}\beta \quad a_{13}\gamma = 0$$
$$a_{21}\alpha \quad (a_{22} - 1)\beta \quad a_{23}\gamma = 0$$
$$a_{31}\alpha \quad a_{32}\beta \quad (a_{33} - 1)\gamma = 0$$

$$\therefore \quad 0\alpha + 0\beta + \gamma = 0 \implies \gamma = 0$$

$$0\alpha - \beta - 2\gamma = 0 \implies$$

$$0\alpha + \beta - \gamma = 0 \implies \beta = \gamma = 0$$

Thus the eigen space belonging to eigen value 1 is given by $\{(t, 0, 0), \ t \in R\}$.

Thus we can take $(1, 0, 0) \in R^3$ as an eigen vector belonging to eigen value 1.

We denote it by e_1.

We next find the eigen space belonging to non real eigen value $+\sqrt{2}\,i$.

$\therefore$ The eigen vector is (z_1, z_2, z_3)

i.e. $(a_{11} - \sqrt{2}\,i)z_1 + a_{12}z_2 + a_{13}z_3 = 0$

$$a_{21}z_1 + (a_{22} - \sqrt{2}\,i)z_2 + a_{23}z_3 = 0$$

$$a_{31}z_1 + a_{32}z_2 + (a_{33} - \sqrt{2}\,i)z_3 = 0$$

$$(1 - \sqrt{2}\,i)z_1 + 0z_2 + z_3 = 0$$

$$0z_1 - \sqrt{2}\,i z_2 - 2z_3 = 0$$

$$0z_1 + z_2 - \sqrt{2}\,i z_3 = 0$$

$$\Rightarrow \quad z_1 = -\frac{z_3}{1-\sqrt{2}\,i} = -\frac{z_3}{\left(-i^2-\sqrt{2}\,i\right)} = \frac{z_3}{i\left(i+\sqrt{2}\,i\right)} = \frac{-i\,z_3}{\sqrt{2}+i}$$

$$z_2 = \frac{+2\,z_3}{-\sqrt{2}\,i} = i\sqrt{2}\,z_3$$

Thus an element of the type $\left(\dfrac{-i\,z_3}{\sqrt{2}+i},\ i\sqrt{2}\,z_3,\ z_3\right)$

where $z_3 \in C$ is an eigen vector belong to non real eigen value $\sqrt{2}\,i$.

We choose $z_3 = 1$ arbitrarily belonging to eigen value $\sqrt{2}\,i$ i.e.

$$\frac{-i}{\sqrt{2}+i},\ i\sqrt{2},\ 1$$

$$\Rightarrow \quad \frac{-i}{\sqrt{2}+i} = \frac{-i}{\sqrt{2}+i} = \frac{\sqrt{2}-i}{\sqrt{2}+i} = \frac{-i\left(\sqrt{2}-i\right)}{3} = \frac{-1-\sqrt{2}\,i}{3}$$

Thus the eigen value belonging eigen vector $\sqrt{2}\,i$ is

$$\phi = \left(\frac{-1}{3}\ \frac{-\sqrt{2}\,i}{3},\ i\sqrt{2},\ 1\right)$$

writing ϕ as $u+i$ we get

$$u = \left(-\tfrac{1}{3}, 0, 1\right), \quad v = \left(-\tfrac{\sqrt{2}}{3},\ \sqrt{2},\ 0\right)$$

Further we note that, we can write space R^3

i.e. E as $E = E_a \oplus E_b$

Where E_a will be generated by $(1,0,0)$

and E_b will be generated by $\left(-\tfrac{1}{3},\ 0,\ 1\right)$ and $\left(-\tfrac{\sqrt{2}}{3},\ \sqrt{2},\ 0\right)$.

Then we can also write the operator T as

$$T = T_a \oplus T_b$$

where T_a is the restriction of T to the Subspace $(1\,\text{dim})$ generated by $(1,0,0)$ and T_b be the restriction of T to the subspace $(2\,\text{dim})$ generated by $\{v,u\}$.

$\therefore$ Matrix of T_a with respect to basis $(1,0,0)$ is $[1]$.

Further the matrix of T_b with respect to basis $\{v,u\}$ is $\begin{bmatrix} 0 & \sqrt{2} \\ -\sqrt{2} & 0 \end{bmatrix}$ as the eigen value is $\sqrt{2}\,i$.

Thus if we consider the basis system $\{e,v,u\}$ the matrix of operator T will be given by,

$$B = \begin{vmatrix} 1 & 0 & 0 \\ 0 & 0 & -\sqrt{2} \\ 0 & \sqrt{2} & 0 \end{vmatrix}$$

If y_1, y_2, y_3 is the coordinate system in the basis then the corresponding system is

$$\begin{bmatrix} y_1' \\ y_2' \\ y_3' \end{bmatrix} = \begin{bmatrix} 1 & 0 & 0 \\ 0 & 0 & -\sqrt{2} \\ 0 & \sqrt{2} & 0 \end{bmatrix} \begin{bmatrix} y_1 \\ y_2 \\ y_3 \end{bmatrix}$$

Since (y_1, y_2, y_3) are the coordinates of a point in basis system (e_1, v, u) we have

$$y_1 e_1 + y_2 v + y_3 u = x_1(1,0,0) + x_2(0,1,0) + x_3(0,0,1)$$

$$y_1(1,0,0) + y_2\left(-\tfrac{\sqrt{2}}{3}, \sqrt{2}, 0\right) + y_3\left(-\tfrac{1}{3}, 0, 1\right) =$$

$$x_1(1,0,0) + x_2(0,1,0) + x_3(0,0,1)$$

$$y_1 - \tfrac{\sqrt{2}}{3} y_2 - \tfrac{y_3}{3} = x_1$$

$$\sqrt{2}\, y_2 = x_2$$

$$y_3 = x_3$$

Thus after solving the system $y' = B\, y$

We can get $x_1(t)$, $x_2(t)$, $x_3(t)$ by using the above relations.

$y' = B\, y$ we have

$$y_1' = y_1$$

$$y_2' = -\sqrt{2}\, y_3$$

$$y_3' = \sqrt{2}\, y_2$$

i.e. $y_1 = a\, e^{t}$ for some constant 'a'.

$$y_2 = a\, e^{-\sqrt{2}\, t}$$

$$y_3 = a\, e^{\sqrt{2}\, t}$$

2) The given matrix is

$$A = \begin{bmatrix} 0 & 0 & 15 \\ 1 & 0 & -17 \\ 0 & 1 & 7 \end{bmatrix}$$

$\therefore$ The eigen value given by characteristic polynomial is given as

$$\begin{vmatrix} -\lambda & 0 & 15 \\ 1 & -\lambda & -17 \\ 0 & 1 & 7-\lambda \end{vmatrix} = 0$$

$\therefore$ We consider the real eigen value 3 and the eigen space corresponding to eigen value 3 given by

$(\alpha, \beta, \gamma) \in R$

$\therefore (a_{11} - 3)\alpha + a_{12}\beta + a_{13}\gamma = 0$

$$a_{21}\alpha + (a_{22}-3)\beta + a_{23}\gamma = 0$$

$$a_{31}\alpha + a_{32}\beta + (a_{33}-3)\gamma = 0$$

(i) $-3\alpha + 15\gamma = 0 \quad \Rightarrow \alpha = 5\gamma$

(ii) $\alpha - 3\beta - 17\gamma = 0$

(iii) $\beta + 4\gamma = 0 \quad \Rightarrow \beta = -4\gamma$

(ii) $\Rightarrow \quad 5\gamma + 12\gamma - 17\gamma = 0$

Where γ can be chosen arbitrarily so let

$$\gamma = t \quad , \quad t \in R$$

$\therefore \qquad$ The eigen space belonging to eigen value 3 is

$$\{5t, -4t, t\} \quad , \quad t \in R$$

Now we consider the non real eigen value $-2+i$, therefore the eigen space is $\{z_1, z_2, z_3\} \quad , \quad z \in R$ subspace of c^3.

$$\begin{vmatrix} (a_{11}-2+i) & a_{12} & a_{13} \\ a_{21} & (a_{22}-2+i) & a_{23} \\ a_{31} & a_{32} & (a_{33}-2+i) \end{vmatrix} \begin{vmatrix} z_1 \\ z_2 \\ z_3 \end{vmatrix} = \begin{vmatrix} 0 \\ 0 \\ 0 \end{vmatrix}$$

$\therefore \quad -2 + i z_1 + 15 z_3 = 0$

$$z_1 - 2 + i z_2 - 17 z_3 = 0$$

$$z_2 + (7-(2+1))z_3 = 0 \quad \Rightarrow \quad z_2 + (5-i)z_3 = 0$$

$\therefore z_1 = \frac{15}{2+i} z_3 \quad , \quad z_2 = -(5-i)z_3$

Thus the element of the type

$$\left(\frac{15}{2+i} z_3 , -(5-i)z_3 , z_3\right)$$

Where $z_3 \in C$ is an eigen vector belonging to eigen value $2+i$.

We choose $z_3 = 1$ arbitrarily then we have

$\left(\frac{15}{2+i} , -(5-i) , 1\right)$

$\Rightarrow \quad \frac{15}{2+i} \frac{(2-i)}{(2-i)} = \frac{15(2-i)}{5} = 3(2-i) = 6-3i$

We have, $\quad \{6-3i, -(5-i), 1\}$

$\quad\quad \phi = or \{6-3i, -5+i, 1\}$

$\therefore u = (6,-5,1), \quad v = \{-3,1,0\}$

We can write $\quad\quad E = E_a + E_b$

E_a will be generated by $(5,-4,1)$

$E_b \quad$ will be generated by $(6,-5,1)$ and $(-3,1,0)$

$\therefore \quad$ operator $T = T_a \oplus T_b$

$\therefore$ Matrix of T_a with respect to basis $(5,-4,1)$ is $[1]$

Matrix T_b with respect to basis $\{v,u\}$ is $\begin{bmatrix} 2 & -1 \\ 1 & 2 \end{bmatrix}$

Thus if we consider the basis system $\{e,v,u\}$ the matrix of operator T will be given by

$$B = \begin{vmatrix} 5 & -4 & 1 \\ -4 & 2 & -1 \\ 1 & 1 & 2 \end{vmatrix}$$

$\therefore$ If $\quad y_1, y_2, y_3$ are the coordinate system in this basis then the corresponding system is

$$\begin{bmatrix} y_1' \\ y_2' \\ y_3' \end{bmatrix} = \begin{vmatrix} 5 & -4 & 1 \\ -4 & 2 & -1 \\ 1 & 1 & 2 \end{vmatrix} \begin{bmatrix} y_1 \\ y_2 \\ y_3 \end{bmatrix}$$

Since (y_1, y_2, y_3) are the coordinates of a point in this basis system

$\{e,v,u\}$ we have.

$$y_1 e_1 + y_2 v + y_3 u = x_1(1,0,0) + x_2(0,1,0) + x_3(0,0,1)$$

$$y_1(5,-4,1) + y_2(-3,1,0) + y_3(6,-5,1) =$$

$$x_1(1,0,0) + x_2(0,1,0) + x_3(0,0,1)$$

$$\therefore \quad 5y_1 - 3y_2 + 6y_3 = x_1$$

$$-4y_1 + y_2 - 5y_3 = x_2$$

$$y_1 + y_3 = x_3$$

TOPOLOGY ON R^n :

Let $x, y \in R^n$. Let us write $x = (x_1, x_2 \ldots x_n)$, $\quad y = (y_1, y_2 \ldots y_n)$

We define the scalar product (inner product) of x, y as

$$\langle x, y \rangle = \sum_{i=1}^{n} x_i y_1$$

Obviously $\quad \langle x, x \rangle \geq 0$, $\langle x, y \rangle = \langle y, x \rangle$

Also, $\quad \langle x+y, z \rangle = \langle x, z \rangle + \langle y, z \rangle$ for $x, y, z \in R^n$

If α is any real number then,

$$\langle \alpha x, y \rangle = \alpha \langle x, y \rangle = \langle x, \alpha y \rangle$$

We define,

$$|x| = \sqrt{\langle x, x \rangle}$$

1·1 is known as Euclidean norm.

Prove $\Rightarrow$ For $x, y \in R^n$ $\quad \langle x, y \rangle \leq |x| \cdot |y|$

So obviously if $x = 0$ or $y = 0$ $\in R^n$ we give the equality.

So let $x \neq 0$ and $y \neq 0$ in R^n.

COMPLEX VECTOR SPACES

For any $\alpha \in R$, consider $\langle x+\alpha y, \ x+\alpha y \rangle$

$$\langle x+\alpha y, \ x+\alpha y \rangle \geq 0$$

i.e., $\langle x,x \rangle + \alpha^2 \langle y,y \rangle + 2\alpha \langle x,y \rangle \geq 0$

In particular for $\quad \alpha = \dfrac{-\langle x,y \rangle}{\langle y,y \rangle}$, we get,

$$\langle x,x \rangle + \frac{\langle x,y \rangle^2}{\langle y,y \rangle^2} \langle y,y \rangle - 2\frac{\langle x,y \rangle}{\langle y,y \rangle} \langle x,y \rangle \geq 0$$

$$\Rightarrow \langle x,x \rangle \langle y,y \rangle + \langle x,y \rangle^2 - 2\langle x,y \rangle^2 \geq 0$$

i.e. $\quad \langle x,x \rangle \langle y,y \rangle \geq \langle x,y \rangle^2$

$$\Rightarrow |x|^2 \cdot |y|^2 \geq \langle x,y \rangle^2$$

$$\Rightarrow |x| \cdot |y| \geq \langle x,y \rangle$$

We consider the Euclidean norm (1.1) on R^n namely,

$$|x| = \sqrt{\langle x,x \rangle} = \sqrt{\sum_{i=1}^{n} x_i^2}$$

Obviously,

1) $\quad |x| \geq 0$; $\quad |x| = 0$ If and only if $x = 0$.

for any real $\alpha \in R$,

2) $\quad |\alpha x| = |\alpha||x|$

3) $\quad$ For $x, y \in R^n$, $\quad |x+y| \leq |x| + |y|$ (prove)

Proof. Consider,

$$\langle x+y, \ x+y \rangle = \langle x,x \rangle + \langle y,y \rangle + 2\langle x,y \rangle$$

$$\leq \langle x,x \rangle + \langle y,y \rangle + 2|x| \cdot |y|$$

$$\leq |x|^2 + |y|^2 + 2|x||y|$$
$$\leq (|x|+|y|)^2$$

i.e. $|x+y|^2 \leq (|x|+|y|)^2$

$\therefore$ $|x+y| \leq |x|+|y|$

We can define distance d on R^n as

$$d(x,y) = |x-y|$$

Obviously, i) $d(x,y) \geq 0$

ii) $d(x,y) = 0$ if and only if $x=y$

iii) $d(x,y) = d(y,x)$

iv) $d(x,z) \leq d(x,y) + d(y,z)$

If $\varepsilon > 0$, we define ε–nbd. of $x \in R^n$ as $\{ y \in R^n$ such that $|x-y| < \varepsilon \}$.

A subset A of R^n is said to be nbd of x in R^n if $x \in A$ and some ε–nbd. of x is contained in A for some $\varepsilon > 0$. An ε–nbd of x is denoted by $B_\varepsilon(x)$.

A set $X \subseteq R^n$ is open if it is a nbd of every $x \in X$ (every point is interior point).

A sequence $\{a_n\}$ in R^n is said to converge iff $\exists$ a point $a \in R^n$ such that $|a_n - a| \to 0$ as $n \to \infty$

A sequence $\{a_n\}$ in R^n is said to be Cauchy/ fundamental sequence if for a given $\varepsilon > 0$ $\exists$ N such that for all $m,n \geq N$ $|a_n - a_m| < \varepsilon$

COMPLEX VECTOR SPACES

If a sequence is convergent then it is Cauchy. A metric space is said to be complete if every Cauchy sequence converges.

A subset y if R^n is closed if every sequence of points in y,

i.e. convergent has its limit in y.

Also we can say a subset y of R^n is closed if its compliment is open.

Let $X \subseteq R^n$ be any subset. A map $f : X \to R^m$ is continuous if it takes convergent sequence to convergent sequence.

This mean for every sequence $\{x_k\}$ in R^m with $\lim\limits_{k \to \infty} x_k = y \in X$

it is true that $\lim\limits_{k \to \infty} f(x_k) = f(y)$.

A subset $X \subseteq R^n$ is bdd. If there exists $a > 0$ such that $X \subseteq B_a(0)$.

A subset X is compact if every sequence in X has a subsequence converging to a point in X.

Bolzano-Weierstrass theorem states that a subset of is compact iff it is closed and bounded.

Every continuous map $f : K \to R$ defined on a compact set takes a maximum value and a minimum value.

NORM:

Let $x \in R^n$, $n \geq 1$

A function $N : R^n \to R$ denoted as $N(x)$ is known as a Norm on R^n provided it satisfies the following properties.

i) $N(x) \geq 0$

ii) $N(x) = 0$ *iff* $x = 0$

iii) $N(\alpha x) = |\alpha| \cdot N(x)$ *for* $x \in R$

iv) $N(x+y) \leq N(x) + N(y)$, $x\,y \in R^n$

105

Ex 8(i):   The Euclidean norm namely if $x = (x_1, x_2, \ldots x_n)$ in the standard basis then

$$|x| = \sqrt{\sum_{i=1}^{n} x_i^2}$$

(ii)   In the standard basis with $x = (x_1, \ldots x_n)$

Let $|x|_{max} = \max \{|x_i|, \quad 1 \le i \le n\}$

Here x_i can be taken as sum of $x_1, x_2, \ldots x_n$ to prove all the properties of norm.

It is known as Max Norm.

(iii)   If $x = (x_1, \ldots x_n)$

We let $|x|_{sum} = |x_1| + |x_2| + \ldots + |x_n| = \sum_{i=1}^{n} |x_i|$

It is known as Sum Norm.

Every norm gives rise to metric and metric give rise to topology.

(iv)   If $\{f_1 \ldots f_n\}$ is a basis of R^n denoted by β, we can define num of $x \in R^n$ with respect to this basis,

Thus if $x = x_1 f_1 + x_2 f_2 + \ldots +$

Then we can define Euclidean β-Norm as

$$|x|_\beta = \sqrt{\sum_{i=1}^{n} x_i^2}$$

We can also define max and sum norm of x with respect to the basis β.

EQUIVALENCE OF NORMS ON R^n

Let $N : R^n \to R$ be any norm on R^n. There exists constants A and $B > 0$ such that

COMPLEX VECTOR SPACES

$$A|x| \leq N(x) \leq B|x|$$

where $1 \cdot 1$ denotes the Euclidean Norm.

Proof: We first consider the max norm namely $1 \cdot 1$ max.

If $x = (x_1, x_2, \ldots x_n)$ then we have

$$\left(\max\{|x_i| \quad 1 \leq i \leq n\}\right)^2 \leq \sum_{i=1}^{n} x_i^2$$

$$\leq \sum_{i=1}^{n}\left(\max\{|x_i| \quad 1 \leq i \leq n\}\right)^2$$

i.e., $\left(|x|_{max}\right)^2 \leq |x|^2 \leq n\left(|x|_{max}\right)^2$

Taking square roots we get

$$|x|_{max} \leq |x| \leq \sqrt{n} \ |x|_{max}$$

i.e., $\dfrac{1}{\sqrt{n}} |x| \leq |x|_{max} \leq |x|$

Thus, the max norm satisfies the inequality with

$$A = \frac{1}{\sqrt{n}} \quad \text{and} \quad B = 1$$

Let N be any norm on R^n,

We show that N is uniformly continuous on R^n.

Let $x_1 e_1 + \ldots + x_n e_n$ where $\{e_1 \ldots e_n\}$ is the standard basis, then we have,

$$N(x) = N(x_1 e_1 + \ldots x_n e_n)$$

$$\leq N(x_1 e_1) + N(x_2 e_2) + \ldots + N(x_n e_n)$$

As $x_1, x_2, \ldots x_n \in R$ and $e_1, e_2 \ldots e_n \in R^n$ we have,

$$N(x_1 e_1) = |x_1| \cdot N(e_1), N(x_2 e_2) = |x_2| N(e_2) \ldots N(x_n e_n) = |x_n| \cdot (e_n)$$

$$\therefore N(x) \le |x_1| N(e_1) + |x_2| N(e_2) + \ldots + |x_n| \cdot N(e_n)$$

Let M be max m of $N(e_1),\ N(e_2)\ldots N(e_n)$

i.e., $\quad M = \max\{N(e_1),\ N(e_2)\ldots N(e_n)\}$

$$N(x) \le |x_1| M + |x_2| M + \ldots + |x_n| M$$

as, $\quad |x_i| \le |x|_{max} \qquad 1 \le i \le n$ we get,

$$N(x) \le |x|_{max} M + |x|_{max} M + \ldots + |x|_{max} M$$

$$\le n|x|_{max} M \le n M |x|, \quad \text{(as } |x|_{max} \le |x|)$$

Thus we have,

$$N(x) \le n M |x| \qquad \text{for all } x \in R^n$$

Here n and m are independent of $x \in R^n$

Let $\quad y \in R^n \qquad N(y) \le n M |y|$

We can write,

$$x = x - y + y$$

$$\therefore \ N(x) = N(x - y + y) \le N(x - y) + N(y)$$

(By triangle inequality)

i.e., $\quad N(x) - N(y) \le N(x - y)$

Interchanging x and y we get

$$N(y) - N(x) \le N(y - x) = N(x - y)$$

So, $\quad N(x) - N(y) \le N(x - y)$

$$N(y) - N(x) \le N(x - y)$$

$$\therefore \ |N(x) - N(y)| \le N(x - y)$$

where $1 \cdot 1$ on the left hand side is the usual absolute value in R

But $\quad N(x) \leq n M |x| \qquad \forall\; x \in R^n$

$\therefore \qquad N(x-y) \leq M |x-y|$

Thus for $\quad x, y \in R^n$, $\quad |N(x) - N(y)| \leq n M |x-y|$

Obviously, as $|x-y| \to 0$, $|N(x) - N(y)| \to 0$

$\therefore$ N is uniformly continuous.

Thus, N is continuous.

Consider the set $\{x \in R^n \;such that\; |x| = 1\}$

$\therefore$ It is a compact subset of R^n.

Since N is continuous and the set $\{x \in R^n \;such that\; |x| = 1\}$ is compact, N will attain its minimum say A and maximum say B (exist) on it.

Obviously $A, B \geq 0$

Since $0 \notin \{x \in R^n \;such that\; |x| = 1\}$ it follows that $N(x) > 0$ for all x with $|x| = 1$.

As this set is compact (hence closed) we have $A > 0$, $B > 0$.

The inequality

$A|x| \leq N(x) \leq B|x|$ is obviously true for $x = 0$.

So let us assume that $x \neq 0$.

$\therefore \qquad |x| \neq 0$

Consider, $\quad y = \dfrac{x}{|x|}$

So obviously $|y| = 1$.

Therefore , we have

$$A \leq N(y) \leq B$$

i.e., $\quad A \leq \dfrac{N(x)}{|x|} \leq B$

i.e., $\quad A|x| \leq N(x) \leq B|x|$

Thus we find +ve constants A and B such that $A|x| \leq N(x) \leq B|x|$

and the constants A and B depends on N.

If E is a subspace of R^n, we can define norm on E to
be a function $N : E \to R$ such that,

(1) $\quad N(x) \geq 0 \qquad \forall \ x \in E$

(2) $\quad N(x) = 0 \quad iff \quad x = 0$

(3) $\quad N(\alpha x) = |\alpha| \cdot N(x) \quad for \quad x \in R$

(4) $\quad N(x + y) \leq N(x) + N(y) \ , \quad x \, y \in E$

Proof. If E is a subspace of R^n , then restriction of any norm on R^n to E, gives a norm on E.

Conversely we can show that any norm on E is a restriction of some norm on R^n.

Let N be a norm on E.

As E is a subspace of R^n. let $\{e_1 \ldots e_n\}$ be a basis of E.

As $\{e_1 \ldots e_n\}$ is linearly independent in R^n.

We can find $n - r$ members say $f_1, f_2 \ldots f_n$ in R^n such that $\{e_1 \ldots e_n, \ f_1 \ldots f_{n-r}\}$ is a basis of E.

Let F be the subspace of R^n generated by $(f_1, f_2 \ldots f_n)$

Obviously $R^n = E \oplus F$.

COMPLEX VECTOR SPACES

As F is a subspace of R^n with basis $\{f_1 ... f_{n-r}\}$.

We define a Euclidean norm $1 \cdot 1$ on F with respect to this basis.

For any $x \in R^n$, we can find unique $y \in E$ and $z \in F$ such that

$$x = y + z$$

We define N' on R^n as,

$$N'(x) = N(y) + |z|$$

We show that N' is a norm on R^n.

Obviously, $N'(x) \geq 0$, $x \in R^n$

If $x = 0$ in R^n,

We have $\quad x = 0 + 0 \qquad$ (because it is direct sum)

$$\downarrow \ \downarrow$$

$$E \quad F$$

$\therefore \ N' = N(0) + |0|$

$$= 0$$

Conversely: if $N'(x) = 0 \quad$ then

$$N'(x) = N(y) + |z|$$

$\Rightarrow \quad N(y) = 0 \quad$ and $\quad |z| = 0$

$\Rightarrow \quad y = 0 \quad$ and $\quad z = 0$

$\Rightarrow \quad x = 0$

$\therefore \quad N'(x) = 0 \quad$ iff $\quad x = 0$

Now let $\alpha \in R$, consider $\ N' \ (\alpha x)$

$x = y + z \quad , \quad y \in E , \ z \in F$

$\Rightarrow \quad \alpha x = \alpha y + \alpha z \qquad\qquad N'(\alpha x) = N(\alpha y) + |\alpha z|$

111

$$= |\alpha| N(y) + |\alpha| \cdot |z|$$

$$= |\alpha| \left(N(y) + |z| \right)$$

$$= |\alpha| \; N'(x)$$

Let $\quad x_1, x_2 \in R^n \quad$ and $\quad x_1 = y_1 + z_1$

$$x_2 = y_2 + z_2 \;, \quad y_1 \; y_2 \in E$$

$$z_1 \; z_2 \in F$$

$$\therefore \quad x_1 + x_2 = \left(y_1 + y_2 \right) + \left(z_1 + z_2 \right)$$

$$N'\left(x_1 + x_2 \right) = N\left(y_1 + y_2 \right) + |z_1 + z_2|$$

$$\leq N(y_1) + N(y_2) + |z_1| + |z_2|$$

$$= N(y_1) + |z_1| + N(y_2) + |z_2|$$

$$= N'\left(x_1 \right) + N'\left(x_2 \right)$$

Thus $\quad N'$ is a norm on $\quad R^n$.

We now show that $\quad N$ is a restriction of $\quad N'$ to $\quad E$.

i.e., $\quad N' \big|_E = N$

$\therefore \qquad$ for any $\quad x \in E$, we can write $\quad x = x + 0$

$$\downarrow \quad \downarrow$$

$$E \quad F$$

$$\therefore \quad N'(x) = N(x) + |0|$$

$$= N(x)$$

Theorem: Let (E, N) be any normal vector space. A sequence $\{x_k\}$ in E converges to y iff $\lim\limits_{k \to \infty} N\left(x_k - y \right) = 0$.

Proof. We know that, in $E \subseteq R^n$ for some n

COMPLEX VECTOR SPACES

$\lim\limits_{k\to\infty} x_k = y$ $\qquad$ iff $\quad |x_k - y| = 0$

where $1 \cdot 1$ is a Euclidean norm.

We know that there exist +ve constants. A and B such that

$A|x| \subseteq N(x) \le B|x|$

Let $\quad \lim\limits_{k\to\infty} x_k = y$

i.e. $\quad \lim\limits_{k\to\infty} |x_k - y| = 0$

$\therefore \quad N(x_k - y) \le B|x_k - y|$

$\therefore \quad \lim\limits_{k\to\infty} |x_k - y| = 0 \;\Rightarrow\; \lim\limits_{k\to\infty} N(x_k - y) = 0$

Conversely, let $\; N(x_k - y) = 0 \quad$ as $\; k \to \infty$

As, $\quad A|x_k - y| \le N(x_k - y)$

$N(x_k - y) \to 0 \Rightarrow A|x_k - y| \to 0 \quad$ as $\; k \to \infty$

$\Rightarrow |x_k - y| \to 0$

Thus if a seq. $\{x_k\}$ converges in E in some norms, then it converges to the same limit in every other norm.

Proof: Let (E, N) be a normed vector space. Then the unit ball

$D = \{x \in E \; such that \; N(x) \le 1\}$ is compact.

$\Rightarrow \qquad$ The set D is a subset of the set $\{x \in E \; such that \; |x| \le \frac{1}{A}\}$

But the set $\{x \in E \; such that \; |x| \le \frac{1}{A}\}$ is compact.

(being closed and bdd.).

However as N is continuous and $D \subseteq N^{-1}\{(-\infty, 1)\}$, it follows that D is closed subset of a compact set

$\therefore \quad D$ is compact.

Thus a unit ball in any norm is compact.

113

DYNAMICAL SYSTEM – A SHORT COURSE

Proof: Let (E, N) be a normed vector space. Then a sequence $\{x_k\}$ in E converges to an element in E iff   For every $\varepsilon > 0$ $\exists$ an integer $n_0 > 0$ such that if $p > n \geq n_0$ then $N(x_p - x_n < \varepsilon)$.

We use the fact that R^n is complete and every subspace of R^n is a closed set.

We know that $\{x_k\}$ in E converges iff $\exists$ y in E such that $|x_k - y| \to 0$ as $k \to \infty$.

But because of equivalence of norm $N(x_p - x_n) < \varepsilon$ for, $p > n \geq n_0$ is equivalent to the fact that the sequence $\{x_k\}$ is Cauchy in R^n with respect to the Euclidean norm.

As R^n is complete, the Cauchy sequence $\{x_k\}$ will converge in R^n.

Thus $\exists$ $y \in R^n$ such that $\lim\limits_{k \to \infty} x_k = y$

However as $\{x_k\}$ is in E and E is a closed subspace of R^n it follows that $y \in E$.

$\therefore$ $x_k \to y$ in E.

Definition: A series $\sum a_k$ of elements in (E, N) is said to converge absolutely iff the series $\sum N(a_n)$ is convergent in R.

Find out the sets $\{x \in R^2 \; such that \; |x|_{\max} = 1\}$ and $\{x \in R^2 \; such that \; |x|_{sum} = 1\}$ where the standard basis is used.

In R^2, $\{x \in R^2 \; such that \; |x|_{\max} = 1\}$ is given by the square joining the vertices $(1,1)(-1,1)(1,-1)$ and $(-1,-1)$

In R^2 $\{x \in R^2 \; such that \; |x|_{sum} = 1\}$ is given by the square with vertices

114

COMPLEX VECTOR SPACES

$(1,0)(0,1)(-1,0)$ and $(0,-1)$

Find the largest A and smallest B such that $A\,|x| \le |x|_{sum} \le B|x|$ in R^n , where standard basis is used for all the norms.

If A and B are the +ve constants such that $A|x| \le N(x) \le B|x|$

For any norm $N(x)$ then consider the set $\{x \ \ such \ that \ |x| = 1\}$

For $|x| = 1$ we must have, $A \le N(x) \le B$.

Thus, we need A to be largest and B to be smallest such that $A \le N(x) \le B$ for $|x| = 1$.

But N is a continuous function and

$\{x \ \ such \ that \ |x| = 1\}$ is a compact set.

It follows that, $A = \min\ N(x)\ such\ that\ |x| = 1$

And $B = \max\ N(x)\ such\ that\ |x| = 1$

Thus the largest A and smallest B such that

$$A|x| \le |x|_{sum} \le B|x|$$

are given by min of $|x|_{sum}$ and max of $|x|_{sum}$ on $|x| = 1$ respectively.

In R^2, $|x| = 1$ is the (circle)

$$x^2 + y^2 = 1 \quad where \quad \overline{x} \equiv (x, y)$$

Consider the first quadrant (because of the symmetry of). We have to find out minimum and maximum of $x + y$ such that $x^2 + y^2 = 1$, $x \ge 0$, $y \ge 0$

The arc of the circle $x^2 + y^2 = 1$ in the 1st quadrant lies above the line $x + y = 1$, meeting it at the points $(1,0)$ and $(0,1)$.

Obviously, the minimum of $x + y$ on $x^2 + y^2 = 1$ in the 1st quadrant is 1.

DYNAMICAL SYSTEM – A SHORT COURSE

Since the arc is symmetric about the line $x = y$, it follows that the farthest point of the arc from the line $x + y = 1$ will be given by $x = y$ and $x^2 + y^2 = 1$ i.e.,

$$x = \tfrac{1}{\sqrt{2}}, \ y = \tfrac{1}{\sqrt{2}}$$

For $x = \tfrac{1}{\sqrt{2}}$ and $y = \tfrac{1}{\sqrt{2}}$ $\quad |(x,y)|_{sum} = \tfrac{1}{\sqrt{2}} + \tfrac{1}{\sqrt{2}} = \sqrt{2}$

Thus in the first quadrant

$A = 1$ and $B = \sqrt{2}$

Because of the symmetry of $x^2 + y^2 = 1$ about the origin, it follows that

min $\left\{ |x| + |y| > x^2 + y^2 = 1 \right\} = 1$

max $\left\{ |x| + |y| \ such \ that \ x^2 + y^2 = 1 \right\} = \sqrt{2}$

$\therefore \ A = 1$, $B = \sqrt{2}$ in R^2

In R^3 we get $A = 1$, $B = \sqrt{3}$ (attained at the point $\left(\tfrac{1}{\sqrt{3}} , \tfrac{1}{\sqrt{3}} , \tfrac{1}{\sqrt{3}} \right)$).

In general in R^n , $A = 1$, $B = \sqrt{n}$

which of the following formulas define norms on R^2 (let (x, y) be the coordinates of a point in R^2).

(a) $\left(x^2 + xy + y^2 \right)^{\frac{1}{2}}$

(b) $\left(x^2 - zxy + y^2 \right)^{\frac{1}{2}}$

(c) $\left(|x| + |y| \right)^2$

(d) $\tfrac{1}{3}\left(|x| + |y| \right) + \tfrac{2}{3}\left(x^2 + y^2 \right)^{\frac{1}{2}}$

NORM OF OPERATORS

Let A be an operator on R^n , since A can be identified with an $n \times n$ real matrix and as every $n \times n$ real matrix can be considered as an element of

COMPLEX VECTOR SPACES

R^{n^2} , we can identify any $A \in L(R^n)$ with an element of R^{n^2} and as such we can define a norm of A.

Since on R^n we can define a number of norms and all these are equivalent, it follows that for an operator $A \in L(R^n)$

We can define a number of norms.

One of the most commonly used norm of an operator is the uniform Norm. $(\|\cdot\|)$

Let $|\cdot|$ be a norm on R^n and $T \in L(R^n)$. we define uniform norm of T with respect to norm on R^n as ,

$\|T\| = \max \{|T(x)| : |x| \le 1\}$

Obviously, the value of uniform norm $\|T\|$ will depend on the norm $|\cdot|$ that is being considered in the definition.

Since a linear operator is a continuous function, $\|T\|$ is well defined for each $T \in L(R^n)$.

PROPERTIES

1) If $T = I$ the identity operator, then, $\|T\| = 1$.

2) If $\|T\| = k$ for some $R \ge 0$ Then,

$\quad |T(x)| \le k|x|$ for $x \in R^n$

Or $|T(x)| \le \|T\| \, |x|$

If $x = 0$, then $T(x) = 0$, the above inequality is indeed an equality.

We now Consider the case $x \ne 0$,

$\therefore \quad |x| \ne 0$

Consider , $y = \frac{x}{|x|}$ and $|y| = 1$

From the definition norm, we have

117

$$|T(y)| \le \|T\| = R$$

As $|T(y)| = \left|\frac{T(x)}{|x|}\right| = \left|\frac{T(x)}{|x|}\right|$

We have, $\left|\frac{T(x)}{|x|}\right| \le k$

$\therefore \quad |T(x)| \le k|x|$

Thus for all $x \in R^n$, $|T(x)| \le \|T\| \cdot |x|$

(3) NORM OF COMPOSITION

If $S, T \in L(R^n)$ then

$$\|ST\| \le \|S\| \cdot \|T\|$$

for $x \in R^n$, consider $ST(x)$

$\therefore \quad ST(x) = S(T(x))$

$|ST(x)| = |S(T(x))| \le \|S\| \cdot |T(x)|$

$\qquad \le \|S\| \cdot \|T\| \cdot |x| \qquad$ {by property}

Thus, for any $x \in R^n$

$$|ST(x)| \le \|S\| \cdot \|T\| \cdot |x|$$

In particular for $|x| \le 1$, we get

$$|ST(x)| \le \|S\| \cdot \|T\| \cdot 1$$

$\max\{|ST(x)| \ such that \ |x| \le 1\} \le \|S\| \cdot \|T\|$

i.e. $|ST| \le \|S\| \cdot \|T\|$

(4) For any $T \in L(R^n)$ for $j \ge 1$

$$\|T^j\| \le \|T\|^j$$

Repeated application of the result.

$|ST| \le \|S\| \cdot \|T\|$ gives $\|T^j\| \le \|T\|^j$

COMPLEX VECTOR SPACES

Definition: Let $T \in L(R^n)$

By exp (T) i.e. e^T is an operator in $L(R^n)$ defined as,

$$e^T = I + T + \frac{T^2}{2!} + \frac{T^3}{3!} + ...$$

where the limit of the series is with respect to any norms in $L(R^n)$ (the definition is independent of the norm on $L(R^n)$ as all norms on $L(R^n)$ are equivalent).

Theorem: For any operator T in $L(R^n)$ the series $\displaystyle\sum_{k=0}^{\infty} \frac{T^k}{k!}$ is absolutely convergent i.e., the operator e^T is well defined.

Further we show that $\left\| e^T \right\| \leq e^{\|T\|}$

Proof: Let $\|T\| = a$ for some real number $a \geq 0$.

Consider $\quad \dfrac{\left\| T^k \right\|}{k!} \leq \dfrac{\|T\|^k}{k!} = \dfrac{a^k}{k!}$

Thus every term of series $\displaystyle\sum_{k=0}^{\infty} \frac{\left\| T^k \right\|}{k!}$ is dominated by the

corresponding term $\dfrac{a^k}{k!}$ of the series $\displaystyle\sum_{k=0}^{\infty} \frac{a^k}{k!}$.

But the series $\displaystyle\sum_{k=0}^{\infty} \frac{a^k}{k!}$ is convergent with sum given by e^a,

Therefore by comparison theorem, the series $\displaystyle\sum_{k=0}^{\infty} \frac{\left\| T^k \right\|}{k!}$ is convergent.

i.e., the series $\displaystyle\sum_{k=0}^{\infty} \frac{T^k}{k!}$ is absolutely convergent.

$\therefore$ The series $\displaystyle\sum_{k=0}^{\infty} \frac{T^k}{k!}$ is convergent.

DYNAMICAL SYSTEM – A SHORT COURSE

To find $\left\| e^T \right\|$ we consider $e^T(x)$, for $x \in k$

By definition,

$$e^T(x) = \lim_{n \to \infty} \left(I + T + \tfrac{T^2}{2!} + \tfrac{T^3}{3!} + \dots \tfrac{T^n}{n!} \right)$$

$$= \lim_{n \to \infty} \left| \left(I + T + \tfrac{T^2}{2!} + \dots + \tfrac{T^n}{n!} \right)(x) \right| \quad \text{(by continuity of norm)}$$

$$\leq \lim_{n \to \infty} \left\| I + T + \tfrac{T^2}{2!} + \dots + \tfrac{T^n}{n!} \right\| \cdot |x|$$

$$\leq \lim_{n \to \infty} \left(\|I\| + \tfrac{\|T\|}{1!} + \tfrac{\|T^2\|}{2!} + \dots + \tfrac{\|T^n\|}{n!} \right) \cdot |x|$$

$$\leq \lim_{n \to \infty} \left(1 + \|T\| + \tfrac{\|T\|^2}{2!} + \dots + \tfrac{\|T\|^n}{n!} \right) \cdot |x|$$

Hence for $|x| \leq 1$, we get

$$\left| e^T(x) \right| \leq \lim_{n \to \infty} \left(1 + \tfrac{\|T\|}{1!} + \tfrac{\|T\|^2}{2!} + \dots + \tfrac{\|T\|^n}{n!} \right)$$

$$= e^{\|T\|}$$

Thus for $|x| \leq 1$, $\left| e^T(x) \right| \leq e^{\|T\|}$

$\therefore \quad \max \left\{ \left| e^T(x) \right| \quad such\,that \quad |x| \leq 1 \right\} \leq e^{\|T\|}$.

$\therefore \quad \left\| e^T \right\| \leq e^{\|T\|} \quad$ (for uniform norm).

Lemma: Let $\sum\limits_{j=0}^{\infty} A_j = A$ and $\sum\limits_{k=0}^{\infty} B_k = B$ be absolutely convergent series of

operators on R^n. Then, $A \cdot B = C$, where $C = \sum\limits_{l=0}^{\infty} C_l$

where $\quad C_l = \sum\limits_{j+k=l} A_j \cdot B_k$

Proof: Let $\{\alpha_n\}, \{\beta_n\}, \{\gamma_n\}$ be the sequences of partial sums of the series

COMPLEX VECTOR SPACES

$$\sum_{j=0}^{\infty} A_j,\ \sum_{j=0}^{\infty} B_j,\ \sum_{j=0}^{\infty} C_j \quad \text{respectively.}$$

We have $A = \lim_{n \to \infty} \alpha_n$, $B = \lim_{n \to \infty} \beta_n$, $C = \lim_{n \to \infty} \gamma_n$

$$\therefore\ \lim_{n \to \infty} \alpha_n \beta_n = AB$$

As the sequence $\{\gamma_n\}$ converges to C, its subsequence $\{\gamma_{2n}\}$ will also converge to C.

Thus to show that $C = A \cdot B$

We need to show that

$$|\gamma_{2n} - \alpha_n \beta_n| \to 0 \quad \text{as} \quad n \to \infty$$

We note that,

$$\gamma_{2n} - \alpha_n \beta_n = \sum{}^{1} A_j B_k + \sum{}^{11} A_j B_k$$

Then $0 \le j + k \le 2n$, $\sum^{1}$ is such that

$0 \le j + k \le 2n$, $0 \le j \le n$, $n+1 \le k \le 2n$ $\sum^{1}$ is such that

$0 \le j + k \le 2n$, $n+1 \le j \le 2n$ $0 \le k \le n$

$$\therefore\ |\gamma_{2n} - \alpha_n \beta_n| \le \sum_{j=0}^{\infty}|A_j| \cdot \sum_{k=n}^{2n}|B_k| + \sum_{j=n}^{2n}|A_j| \cdot \sum_{k=0}^{\infty}|B_k|$$

But as the series $\sum A_j$ and $\sum B_j$ are absolutely convergent,

$\sum_{j=0}^{\infty}|A_j|$ and $\sum_{j=0}^{\infty}|B_j|$ are finite numbers.

Further as $\sum|A_j|$ is convergent,

We must have, $\lim_{n \to \infty} \sum_{j=n}^{2n}|A_j| = 0$ and $\lim_{n \to \infty} \sum_{j=n}^{2n}|_j = 0|$

$$\therefore\ \lim_{n \to \infty} |\gamma_{2n} - \alpha_n \beta_n| = 0$$

$$\therefore\ \lim \gamma_{2n} = \lim \alpha_n \beta_n$$

$C = A \cdot B$

Prop: Let P, S, T denotes operators on R^n , then

(a) If $Q = PTP^{-1}$, then $e^Q = Pe^T P^{-1}$

(b) If $ST = TS$ then $e^{S+T} = e^S \cdot e^T$

(c) $e^{-S} = \left(e^S\right)^{-1}$

(d) If $n = 2$ and $T = \begin{bmatrix} a & -b \\ b & a \end{bmatrix}$ then,

$$e^T = e^a \begin{bmatrix} \cos b & -\sin b \\ \sin b & \cos b \end{bmatrix}$$

Proof: a) By definition,

$$e^Q = I + Q + \frac{Q^2}{2!} + \dots$$

$$= \lim_{n \to \infty} \left(\sum_{k=0}^{n} \frac{Q^k}{k!} \right) = \lim_{n \to 0} \left(\sum_{k=0}^{n} \frac{\left(PTp^{-1}\right)^k}{k!} \right)$$

But $\left(PTp^{-1}\right)^k = PT^k p^{-1}$, $k = 0, 1, 2 \dots$

$$\therefore \quad e^Q = \lim_{n \to 0} \left(\sum_{k=0}^{n} \frac{PT^k p^{-1}}{k!} \right)$$

$$= \lim_{n \to 0} \left(P \left(\sum_{k=0}^{n} \frac{T^k}{k!} \right) p^{-1} \right)$$

$$= P \left(\lim_{n \to 0} \sum_{k=0}^{n} \frac{T^k}{k!} \right) p^{-1}$$

$$= P e^T p^{-1}$$

$\therefore$ If $Q = PTp^{-1}$ then $e^Q = Pe^T p^{-1}$.

(b) we are given that $ST = TS$

Suppose

COMPLEX VECTOR SPACES

$$(S+T)^n = S^n + n \cdot S^{n-1} \cdot T + \frac{n(n-1)}{2!} S^{n-2} \cdot T^2 + \ldots + n \cdot S \cdot T^{n-1} + T^n$$

$$= S^n + \frac{n!}{(n-1)!} S^{n-1} \cdot T + \frac{n!}{(n-2)!2!} S^{n-2} T^2 + \frac{n!}{(n-3)!3!} S^{n-3} T^3 + \ldots + \frac{n!}{(n-1)!} S'T^{n-1} + 1 \cdot T^n$$

$$= \frac{n!}{n!0!} S^n T^0 + \frac{n!}{(n-1)!1!} S^{n-1}T^1 + \frac{n!}{(n-2)!2!} S^{n-2}T^2 + \ldots + \frac{n!}{(n-1)!} S^1 T^{n-1} + \frac{n!}{n!0!} S^0 T^n$$

$$= \sum_{i+j=n} \frac{n!}{i!\,j!} S^i \cdot T^j = n! \sum_{i+j=n} \left(\frac{S^i}{i!}\right) \cdot \left(\frac{T^j}{j!}\right)$$

Thus, $\dfrac{(S+T)^n}{n!} = \displaystyle\sum_{i+j=n} \frac{S^i}{i!} \cdot \frac{T^j}{j!}$

Since the series e^S and e^T are absolutely convergent series, we have

$$e^{S+T} = e^S \cdot e^T = e^T \cdot e^S$$

(c)   we can write

$$0 = S - S$$

Taking exponential

$$I = e^0 = e^{S-S}$$

As $S(-s) = (-s)(S)$

We have, $e^{S+(-s)} = e^S \cdot e^{-s}$

$\therefore\ 1 = e^S e^{-s}$

$\therefore\ e^{-s} = \left(e^S\right)^{-1}$

(d) Given $T = \begin{bmatrix} a & -b \\ b & a \end{bmatrix}$ and we show that

$$e^T = e^a \begin{bmatrix} \cos b & -\sin b \\ \sin b & \cos b \end{bmatrix}$$

We know that, the operator $\begin{bmatrix} a & -b \\ b & a \end{bmatrix}$ operating on element $\begin{bmatrix} x \\ y \end{bmatrix}$ of R^2 is equivalent to the multiplication of the complex number $x+iy$ by the complex number $a+ib$ where the point (x,y) of R^2 is given by the complex number $x+iy$.

However, the expression for e^T where $T = \begin{bmatrix} a & -b \\ b & a \end{bmatrix}$ involves addition, multiplication and limiting. Thus we need to show that the representation $\begin{bmatrix} a & -b \\ b & a \end{bmatrix} \to a+ib$ preserved under addition, multiplication and limiting process.

Consider $T_1 = \begin{bmatrix} a_1 & -b_1 \\ b_1 & a_1 \end{bmatrix}$, $T_2 = \begin{bmatrix} a_2 & -b_2 \\ b_2 & a_2 \end{bmatrix}$

$$\downarrow \qquad\qquad \downarrow$$

$$a_1 + i b_1 \qquad\qquad a_2 + i b_2$$

$$\therefore \; T_1 + T_2 = \begin{bmatrix} a_1 + a_2 & -(b_1 + b_2) \\ b_1 + b_2 & a_1 + a_2 \end{bmatrix}$$

$$\downarrow$$

$$(a_1 + a_2) + i(b_1 + b_2)$$

$$\parallel$$

$$(a_1 + i b_1) + (a_2 + i b_2)$$

$$\therefore \; T_1 \cdot T_2 = \begin{bmatrix} a_1 & -b_1 \\ b_1 & a_1 \end{bmatrix}\begin{bmatrix} a_2 & -b_2 \\ b_2 & a_2 \end{bmatrix} = \begin{bmatrix} a_1 a_2 - b_1 b_2 & -a_1 b_2 - b_1 a_2 \\ a_2 b_1 - a_1 b_2 & -b_1 b_2 + a_1 a_2 \end{bmatrix}$$

$$\downarrow \qquad\qquad \downarrow \qquad\qquad \downarrow$$

$$(a_1 + i b_1) \quad (a_2 + i b_2) \quad (a_1 a_2 - b_1 b_2) + i(b_1 a_2 + a_1 b_2)$$

$$= (a_1 a_2 - b_1 b_2) + i(a_1 b_2 + b_1 a_2)$$

Now, we know that a sequence $\{a_n + i b_n\}$ of complex numbers converges to a complex numbers $a+ib$ iff $a_n \to a$ and $b_n \to b$.

COMPLEX VECTOR SPACES

Identifying the operator $\begin{bmatrix} a_n & -b_n \\ b_n & a_n \end{bmatrix}$ to an element $(a_n, -b_n, b_n, a_n)$ of R^4; it

follows that the sequence $\begin{bmatrix} a_n & -b_n \\ b_n & a_n \end{bmatrix}$ of operators will converge to an

operator $\begin{bmatrix} a & -b \\ b & a \end{bmatrix}$ iff $a_n \to a \quad b_n \to b$.

Thus the representation $\begin{bmatrix} a & -b \\ b & a \end{bmatrix} \leftrightarrow a + ib$ preserves addition, multiplication

and limiting process.

Thus if $T = \begin{bmatrix} a & -b \\ b & a \end{bmatrix}$ then the operator e^T will be represented by the

complex number $e^{a+ib} = e^a \cdot (\cos b + i \sin b)$.

But the complex number $\cos b + i \sin b$ corresponds to the operator

$\begin{bmatrix} \cos b & -\sin b \\ \sin b & \cos b \end{bmatrix}$.

Thus if $T = \begin{bmatrix} a & -b \\ b & a \end{bmatrix}$ then $e^T = e^a \begin{bmatrix} \cos b & -\sin b \\ \sin b & \cos b \end{bmatrix}$

Ex 9: Let $T = \begin{bmatrix} a & 0 \\ 0 & b \end{bmatrix}$

By definition

$$e^T = \lim_{n \to \infty} \sum_{k=0}^{n} \frac{T^k}{k!} = \lim_{n \to \infty} \sum_{k=0}^{n} \frac{\begin{bmatrix} a^k & 0 \\ 0 & b^k \end{bmatrix}}{k!}$$

$$= \lim_{n \to 0} \begin{bmatrix} 1 + a + \frac{a^2}{2!} + \frac{a^3}{3!} + \dots + \frac{a^n}{n!} & 0 \\ 0 & 1 + b + \frac{b^2}{2!} + \frac{b^3}{3!} + \dots + \frac{b^n}{n!} \end{bmatrix}$$

$$= \begin{bmatrix} e^a & 0 \\ 0 & e^b \end{bmatrix}$$

DYNAMICAL SYSTEM – A SHORT COURSE

2) Let $T = \begin{bmatrix} 2 & 0 \\ 3 & 2 \end{bmatrix} = \begin{bmatrix} 2 & 0 \\ 0 & 2 \end{bmatrix} + \begin{bmatrix} 0 & 0 \\ 3 & 0 \end{bmatrix}$

As $\begin{bmatrix} 2 & 0 \\ 0 & 2 \end{bmatrix} = 2 \cdot I$,

The matrices $\begin{bmatrix} 2 & 0 \\ 0 & 2 \end{bmatrix}$ and $\begin{bmatrix} 0 & 0 \\ 3 & 0 \end{bmatrix}$ commute with each other

$\therefore \quad e^T = e^{\begin{bmatrix} 2 & 0 \\ 0 & 2 \end{bmatrix}} \cdot e^{\begin{bmatrix} 0 & 0 \\ 3 & 0 \end{bmatrix}}$

But $e^{\begin{bmatrix} 2 & 0 \\ 0 & 2 \end{bmatrix}} = \begin{bmatrix} e^2 & 0 \\ 0 & e^2 \end{bmatrix}$

$\begin{bmatrix} 0 & 0 \\ 3 & 0 \end{bmatrix}^2 = \begin{bmatrix} 0 & 0 \\ 3 & 0 \end{bmatrix}\begin{bmatrix} 0 & 0 \\ 3 & 0 \end{bmatrix} = \begin{bmatrix} 0 & 0 \\ 0 & 0 \end{bmatrix}$

It follows that,

$e^{\begin{bmatrix} 0 & 0 \\ 3 & 0 \end{bmatrix}} = I + \begin{bmatrix} 0 & 0 \\ 3 & 0 \end{bmatrix} = \begin{bmatrix} 1 & 0 \\ 3 & 1 \end{bmatrix}$

$e^T = \begin{bmatrix} e^2 & 0 \\ 0 & e^2 \end{bmatrix} \cdot \begin{bmatrix} 1 & 0 \\ 3 & 1 \end{bmatrix} = e^2 \begin{bmatrix} 1 & 0 \\ 3 & 1 \end{bmatrix}$

1) Let N be any norm on $L(R^n)$.

Prove that $\exists$ a constant k such that

$N(ST) \le k\, N(S) \cdot N(T)$

Solution: We have $\|ST\| \le \|S\| \cdot \|T\|$ where $\|\cdot\|$ is uniform norm.

Since all norms on $L(R^n)$ are equivalent, we can find positive constants A and B such that

$A\|S\| \le N(S) \le B\|S\|$

Similarly, $A\|T\| \le N(T) \le B\|T\|$

By equivalence of norms on $L(R^n)$.

COMPLEX VECTOR SPACES

We can find +ve constants A and B such that

for any $S \in L(R^n)$.

$A\|S\| \le N(S) \le B\|S\|$

Thus, for the composition ST we have

$A\|ST\| \le N(ST) \le B\|ST\|$

i.e., $N(ST) \le B\|ST\|$

But for uniform norm, $\|ST\| \le \|S\| \cdot \|T\|$

$\therefore \quad N(ST) \le B\|S\| \cdot \|T\|$

But as $\|S\| \le \frac{1}{A} N(S)$, we get

$N(ST) \le B\|S\| \cdot \|T\| \le B \frac{N(S)}{A} \cdot \frac{N(T)}{A}$

$$= \frac{B}{A^2} N(S) N(T)$$

We choose $k = \frac{B}{A^2}$ to get

$N(ST) \le k \, N(S) N(T)$

2) Let $T : R^n \to R^m$ be a linear transformation.

Show that T is uniformly continuous.

Solution: Let $\{e_1, e_2 \dots e_n\}$ be a basis of R^n.

Then for every $x \in R^n$, we can write

$x = x_1 e_1 + x_2 e_2 + \dots + x_n e_n$ { for real numbers $x_1 \, x_2 \dots x_n$ }

$\therefore \ T(x) = x_1 T(e_1) + x_2 T(e_2) + \dots + x_n T(e_n)$

Consider any norm in R^m and denote it by $|\cdot|$

Then we have,

$|T(x)| \le |x_1| \cdot |T(e_1)| + |T(e_2)| + \dots + |T(e_n)| \cdot |x_n|$

In particular for Euclidean norm in R^m.

$|T(x)| \le |x_1| \cdot |T(e_1)| + |x_2| \cdot |T(e_2)| + \dots + |x_n| \cdot |T(e_n)|$

Since, $|x_i| \le |x|$ for $1 \le i \le n$,

where $|x|$ is the Euclidean norm of x , we get

$|T(x)| \le |x_1| \{|T(e_1)| + |T(e_2)| + ... + |T(e_n)|\}$

Let $M = \max \{|T(e_1)| + |T(e_2)| + ... + |T(e_n)|\}$

$\therefore |T(x)| \le m \cdot M |x|$

Thus for every $x \in R^n$, we have

$|T(x)| \le m \cdot M |x|$

So, for $x, y \in R^n$, we get

$|T(x-y)| \le m \cdot M |x-y|$

$|T(x) - T(y)| \le m \cdot M |x-y|$

For a given $\varepsilon > 0$, choose $\delta = \dfrac{\varepsilon}{m\,M}$

$\therefore |x-y| < \delta \Rightarrow |x-y| < \dfrac{\varepsilon}{m\,M}$

i.e., $m\,M |x-y| < \varepsilon$

$\therefore |T(x) - T(y)| \le m\,M |x-y| < \varepsilon$

$\therefore T$ is uniformly continuous.

3) Show that if $x \in R^n$ is an eigen vector of T belonging to a real eigen value α of T then x is also an eigen vector of e^T belonging to eigen value e^α .

Solution: Let x be an eigen vector of T belonging to eigen value α . By the definition,

$Tx = \alpha T x = \alpha x$

$\therefore T^2 x = T(\alpha x) = \alpha T(x) = \alpha^2 x$

i.e., $\quad T^j x = \alpha^j x \qquad , \quad j = 0, 1, 2 .....$

$\therefore \quad \dfrac{T^j x}{j!} = \dfrac{\alpha^j x}{j!}$

Hence,

$\left(1 + T + \dfrac{T^2}{2!} + ... \dfrac{T^n}{n!}\right) x = 1x + Tx + \dfrac{T^2}{2!} x + ... \dfrac{T^n}{n!} x$

$$= 1 \cdot x + \alpha\, x + \dfrac{\alpha^2}{2!} x + ... + \dfrac{\alpha^n}{n!} x$$

$$= \left(1 + \alpha + \dfrac{\alpha^2}{2!} + ... + \dfrac{\alpha^n}{n!}\right) x$$

As, $\quad e^T = \lim\limits_{n \to \infty}\left(I + T + \dfrac{T^2}{2!} + ... \dfrac{T^n}{n!}\right)$

$$e^T(x) = \left(\lim\limits_{n \to \infty}\left(1 + T + \dfrac{T^2}{2!} + ... \dfrac{T^n}{n!}\right)\right) x$$

$$= \lim\limits_{n \to \infty}\left(\left(1 + T + \dfrac{T^2}{2!} + ... \dfrac{T^n}{n!}\right) x\right)$$

$$= \lim\limits_{n \to \infty}\left(\left(1 + \alpha + \dfrac{\alpha^2}{2!} + ... \dfrac{\alpha^n}{n!}\right) x\right)$$

$$= \left(\lim\limits_{n \to \infty}\left(1 + \alpha + \dfrac{\alpha^2}{2!} + ... \dfrac{\alpha^n}{n!}\right) x\right)$$

$$= e^\alpha x$$

4) If $T \in L(R^n)$ show that,

$\alpha = \max \ \{|T(x)| \cdot \ \text{such that} \ |x| \le 1\}$

$\beta = \max \ \{|T(x)| \cdot \ \text{such that} \ |x| = 1\}$

$\gamma = \sup \ \left\{\dfrac{|T(x)|}{|x|}, \ x \ne 0 \ , \ x \in R^n\right\}$

Solution: Let us denote these expressions by α, β, γ respectively.

i.e., $\quad \alpha = \max \left\{ |T(x)| \cdot \text{ such that } |x| \le 1 \right\}$

$\qquad \beta = \max \left\{ |T(x)| \cdot \text{ such that } |x| = 1 \right\}$

$\qquad \gamma = \sup \left\{ \dfrac{|T(x)|}{|x|}, x \ne 0, x \in R^n \right\}$

Obviously, $\quad \beta \le \alpha$

Since, on $\quad |x| = 1$, $\quad |T(x)| = \dfrac{|T(x)|}{|x|}$

we have end as $\quad \left\{ x \in R^n \text{ such that } |x| = 1 \right\} \le \left\{ x \in R^n \text{ such that } |x| \ne 0 \right\}$

then we have, $\quad \beta \le \gamma$.

For any $x \in R^n$ such that $x \ne 0$,

Consider $\quad y = \dfrac{x}{|x|} \quad$ then $\quad |y| = 1$

Further, $\quad T(y) = \dfrac{T(x)}{|x|} \quad \therefore |T(y)| = \dfrac{|T(x)|}{|x|}$

Thus for each number $\quad \dfrac{|T(x)|}{|x|}$, $x \ne 0$

We have an equal value number $|T(y)|$ with $|y| = 1$.

$\therefore \beta \ge \gamma$. But we already have $\beta \le \gamma$.

$\therefore \quad \beta = \gamma$

Thus we have $\quad \alpha \ge \beta = \gamma$

We show that $\quad \alpha \le \gamma$

Since for $x = 0$, $T(x) = 0$, $|T(x)| = 0$ and as $|T(x)| \ge 0$, $x \in R^n$, we can consider

$\alpha = \max \left\{ |T(x)| \text{ such that } x \ne 0, |x| \le 1 \right\}$

COMPLEX VECTOR SPACES

For $x \neq 0$, $|x| \leq 1$

$$\frac{|T(x)|}{|x|} \geq |T(x)| \qquad \because \left\{ \frac{1}{|x|} \geq 1 \right\}$$

Thus,

$$\max \left\{ |T(x)| \quad such\,that,\; x \neq 0, |x| \leq 1 \right\} = \alpha$$

$$\leq \max \left\{ \frac{|T(x)|}{|x|},\; x \neq 0,\; |x| \leq 1 \right\}$$

$$\leq \sup \left\{ \frac{|T(x)|}{|x|},\; x \neq 0,\; x \in R^n \right\} = \gamma$$

$$\therefore \quad \alpha \leq \gamma$$

Thus we have $\quad \alpha = \beta = \gamma$

Thus we can use,

$$\max \left\{ |T(x)| \quad such\,that \; |x| = 1 \right\}$$

a) the definition of $\|T\|$.

5) Find the uniform norm of $\begin{bmatrix} 3 & 0 \\ 0 & -4 \end{bmatrix} \begin{bmatrix} 0 & 1 \\ -1 & 0 \end{bmatrix}$

Solution. Let us denote any point $\bar{x}$ of R^2 by (x,y).

And we consider $|\bar{x}|$ to the Euclidean norm.

$\therefore$ Thus $|\bar{x}| = 1$ is given by $\left\{ (x,y) \in R^2 \quad such\,that \; x^2 + y^2 = 1 \right\}$

Thus, $(x,y) \in R^2$ with $x^2 + y^2 = 1$.

Consider $T(x,y)$ i.e. $\begin{bmatrix} 3 & 0 \\ 0 & -4 \end{bmatrix} \begin{bmatrix} x \\ y \end{bmatrix} = \begin{bmatrix} 3x \\ -4y \end{bmatrix}$

$\therefore \quad |T(x,y)| = \sqrt{9x^2 + 16y^2}$ with $\quad x^2 + y^2 = 1$

i.e. $|T(x,y)| = \sqrt{9x^2 + 16(1-x^2)} = \sqrt{16 - 7x^2}$, $\quad -1 < x < 1$

$\therefore$ Since $\quad \|T\| = \max \ \left\{ |T(x,y)| \ \text{such that} \ x^2 + y^2 = 1 \right\}$

it follows that,

$\|T\| = 4$

Now let $\quad T(x,y) \quad$ i.e. $\quad \begin{bmatrix} 0 & 1 \\ -1 & 0 \end{bmatrix} \begin{bmatrix} x \\ y \end{bmatrix} = \begin{bmatrix} x \\ -y \end{bmatrix}$

$|T(x,y)| = \sqrt{y^2 + (1 - y^2)} = \sqrt{y^2 + 1 - y^2} = \sqrt{1} = 1$

CHAPTER 3
HOMOGENEOUS LINEAR SYSTEM

We consider a system of homogeneous linear differential equation given by

$$\frac{dx}{dt} = Ax \quad \text{where} \quad x \in R^n \quad A \in L(R^n)$$

Since an operator of the type e^{tA} operating on an element on R^n is expected to be a solution of this system, we first find $\frac{d}{dt} e^{tA}$.

We consider $t \to e^{tA}$ as a map from $R \to L(R^n) = R^{n^2}$

Theorem 1: Prove that $\dfrac{d}{dt} e^{tA} = e^{tA} \cdot A = A \cdot e^{tA}$

Proof: By definition ,

$$\frac{d}{dt} e^{tA} = \lim_{\to 0} \left(\frac{e^{(t+h)A} - e^{tA}}{h} \right)$$

$$= \lim_{h \to 0} \frac{e^{tA} \cdot e^{hA} - e^{tA}}{h} \quad \begin{cases} \because (t+h)A = tA + hA \;\; and \\ ta \cdot hA = hA \cdot tA \end{cases}$$

$$= \lim_{h \to 0} \frac{e^{tA} \left(e^{hA} - I \right)}{h}$$

$$= e^{tA} \lim_{h \to 0} \left(\frac{e^{hA} - I}{h} \right)$$

By definition of e^{hA}, we have

$$e^{hA} - I = I + hA + \frac{h^2 A^2}{2!} + \dots - I$$

$$\therefore \quad \lim_{h \to 0} \frac{e^{hA} - I}{h} = A$$

$$\therefore \quad \frac{d}{dt}e^{tA} = e^{tA} \cdot A$$

$$e^{tA} \cdot A = \left(\lim_{n \to \infty} \left(\sum_{k=0}^{n} \frac{(tA)^k}{k!} \right) \right) \cdot A$$

$$= \lim_{n \to \infty} \left(\sum_{k=0}^{n} \frac{(tA)^k}{k!} \cdot A \right)$$

$$= \lim_{n \to \infty} \left(\sum_{k=0}^{n} A \cdot \frac{(tA)^k}{k!} \right)$$

$$= A \lim_{n \to \infty} \left(\sum_{k=0}^{n} \frac{(tA)^k}{k!} \right)$$

$$= A \cdot e^{tA}$$

Thus $\quad \dfrac{d}{dt}e^{tA} = e^{tA} \cdot A = A \cdot e^{tA}$

Theorem 2: Let A be an operator on R^n. Then the solution to the initial value problem $x' = Ax$, $x(0) = k \in R^n$ is $e^{tA} \cdot k$ and there are no other solution.

Proof. Consider the function

$$x(t) = e^{tA} \cdot k$$

$$\therefore \quad x'(t) = \frac{d}{dt}\left(e^{tA} \cdot k \right) = \left(\frac{d}{dt}e^{tA} \right)k$$

$$= A \cdot e^{tA} \ k = A \ x(t)$$

i.e. $x'(t) = A \cdot x(t)$

Thus

HOMOGENEOUS LINEAR SYSTEM

$e^{tA} \cdot k$ is a solution of the homogenous differential equation

$$x'(t) = A \cdot x(t)$$

Further $x(t) = e^{tA} \cdot k$

$$x(0) = e^0 \cdot k = 1 \cdot k = k$$

Thus $e^{tA} \cdot k$ is a solution of the given initial value problem.

If possible,

let $y(t)$ be a solution of the initial value problem $x' = Ax$, $x(0) = t$

i.e., $y'(t) = A y(t)$, $y(0) = k$

Consider a new function $z(t) = e^{-tA} \cdot y(t)$

Obviously, $z(t)$ is differentiable.

$$\therefore \quad z'(t) = -A e^{-tA} \cdot y(t) + e^{-tA} \cdot y'(t)$$

$$= -A e^{-tA} \cdot y(t) + e^{-tA} \cdot A y(t)$$

$$= -A e^{-tA} \cdot y(t) + A e^{-tA} \cdot y(t)$$

$$= 0$$

As $z'(t) = 0$, we must have

$z(t)$ = constant.

i.e. $e^{-tA} \cdot y(t) =$ constant.

Since this holds for all t, put $t = 0$ to get $y(0) =$ constant

i.e., constant=k

Thus $e^{-tA} \cdot y(t) = k$

$$\therefore \quad y(t) = e^{tA} \cdot k$$

Thus the solution is uniquely given by $e^{tA} \cdot k$.

Ex 1: Consider the system,

$$x_1' = a x_1$$

$$x_2' = b x_1 + a x_2; \qquad x_1(0) = 1 \ , \ x_2(0) = 2$$

Here the matrix for the operator A is

$$\begin{bmatrix} a & 0 \\ b & a \end{bmatrix} \quad \text{and} \quad k = \begin{bmatrix} 1 \\ 2 \end{bmatrix}$$

$\therefore$ The solution is given by

$$x(t) = e^{tA} \cdot k$$

$$\therefore \ tA = \begin{bmatrix} ta & 0 \\ tb & ta \end{bmatrix} = \begin{bmatrix} ta & 0 \\ 0 & ta \end{bmatrix} \begin{bmatrix} 0 & 0 \\ tb & 0 \end{bmatrix}$$

As $\begin{bmatrix} ta & 0 \\ 0 & ta \end{bmatrix} = ta \cdot 1$, the two matrices on the RHS commute with each other.

$$\therefore \ e^{tA} = e^{\begin{bmatrix} ta & 0 \\ 0 & ta \end{bmatrix}} \cdot e^{\begin{bmatrix} 0 & 0 \\ tb & 0 \end{bmatrix}}$$

$$= \begin{bmatrix} e^{ta} & 0 \\ 0 & e^{ta} \end{bmatrix} \cdot e^{\begin{bmatrix} 0 & 0 \\ tb & 0 \end{bmatrix}}$$

As, $\begin{bmatrix} 0 & 0 \\ tb & 0 \end{bmatrix}^2 = \begin{bmatrix} 0 & 0 \\ 0 & 0 \end{bmatrix}$

$$\therefore \ e^{\begin{bmatrix} 0 & 0 \\ tb & 0 \end{bmatrix}} = I + \begin{bmatrix} 0 & 0 \\ tb & 0 \end{bmatrix} = \begin{bmatrix} 1 & 0 \\ tb & 1 \end{bmatrix}$$

$$\therefore \ e^{tn} = \begin{bmatrix} e^{tA} & 0 \\ 0 & e^{tA} \end{bmatrix} \begin{bmatrix} 1 & 0 \\ tb & 1 \end{bmatrix}$$

$$= e^{tA} \cdot I \begin{bmatrix} 1 & 0 \\ tb & 1 \end{bmatrix} = e^{ta} \begin{bmatrix} 1 & 0 \\ tb & 1 \end{bmatrix}$$

The solution is

$$x(t) = e^{tA} \cdot k = e^{ta} \cdot \begin{bmatrix} 1 & 0 \\ tb & 1 \end{bmatrix} \begin{bmatrix} 1 \\ 2 \end{bmatrix}$$

$$= e^{ta} \begin{bmatrix} 1 \\ tb+2 \end{bmatrix} = \begin{bmatrix} e^{ta} \\ (2+tb)e^{ta} \end{bmatrix}$$

Thus the solution is

$$x(t) = \begin{bmatrix} x_1(t) \\ x_2(t) \end{bmatrix} \text{ with } x_1(t) = e^{ta}, \ x_2(t) = (2+tb)e^{ta}$$

QUALITATIVE BEHAVIOUR OF SOLUTION

Consider $x' = Ax$, $A \in L(R^2)$, $x \in R^2$

Case 1: A is diagonalizable.

We shall assume the diagonalized form as

$$B = \begin{bmatrix} \lambda & 0 \\ 0 & \mu \end{bmatrix}$$

And the equivalent system given by

$$y' = By$$

Case (a):- $\lambda \neq \mu < 0$

Case (b):- $\lambda = \mu < 0$

Case (c):- $\lambda < 0 < \mu$

Case (d):- $0 < \lambda < \mu$

Case (e):- $0 < \lambda = \mu$

By phase portrait we mean plot of $y_1(t)$ and $y_2(t)$ in the (y_1, y_2) plane for different value of t .

DYNAMICAL SYSTEM – A SHORT COURSE

Consider the **Case (b)**: $\quad \lambda = \mu < 0$

Now, here $\quad y_1'(t) = \lambda\, y_1(t)$

$$y_2'(t) = \lambda\, y_2(t)$$

i.e., $\quad y_1(t) = y_1(0)e^{\lambda t}$

$$y_2(t) = y_2(0)e^{\mu t}$$

In this case both $y_1(t)$ as well as $y_2(t) \to 0$ as $t \to \infty$

Now slope of the curve $(y_1(t),\, y_2(t))$ is given by,

$$\frac{y_2'(t)}{y_1'(t)} = \frac{y_2'(0)}{y_1'(0)}, \quad \text{a constant}$$

Thus the phase portrait in this case consist of straight lines moving to origin.

In this case we say that origin is the focus of the differential equation or the differential system has focus at origin.

Case (c) . $\lambda < 0 < \mu$

In this case the solution are given by

$$y_1(t) = y_1(0)e^{\lambda t}$$

$$y_2(t) = y_2(0)e^{\mu t}$$

$$\therefore \; |y_1(t)| \to 0 \;, \; |y_2(t)| \to \infty \text{ as } t \to \infty$$

Slope is given by,

$$\frac{y_2'(t)}{y_1'(t)} = \frac{y_2(0)\mu e^{\mu t}}{y_1(0)\lambda e^{\lambda t}} = \frac{y_2(0)}{y_1(0)}\,\frac{\mu}{\lambda}\,e^{\lambda t}$$

138

HOMOGENEOUS LINEAR SYSTEM

As $\quad \lambda < 0 < \mu$, $\quad \dfrac{\left|y_2'(t)\right|}{\left|y_1'(t)\right|} \to \infty$ as $\quad t \to \infty$

In this case origin is known as Saddle point.

Case d). Consider the case, $\quad \lambda < \mu < 0$

Now, $\quad y_1(t) = y_1(0)e^{\lambda t}$

$$y_2(t) = y_2(0)e^{\mu t}$$

$\therefore \quad y_1(t) \to 0$, $y_2(t) \to 0 \quad as \quad t \to \infty$

Slope is given by

$$\frac{y_2(0)\mu e^{\mu t}}{y_1(0)\lambda e^{\lambda t}} = \frac{y_2(0)}{y_1(0)} \frac{\mu}{\lambda} e^{(\mu - \lambda)t}$$

As $\lambda < \mu$, we have,

$$\frac{\left|y_2'(t)\right|}{\left|y_1'(t)\right|} \to \infty \text{ as } t \to \infty$$

In this case we say that the origin is a node.

for $\quad \lambda = \mu > 0$, origin is source.

Case 2: i). The operator is not diagonalizable.

Consider the case where, $\quad \beta = \begin{bmatrix} \lambda & 0 \\ 1 & \lambda \end{bmatrix}$, $\quad \lambda < 0$

$\therefore \quad y' = By \quad \Rightarrow \quad y_1'(t) = \lambda\, y_1(t)$

$$y_2'(t) = y_1(t) + \lambda\, y_2(t)$$

$\therefore \quad y_1(t) = y_1(0)e^{\lambda t}$

$\therefore \quad y_2'(t) = y_1(0)e^{\lambda t} + \lambda\, y_2(t)$

i.e., $\quad y_2'(t) - \lambda y_2(t) = y_1(0)e^{\lambda t}$

$$y_2'(t)e^{-\lambda t} - \lambda y_2(t) \cdot e^{-\lambda t} = y_1(0)e^{\lambda t} \cdot e^{-\lambda t}$$

$$\frac{d}{dt}\left(y_2(t)e^{-\lambda t}\right) = y_1(0)$$

$$y_2(t)e^{-\lambda t} = y_1(0) + C$$

$$y_2(t) = y_1(0)t e^{-\lambda t} + C \cdot e^{-\lambda t}$$

i.e., $\quad y_2(t) = y_1(0)t e^{\lambda t} + y_2(0) \cdot e^{-\lambda t}$

Thus we have,

$$y_1(t) = y_1(0)e^{\lambda t}$$

$$y_2(t) = (y_1(0) + y_2(0))e^{\lambda t}$$

Obviously, $\quad y_1(t) \to 0 \quad$ and $\quad y_2(t) \to 0 \quad as \quad t \to \infty$

We have,

$$y_1'(t) = \lambda y_1(0)e^{\lambda t}$$

$$y_2'(t) = y_1(0)e^{\lambda t} + \lambda(y_1(0)t + y_2(0))e^{\lambda t}$$

$$= (y_1(0) + \lambda y_1(0)t + \lambda y_2(0))e^{\lambda t}$$

In this case we say that origin is an improper node.

ii) we consider the case

where $\quad B = \begin{bmatrix} a & -b \\ b & a \end{bmatrix} \quad b \neq 0 \;, \; a < 0$

Suppose $b > 0$.

i.e. $\quad y' = B y$

$$\therefore \quad y(t) = e^{tB} \cdot k \qquad \text{where} \quad k = \begin{bmatrix} y_1(0) \\ y_2(0) \end{bmatrix}$$

$$\therefore \quad tB = \begin{bmatrix} ta & -tb \\ tb & ta \end{bmatrix}$$

$$\therefore \quad e^{tB} = e^{ta}\begin{bmatrix} \cos(tb) & -\sin(tb) \\ \sin(tb) & \cos(tb) \end{bmatrix}$$

$$\therefore \quad \begin{bmatrix} y_1(t) \\ y_2(t) \end{bmatrix} = e^{ta}\begin{bmatrix} \cos(tb) & -\sin(tb) \\ \sin(tb) & \cos(tb) \end{bmatrix}\begin{bmatrix} y_1(0) \\ y_2(0) \end{bmatrix}$$

$$\therefore \quad y_1(t) = e^{ta}\big(\cos tb - \sin tb\big)\big(y_1(0)\big)$$

$$y_2(t) = e^{ta}\big(\sin tb + \cos tb\big)\big(y_2(0)\big)$$

Since the operator $\quad e^{at}\begin{bmatrix} \cos(tb) & -\sin(tb) \\ \sin(tb) & \cos(tb) \end{bmatrix}$

represents scaling and rotation about origin.

It follows that as $a < 0$ and $b > 0$.

The solution curves $(y_1(t), y_2(t))$ will approach the origin revolving around it.

The solution curves spirals towards origin.

The origin in this case is $k^n/_{as}$ Spiral Sink.

iii) We next consider the case where

$$B = \begin{bmatrix} 0 & -b \\ b & 0 \end{bmatrix} , \quad b > 0$$

$$y' = By \implies y(t) = e^{tB} \cdot y(0)$$

$$e^{tB} = e^{\begin{bmatrix} 0 & -tb \\ tb & 0 \end{bmatrix}} = \begin{bmatrix} \cos(tb) & -\sin(tb) \\ \sin(tb) & \cos(tb) \end{bmatrix}$$

$$\begin{bmatrix} y_1(t) \\ y_2(t) \end{bmatrix} = \begin{bmatrix} \cos(tb) & -\sin(tb) \\ \sin(tb) & \cos(tb) \end{bmatrix}\begin{bmatrix} y_1(0) \\ y_2(0) \end{bmatrix}$$

DYNAMICAL SYSTEM – A SHORT COURSE

In this case there is no scaling as $a = 0$, but only rotation about the origin, the rotation is clockwise or anticlockwise depends on ' b '.

origin is k^n as Centre.

We say that origin is Sink of the system $x' = Ax$, provided all the eigen values of A have negative part.

If all the eigen values have positive real part then we say that origin is source of the system.

NON-HOMOGENEOUS SYSTEM

Consider a system of Differential Equation of first order of the type

$$x' = Ax + B(t)$$

where A is an $n \times n$ matrix (a linear operator on R^n) and $B(t)$ be an element of R^n.

We assume that B is a continuous function of time from R to R^n.

We consider an expression of the type $e^{tA} \cdot f(t)$, where $t \to f(t)$ is a differentiable R^n of t from R to R^n and check is a solution of the system

$$x' = Ax + B(t)$$

$$\therefore \frac{d}{dt}\left(e^{tA} \cdot f(t)\right) = A e^{tA} f(t) + e^{tA} f'(t)$$

Thus for $e^{tA} \cdot f(t)$ to be a solution of a system $x' = A(x) + B(t)$ we must have

$$A e^{tA} f(t) + e^{tA} f'(t) = A e^{tA} f(t) + Bt$$

i.e., $A e^{tA} f(t) = B(t)$

Thus an expression of the type

142

HOMOGENEOUS LINEAR SYSTEM

$A e^{tA} f(t)$ will be a solution of system $x' = Ax + B(t)$ if

$A e^{tA} f'(t) = B(t)$, $t \in R$

i.e., $f'(t) = e^{-tA} \cdot B(t)$

Integrating

i.e., $f(t) = \int_0^t e^{-SA} B(s) ds + K$, where $K = f(0) \in R^n$

Conversely, We can show that an expression of type,

$e^{tA} \left(\int_0^t e^{-SA} B(s) ds + K \right)$ operating on this is the solution of system

$x' = Ax + B(t)$

This is because

$$\frac{d}{dt}\left(e^{tA}\left(\int_0^t e^{-SA} B(s) ds + K \right) \right) = e^{tA}\left(\int_0^t e^{-SA} B(s) ds + K \right) + e^{tA} \cdot e^{-tA} B(t)$$

$$= e^{tA}\left(\int_0^t e^{-SA} B(s) ds + K \right) + B(t)$$

Now we show that a solution to the non-homogeneous system

$x' = Ax + B(t)$ is of the above form.

Thus, let $x(t) = e^{tA}\left(\int_0^t e^{-SA} B(s) ds + K \right)$ and

Let $y(t)$ be any other solution of system $x' = Ax + B(t)$

As $y' = Ay + B(t)$

$\therefore$ $x' - y' = Ax - Ay$

i.e. $(x - y)' = A(x - y)$

143

i.e. $x(t) - y(t)$ is a solution of the corresponding homogeneous system

$x' = Ax$.

$\therefore$ $x(t) - y(t) = e^{tA} \cdot K_1$ for some constant $K_1 \in R^n$.

$\therefore$ $y(t) = x(t) - e^{tA} \cdot K_1$

$$= e^{tA} - \left(\int_0^t e^{-SA} B(s)ds + K \right) - e^{tA} \cdot K_1$$

$$= e^{tA} - \left(\int_0^t e^{-SA} B(s)ds + K \right) - e^{tA}(K - K_1)$$

Thus the expression for $y(t)$ is similar to that of $x(t)$. The expression

$$e^{tA} \left(\int_0^t e^{-SA} B(s)ds \right)$$ is also a solution to the non homogeneous system

$x'(t) = Ax(t) + B(t)$

$\therefore$ It is known as a particular solution of non-homogeneous system

$x'(t) = Ax(t) + B(t)$.

Theorem 3: Any solution of non-homogeneous system $x' = Ax + B(t)$ is a sum of a particular solution of the non homogeneous system and a solution of the corresponding homogeneous system.

Proof: We know that any solution of the non homogeneous system $x' = Ax + B(t)$ is of the type

$$x(t) = e^{tA} \left(\int_0^t e^{-SA} B(s)ds + K \right)$$

i.e. $x(t) = e^{tA} \int_0^t e^{-SA} B(s)ds + e^{tA} \cdot K$

HOMOGENEOUS LINEAR SYSTEM

But we know that

$e^{tA} \int\limits_{0}^{t} e^{-SA} B(s)ds$ is a particular solution of the non homogeneous

system while $e^{tA} \cdot K$ is a solution of the corresponding homogeneous
system $x' = A x$.

Ex 2 : Find the general solution to

$$x_1' = -x_2$$

$$x_2' = x_1 + t \quad \text{Here} \quad A = \begin{bmatrix} 0 & -1 \\ 1 & 0 \end{bmatrix}, \ B(t) = \begin{bmatrix} o \\ t \end{bmatrix}$$

Solution. We have,

$$-SA = \begin{bmatrix} 0 & S \\ -S & 0 \end{bmatrix} \quad \text{then} \ \exp(-SA) \ \text{will be}$$

$$e^{-SA} = \begin{bmatrix} \cos(-S) & -\sin(-S) \\ \sin(-S) & \cos(-S) \end{bmatrix} = \begin{bmatrix} \cos S & \sin S \\ -\sin S & \cos S \end{bmatrix}$$

Similarly

$$e^{tA} = \begin{bmatrix} \cos t & -\sin t \\ \sin t & \cos t \end{bmatrix}$$

$$\therefore \ \int\limits_{0}^{t} e^{-SA} B(S)ds = \int\limits_{0}^{t} \begin{bmatrix} \cos S & \sin S \\ -\sin S & \cos S \end{bmatrix} \begin{bmatrix} 0 \\ s \end{bmatrix} ds$$

$$= \int\limits_{0}^{t} \begin{bmatrix} S \sin S \\ S \cos S \end{bmatrix} ds = \begin{bmatrix} \int\limits_{0}^{t} S \sin S \ ds \\ \int\limits_{0}^{t} S \cos S \ ds \end{bmatrix}$$

$$= S(-\cos S)_0^t + \int\limits_{0}^{t} \cos s \ ds$$

145

$$= \begin{bmatrix} \sin t - t\cos t \\ \cos t + t\sin t - 1 \end{bmatrix}$$

A general solution of the given non homogeneous system.

$$\begin{bmatrix} \cos t & -\sin t \\ \sin t & \cos t \end{bmatrix} \left(\begin{bmatrix} \sin t & -t\cos t \\ \cos t & +t\sin t - 1 \end{bmatrix} + \begin{bmatrix} K_1 \\ K_2 \end{bmatrix} \right)$$

$$x_1(t) = -t + K_1 \cos t + (1 - K_2)\sin t$$

$$x_2(t) = 1 - (1 - K_2)\cos t + K_1 \sin t$$

2) $\quad x' = y$

$$y' = -4x + \sin 2t$$

$$\rightarrow \quad A = \begin{bmatrix} 0 & 1 \\ -4 & 0 \end{bmatrix}$$

HIGHER ORDER SYSTEM

Consider an nth order DE

$$y^{(n)} + a_1 y^{(n-1)} + a_2 y^{(n-2)} + \dots - 1 a_n y = 0$$

with initial condition $y(0) = \alpha_1$

$$y'(0) = \alpha_2$$

$$\vdots$$

$$y^{(n-1)}(0) = \alpha_n$$

We introduce new variables namely

$$x_1 \equiv y \quad , \quad x_2 \equiv y' \dots x_n = y^{(n-1)}$$

Thus the given n^{th} order initial value problem is equivalent to the system of n- simultaneous first order difference

$$x_1' = x_2 \quad , \quad x_2' = x_3 \dots x_{n-1}' = x_n \quad , x_n' = -a_1 x_{n-a_2} x_n$$

HOMOGENEOUS LINEAR SYSTEM

i.e., $x_1' = x_2$, $x_2' = x_3 .... x_{n-1}' = x_n$, $x_n' = -a_1 x_1 - a_n ..... - a_n x_1$

with

$$x_1(0) = \alpha_1 \quad , \quad x_2(0) = \alpha_2 .... x_n(0) = \alpha_n$$

If $\begin{pmatrix} x_1 \\ x_2 \\ \vdots \\ x_n \end{pmatrix}$ is denoted by $\overline{x}$ then we get

$\overline{x} = A\overline{x}$ with $\overline{x}(0) = \overline{\alpha}$ where $\overline{\alpha} = (\alpha_1 \alpha_2 ..... \alpha_n)$

and

$$A = \begin{bmatrix} 0 & 1 & 0 & .... \\ 0 & 0 & 1 & .... \\ 0 & 0 & 0 & 1.... \\ \vdots & & & \\ 0 & .... & & ...1 \\ -a_n & -a_{n-1} & ...... a_n \end{bmatrix}$$

Since $\det (A - \lambda I) = 0$ iff $\det (\lambda I - A) = 0$

(if they are equal or opposite to each other).

We can take $\det (\lambda I - A)$ as the characteristic polynomial for the $m \times n$ matrix A.

We show that the characteristic polynomial is given by

$$\lambda^n + a_1 \lambda^{n-1} + ... + a_{n-1} \lambda_0 + a_n = 0$$

The result is true for $n = 2$.

Here the DE is $y'' + a_1 y' + a_2 y = 0$ and the corresponding system is

$$x_1' = x_2 \quad \text{and} \quad x_2' = -a_2 x_1 - a_1 x_2$$

147

and $\quad A = \begin{bmatrix} 0 & 1 \\ -a_2 & -a_1 \end{bmatrix}$

$\det (A - \lambda I) = \begin{vmatrix} -\lambda & 1 \\ -a_2 & -a_1 - \lambda \end{vmatrix}$

$\qquad\qquad = \lambda (a_1 + \lambda) + a_2$

$\qquad\qquad = \lambda^2 + a_1 \lambda + a_2$

$\qquad\qquad = \det (\lambda I - A)$

So we now assume the result to be true for a system of $n-1$ simultaneous equation and prove it to be true for n. we note that

$$(\lambda 1 - A) = \begin{bmatrix} \lambda & 0 ..... \\ 0 & \lambda ..... \\ \vdots \\ 0 ....\lambda \end{bmatrix} - \begin{bmatrix} 0 & 1 & 0 .... \\ 0 & 0 & 1 .... \\ 0 & 0 & 0 & 1.... \\ \vdots \\ 0 .... +1 \\ -a_n & -a_{n-1} ... + a_1 \end{bmatrix}$$

$\therefore \quad \det (\lambda 1 - A) = \lambda$ (corresponding expression for $n-1$ order

$\qquad\qquad\qquad\qquad\qquad\qquad\qquad$ case $+ a_n$)

$\qquad\qquad = \lambda \left(\lambda^{n-1} + a_1 \lambda^{n-2} + ... + a_{n-1} \right) + a_n$

$\qquad\qquad = \lambda^n + a_1 \lambda^{n-1} + ... + a_{n-1} \lambda + a_n$

Theorem 4: Let λ_1, λ_2 be the roots of the polynomial $\lambda^2 + a\lambda + b$. Then every solution of the differential equation $y'' + a y' + b$ is of the following type.

Case (a). λ_1, λ_2 are real and distinct

$$\lambda(t) = c_1 e^{\lambda_1 t} + c_2 e^{\lambda_2 t}$$

(b). $\lambda_1 = \lambda_2 = \lambda$ real , $\lambda(t) = c_1 e^{\lambda t} + c_2 t e^{\lambda t}$

HOMOGENEOUS LINEAR SYSTEM

(c). λ_1, λ_2 are complex with

$$\lambda_1 = a + ib \text{ (i.e. } \lambda_2 = a - ib)$$

$$\lambda(t) = e^{at}\left(c_1 \cos bt + c_2 \sin bt\right)$$

Proof. a) We note that the corresponding system of DE is

$$x_1' = x_2 \ , \quad x_2' = -bx_1 - ax_2 \qquad \begin{pmatrix} x_1 \equiv y \\ x_2 \equiv y' \end{pmatrix}$$

And the origin values of the operator $A = \begin{bmatrix} 0 & 1 \\ -b & -a \end{bmatrix}$ are given by the roots

of the characteristic polynomial $\lambda^2 + a\lambda + b = 0$.

In case where $\lambda_1 \neq \lambda_2$ are real roots of $\lambda^2 + a\lambda + b = 0$

We have A with distinct real roots λ_1 and λ_2.

We know that in this case $x_1(t)$ as well as $x_2(t)$ are linear combinations

of $e^{\lambda_1 t}$ and $e^{\lambda_2 t}$

i.e., $x_1(t) = y(t) = c_1 e^{\lambda_1 t} + c_2 e^{\lambda_2 t}$

b) In case $\lambda_1 = \lambda_2 = \lambda$ (real).

It can be shown that the matrix A can be transformed into the form

$\begin{bmatrix} \lambda & 0 \\ \beta & \lambda \end{bmatrix}$ where $\beta \neq 0$.

In this case the given system is equivalent to new system say

$$y_1'(t) = \lambda y_1(t)$$

$$y_2'(t) = \beta y_1(t) + \lambda y_2(t)$$

i.e. $y_1(t) = y_1(0)e^{\lambda t}$

$$y_2'(t) - \lambda y_2(t) = \beta y_1(0)e^{\lambda t}$$

i.e. $\dfrac{d}{dt}\left(y_2(t)e^{-\lambda t}\right) = \beta y_1(0).$

i.e. $y_2(t)e^{-\lambda t} = \beta y_1(0)t + c$.(After Integrating)

i.e. $y_2(t) = \beta y_1(0)t e^{\lambda t} + c e^{\lambda t}.$

i.e. $y_2(t) = \beta y_1(0)t e^{\lambda t} + y_2(0)e^{\lambda t}.$

c) We now consider the case where

$$\lambda_1 = a + ib \qquad \lambda_2 = a - ib \quad , \quad b \neq 0$$

We know that there exist a basis with respect to

which the matrix A gets transform into the form

$$\begin{bmatrix} a & -b \\ b & a \end{bmatrix}.$$

Thus we can write an equivalent system

$$\therefore \; y'(t) = \begin{bmatrix} a & -b \\ b & a \end{bmatrix} y(t)$$

$$\therefore \; y(t) = \left(\exp + \begin{bmatrix} a & -b \\ b & a \end{bmatrix} \right) y(0)$$

Here $y'(t) = \begin{bmatrix} y_1'(t) \\ y_2'(t) \end{bmatrix}$ and $y_0 = \begin{bmatrix} y_1(0) \\ y_2(0) \end{bmatrix}$

As $t\begin{bmatrix} a & -b \\ b & a \end{bmatrix} = \begin{bmatrix} at & -bt \\ bt & at \end{bmatrix}$

We have $e^{t\begin{bmatrix} a & -b \\ b & a \end{bmatrix}} = e^{\begin{bmatrix} at & -bt \\ bt & at \end{bmatrix}}$

$$= e^{at\begin{bmatrix} \cos bt & -\sin bt \\ \sin bt & \cos bt \end{bmatrix}}$$

Hence solution is of the type

$$e^{at} \cdot \begin{bmatrix} \cos bt & -\sin bt \\ \sin bt & \cos bt \end{bmatrix} \begin{bmatrix} y_1(0) \\ y_2(0) \end{bmatrix}$$

$$e^{at} \cdot \begin{bmatrix} y_1(0)\cos bt - y_2(0)\sin bt \\ y_1(0)\sin bt + y_2(0)\cos bt \end{bmatrix}$$

Hence solution of given 2nd order DE is of type

$$= e^{at} \left(c_1 \cos bt + c_2 \sin bt \right)$$

DECOMPOSITION

Let V be a vector space and T be an operator on V.

We consider the operator $T^j, j = 0,1,2....$ where T^0 denotes the identity operator.

For $j = 0,1,2...$ we get

$N^{(j)} = \ker T^j$ and $M^{(j)} = \text{Im}age\ T^j$ (entire space)

i.e. $M^{(j)}$ is the image of V under the operator T^j.

Obviously, $N^{(j)}$ and $M^{(j)}$ are subspaces of V.

$N^{(0)} = (0)$ ($\because$ kernel of identity operator is 0)

$M^{(0)} = (0)$ ($\because$ Image of V under identity is V itself)

If $x \in N^{(j)}$ then $T^{(j)}x = 0$

$\therefore$ $N^{(j)} \subseteq N^{(j+1)}$ for $j = 0,1,2....$

Similarly, if $y \in M^{j+1}$,

$\exists$ some $x \in V$ such that

$y = T^{j+1}x = T^j(Tx)$

$\therefore$ $y \in M^{(j)}$

$\therefore$ $M^{(j)}$ contains $M^{(j+1)}$

i.e. $M^{(j)} \geq M^{(j+1)}$

We define $N(T) = \bigcup_{j \geq 0} \ker T^j = \bigcup_{j \geq 0} N^{(j)}$

$M(T) = \bigcap_{j \geq 0} M^{(j)}$

Obviously $N(T)$ and $M(T)$ are subspaces of V .

Further we can show that if

$N^{(j)} = N^{(j+1)}$ then $N^{(j)} = N^{(j+1)} = N^{(j+2)} = ...$

We know that, $N^{(j)} \subseteq N^{(j+1)} \subseteq N^{(j+2)} \subseteq ....$

$N^{(j)} = N^{(j+1)}$

$\Rightarrow$ $N^{(j+1)} \subseteq N^{(j)}$

i.e., whenever $x \in N^{(j+1)}$

i.e. whenever $T x^{(j+1)} = 0$, we have $x \in N^{(j)}$ i.e., $T^j x = 0$.

In order to show that $N^{(j+1)} = N^{(j+2)}$

Let $x \in N^{(j+2)}$

i.e. $T^{j+2} x = 0$

i.e. $T^{j+1}(T x) = 0$

$\therefore$ $T x \in \ker T^{j+1} = N^{(j+1)}$

But $N^{(j+1)} = N^{(j)}$

$\Rightarrow$ $Tx \in N^{(j)}$ $\therefore$ $T^j Tx = 0$

i.e. $T^{j+1} x = 0$

$\Rightarrow$ $x \in N^{(j+1)}$

$\therefore$ $N^{(j+2)} = N^{(j+1)}$

HOMOGENEOUS LINEAR SYSTEM

If $N^{(j)} = N^{(j+1)}$ then $N^{(i)} = N^{(j)}$ for all $i \geq j+1$

Similarly, if $M^{(j)} = M^{(j+1)}$ for some j,

then $M^{(j+1)} = M^{(j+2)} = M^{(j+3)} = ...$

$M^{(j)} = M^{(j+1)} \Rightarrow$ image of V under T^j

$\qquad\qquad =$ Image of V under T^{j+1}

Operator T on both sides to get Image of V under $T^{j+1} =$ Image of V

under T^{j+2}.

i.e. $M^{(j+1)} = M^{(j+2)}$

Thus if $M^j = M^{(j+1)}$

Then $M^{(j)} = M^{(i)}$ for all $i \geq j+1$

Since V is finite dimensional vector space and

$$N^{(0)} \subseteq N^{(1)} \subseteq N^{(2)} \subseteq .... \subseteq V$$

It follows that, there exist a +ve integer such that

$$N^{(n)} = N^{(n+1)} = N^{(n+2)} = ...$$

In this case, $N^{(T)} = N^{(n)}$

Similarly, as $V \equiv M^{(0)} \geq M^{(1)} \geq M^{(3)} .... \geq (0)$

We must have for some +ve integer 'm'

$$M^{(m)} = M^{(m+1)} = M^{(m+2)} = ...$$

Hence, $M^{(T)} = M^{(m)}$

Theorem 5: Let V be a vector space (of finite dimension) and an operator
on V. Then $V = N \oplus M$

where $N \equiv N(T) = \bigcup_{j \geq 0} \ker T^j$

$$M = M(T) = \bigcap_{j \geq 0} \quad \text{Image of } V \text{ under } T^j$$

153

Proof: We show that, $V = N + M$ and $N \cap M = (0)$

Let $x \in N \cap M$

Consider $T M$. We know that $M = M^{(m)}$ for some m

and $M^{(m)} = M^{(m+1)} = M^{(m+2)} = \dots$

$\therefore T M = T$ (Image of V under T^m)

$\qquad$ = Image of V under T^{m+1}

$\qquad$ = $M^{(m+1)} = M^{(m)} = M$

Therefore it follows that $\ker(T)|_M = (0)$

Thus T is a linear isomorphism on M.

Similarly, $T M = M \Rightarrow T^2 M = M \Rightarrow T^i M = M$, for all $i \geq 1$

Thus $T^i (i \geq 0)$ is a linear isomorphism on M.

Since $x \in N \cap M$, $x \in N$

$\therefore x \in \ker T^n$ (we can assume that $N = N^{(n)}$)

$\therefore T^n x = 0$

But $x \in M$ $\therefore T^n x = 0 \Rightarrow x = 0$

($\because T^n$ is a linear isomorphism on M)

Thus $x \in N \cap M \Rightarrow x = 0$ so $N \cap M = (0)$

Next to show that $V = N + M$

Obviously , $N + M$ is subspace

i.e. $N + M \subseteq V$

Let $x \in V$. Consider $T^m x$.

$T^m x \in M^{(m)}$ ($\because M^{(m)} = $ Image of V under T^m)

i.e., $T^m x \in M$

HOMOGENEOUS LINEAR SYSTEM

But we know that $T^i M = M$ for $i \geq 0$

In particular $T^m M = M$

As $T^m x \in M$, we can find $y \in M$ such that

$T^m y = T^m x$.

We can write, $x = x - y + y$

We have, $y \in M$,

$T^m(x - y) = T^m x - T^m y = 0$

$\therefore \ x - y \in \ker T^m \subseteq N$

So, $x - y \in N$

Thus every $x \in V$ can be written as

$x = x - y + y$ with $x - y \in N$ and $y \in M$

$\therefore V \subseteq N + M$ i.e. $V = N + M$

$\therefore V = N \oplus M$

Let V be a vector space and T an operator on V .

If V is real vector space we assume that all the eigen values of T are real.

Thus either V is complex V_s or Real V_s , then we consider only those real operators on V whose eigen values are all real.

Let $\alpha_1, \alpha_2, \ldots \alpha_k$ be the distinct eigen values of T of multiplicity say

$r_1, r_2, \ldots r_k$.

i.e., the characteristic polynomial of T is of the type, some constant

$(\lambda - \alpha_1)^{r_1} (\lambda - \alpha_2)^{r_2} \ldots (\lambda - \alpha_k)^{r_k}$

We consider the operators $T - \alpha_j 1$, $1 \leq j \leq k$

By N_j we shall mean the space $N(T - \alpha_j I)$

and by M_j we shall mean the space $M(T - \alpha_j I)$

Thus , $N_j = \bigcup_{i \geq 0} \ker \left(T - \alpha_j I \right)^i$

$M_j = \bigcap_{i \geq o}$ Images of V under $\left(T - \alpha_j I \right)^i$

Theorem 6: $V = N_1 \oplus N_2 \oplus ... \oplus N_k$.

Proof: Considering the operator $T - \alpha_1 I$

The previous theorem implies that,

$V = N_1 \oplus M_1$ where $N_1 \equiv N(T - \alpha_1 I)$, $M_1 \equiv M(T - \alpha_1 I)$

If $V = (0)$ the result is obviously true (as there are no eigen values)

If $\dim V = 1$

Then for the operator T , the characteristic polynomial of T is first (linear) degree polynomial.

Then there is only eigen value of multiplicity 1.

Further, the eigen space (being non trivial) belonging to this eigen value will be the whole space V and thus we have

$V = N$, (and there are no $N_2, N_2...$)

We now assume that the result is true for an operator on any vector space whose dimension on is less than the dim of V (strong induction)

As $V = N_1 \oplus M_1$

and N_1 is non trivial (non zero) subspace of V

$\dim \quad M_1 \;<\; \dim V$

Thus if we show that T restricted to M_1 has eigen values $\alpha_2...\alpha_k$ of multiplicities $r_2...r_k$.

Then by induction hypothesis we can write

HOMOGENEOUS LINEAR SYSTEM

$M_1 = N_2 \oplus N_3 \oplus ...N_k$

Therefore, $V = N_1 \oplus M_1$

$\Rightarrow \quad V = N_1 \oplus N_2 \oplus ... \oplus N_k$

Thus we need to show that the restricted of T to M_1 has only eigen values

$\alpha_2 ... \alpha_k$ with multiplicities $r_2 ... r_k$ and further

$N(T - \alpha_j I)|_{M_1} = N(T - \alpha_j I)$

We obviously know that,

$\ker (T - \alpha_1 I) \cap M_1 = 0$

$(\because \ker (T - \alpha_1 I) \subseteq N_1 \ and \ N_1 \cap M_1 = 0)$

We show that $\ker (T - \alpha_1 I) \cap N_j = (0)$, $2 \le j \le k$

Let $x \in \ker (T - \alpha_1 I)$ and $x \ne 0.$

We show that $x \notin N_j$, $2 \le j \le k$

As $x \in \ker (T - \alpha_1 I)$ we have $(T - \alpha_1 I)x = 0$

i.e., $Tx = \alpha_1 x.$

Thus for $2 \le j \le k$,

$(T - \alpha_j I)x = Tx - \alpha_j Ix \qquad \{\because \ Ix = x\}$

$\qquad\qquad = \alpha_1 x - \alpha_j x$

$\qquad\qquad = (\alpha_1 - \alpha_j)x$

$(T - \alpha_j I)^2 x = (T - \alpha_j I)(T - \alpha_j I)x$

$\qquad\qquad = (T - \alpha_j I)((\alpha_1 - \alpha_j)x)$

$\qquad\qquad = (\alpha_1 - \alpha_j)((T - \alpha_j I)x)$

$\qquad\qquad = (\alpha_1 - \alpha_j)(\alpha_1 - \alpha_j)x$

$$= (\alpha_1 - \alpha_j)^2 x$$

Thus $(T - \alpha_j I)^i x = (\alpha_1 - \alpha_j)^i x \qquad i = 0,1,2...$

As $x \neq 0$, and $\alpha_1 \neq \alpha_j$

So $(T - \alpha_j I)^i x \neq 0$

$\therefore x \notin \ker (T - \alpha_j I)^i$, $i = 0,1,2....$

$\therefore x \notin N_j \qquad 2 \leq j \leq k$

$\therefore \ker (T - \alpha_1 I) \cap N_j = (0) \qquad 2 \leq j \leq k$

Further,

$(T - \alpha_j I) N_j \subseteq N_j$, $2 \leq j \leq k$

For let $x \in N_j \qquad\qquad \therefore x \in \ker (T - \alpha_j I)$ for some j

i.e. $(T - \alpha_j I)^i x = 0$

$\therefore (T - \alpha_j I)(T - \alpha_1 I)x = (T - \alpha_1 I)(T - \alpha_j I)x$

$$= (T - \alpha_j I)(0) = 0.$$

$\therefore (T - \alpha_1 I)x \in N_j$

Thus $(T - \alpha_1 I) N_j \subseteq N_j$

i.e., we can consider,

$T - \alpha_1 I$ as an operator on N_j .

But as $(T - \alpha_1 I) \cap N_j = 0$, we have

$$(T - \alpha_1 I) N_j = N_j$$

$$(T - \alpha_1 I)^2 N_j = N_j$$

and so on.

i.e. $(T - \alpha_1 I)^i N_j = N_j \qquad i = 0, 1, 2....$

Thus $N_j \subseteq$ Image of V under $(T - \alpha_1 I)^i \quad i = 0, 1, 2....$

$\qquad N_j \subseteq M_1 \qquad\qquad 2 \le j \le k$

Thus on the one hand, no nor trivial eigen vector belonging to the eigen value α_1 is in M_1, all the eigen vector belonging to the eigen values $\alpha_2, \alpha_3 ... \alpha_k$ are in M_1.

Therefore if we restrict T to the space M_1 then $\alpha_2 ... \alpha_k$ are the only eigen values of T of multiplicity $r_2 ... r_k$.

Further as no non trivial eigen vector belonging to $\alpha_2 ... \alpha_k$ is in N_1, it follows that

$$N(T - \alpha_j I) \big|_{M_1} = N(T - \alpha_j I)$$

Therefore by Induction hypothesis,

$$M_1 = N_2 \oplus N_3 \oplus ... \oplus N_k$$

But we have $V = N_1 \oplus M_1$

$\therefore \quad V = N_1 \oplus N_2 \oplus N_3 \oplus ... \oplus N_k$

Thus the theorem holds for all finite dimensional spaces.

In the above proof, all the subspaces under consideration namely $N_1, N_2 ... N_k$ as well as $M_1, M_2 ... M_k$ are subspaces of V (as if V is real then this subspaces are real subspaces of V) and if V were real and one of the eigen values were non real, then the corresponding eigen space, would be subspace of V_c and not of V.

NILPOTENT OPERATOR

Let T be an operator on a vector space V. If $\exists$ a +ve integer say k such

DYNAMICAL SYSTEM – A SHORT COURSE

that T^k is zero operator on V then we say that T is a nilpotent operator.

Smallest k such that $T^k = 0$ for a nilpotent operator is known as the order of nilpotent of the operator T.

Let T be an operator on a vector space V. If V is real, we assume that all the eigen values of T are real.

Let $\alpha_1 \ldots \alpha_k$ be the distinct eigen values of T of multiplicity $r_1 \ldots r_k$ respectively.

For $j = 1$ to k we define $E_j = \ker (T - \alpha_j I)^{r_j}$

E_j is known as the generalized eigen space of T belonging to the eigen value r_j .

Theorem 7. Primary Decomposition Theorem

Let T be an operator E where E is a complex vector space or else E is real and has real eigen values. Then E is a direct sum of the generalized eigen spaces of T. The dimension of each generalized eigen space equals the multiplicity of the corresponding eigen value.

Proof: Let $\alpha_1 \ldots \alpha_k$ be the distinct eigen values of T with multiplicities $r_1 \ldots r_k$ respectively.

Thus $E_j = \ker (T - \alpha_j I)^{r_j}$ is the generalized eigen space belonging to the eigen value α_j further,

Let $N_j = \bigcup_{i \geq 0} \ker (T - \alpha_j I)^i$.

If E is a complex vector space then obviously E_j and $N_j, j = 1 \ldots k$ are all complex subspaces of E.

If E is a real vector space then as $\alpha_1 \ldots \alpha_k$ are assumed to be real in this case.

160

HOMOGENEOUS LINEAR SYSTEM

The spaces E_j and N_j will be all real and subspaces of E.

We show that,

$E = E_1 \oplus E_2 \oplus .... \oplus E_k$ and

$\dim E_j = r_j$, $1 \le j \le k$

We know that, $E = N_1 \oplus N_2 \oplus .... \oplus N_k$

We also know that $T \mid_{N_1}$ has only one eigen value namely α_1 of

multiplicity r_1 this is because

$E = N_1 \oplus M_1$ and $T \mid_{M_1}$ has only eigen values $\alpha_1 ... \alpha_k$.

Hence the characteristic polynomial of T restrict constant $(\lambda - \alpha_1)^{r_1}$

Since the degree of the characteristic polynomial of an operator once vector equals the dimension of vector space, it follows that

$\dim N_j = r_j$ $1 \le j \le k$

As $E_j = \ker (T - \alpha_j I)^{r_j}$

$N_j = \bigcup_{i \ge 0} \ker (T - \alpha_j I)^{r_j}$

Obviously $E_j \subseteq N_j$

Since $T \mid_{N_j}$ the characteristic polynomial of T is constant $(\lambda - \alpha_j I)^{r_j}$

$\therefore$ $(\lambda - \alpha_j I)^{r_j}$ must be identically zero on N_j.

i.e., $N_j \subseteq \ker (T - \alpha_j I)^{r_j} = E_j$

$\therefore N_j \subseteq E_j$

$\therefore E_j = N_j$

$\therefore$ $E = E_1 \oplus E_2 \oplus E_3 \oplus .... \oplus E_k$ with $\dim$ $E_j = r_j$

161

Ex 3: Suppose T is an operator on $V \cdot s \ E$ of dim n . Suppose T has only one eigen value say α, so obviously α will be root of character poly. of multiplicity n .

Thus we have $E = E_1$

That is $E = E_1 = \ker (T - \alpha I)^n$

We define $S = \alpha I$ where I is an $n \times n$ identity matrix.

So obviously S is diagonal.

Let $N = T - S$

Thus we have $T = S + N$

Further $SN = NS \quad (\because S = \alpha I)$

Consider N^n but $N = T - \alpha I$

$\therefore N^n = (T - \alpha I)^n$

But as $E = E_1 = \ker (T - \alpha I)^n$

$\therefore N^n = 0$ on $E = E_1$

i.e. N is nilpotent.

Thus if T is an operator with only one eigen value α then we can write T as

$T = S + N$, $S = \alpha I$ is diagonal

$N = T - S$ is nilpotent and

$SN = NS$

Theorem 8: Let $T \in L(E)$ where E is complex if T has non real eigen value. Then $T = S + N$ where $SN = NS$. S is diagonalizable and N is nilpotent.

Proof: We have E as either a complex Vs or a real Vs in which case all eigen values of T are real.

HOMOGENEOUS LINEAR SYSTEM

Let $\alpha_1 ... \alpha_k$ be the eigen values with multiplicity $r_1 ... r_k$.

By the primary decomposition theorem,

$E = E_1 \oplus E_2 \oplus .... \oplus E_k$

Where E_j is generalized eigen space belonging to the eigen value α_j.

Since $E_j = N_j = \bigcup_{i \geq 0} \ker (T - \alpha_j I)^i$,

It follows that $T E_j \subseteq E_j$

Because if $x \in E_j = N_j$ then

$(T - \alpha_j I)^i x = 0$ for some $i \geq 0$

Consider $(T - \alpha_j I)^i \equiv x = T(T - \alpha_j I)^i x = T \cdot 0 = 0$

$\therefore x \in E_j \implies T_x \in E_j$

Thus if we denote by T_j the restriction of T to E_j, it is an operator on E_j and as

$E = E_1 \oplus E_2 \oplus .... \oplus E_k$

we have $T = T_1 \oplus T_2 \oplus .... \oplus T_k$.

Consider T_j, $1 \leq j \leq k$

Since $T_j = T \Big|_{E_j} = T \Big|_{N_j}$, T_j has only one eigen value α_j of

multiplicity r_j.

So we define $S_j = \alpha_j I$ (where I is an identity matrix of

$$\dim r_j \times r_j)$$

We further define, $N_j = T_j - S_j$

We have $T_j = S_j + N_j$

S_j is diagonal, N_j is nilpotent and

$S_j N_j = N_j S_j$

We define,

$S = S_1 \oplus S_2 \oplus .... \oplus S_k$

$N = N_1 \oplus N_2 \oplus .... \oplus N_k$

As $T = T_1 \oplus T_2 \oplus .... \oplus T_k$ and $T_j = S_j + N_j$.

It follows that $T = S + N$

As $S = \oplus \sum_{j=1}^{k} S_j$ and each S_j is diagonal matrix in E_j it follows that S

is diagonal in the basis consisting of independent members of $E_1, E_2 .... E_k$.

In a standard basis system, S may not be diagonal.

Thus S is diagonalizable.

We now show that N is nilpotent.

We note that $N_j^{r_j} = 0$.

Since $N = \oplus \sum_{j=1}^{k} N_j$, $N^i = N_1^i \oplus N_2^i \oplus ... \oplus N_k^i$

Thus if $r = \max \{r_1 ... r_k\}$

We have $N^r = 0$ $\left(\because N_i^r = 0 \quad 1 \le i \le k \right)$

As $S = \oplus \sum_{j=1}^{k} S_j$ and $N = \oplus \sum_{j=1}^{k} N_j$

$\therefore \quad S N = \oplus \sum_{j=1}^{k} S_j N_j$

$= \oplus \sum_{j=1}^{k} N_j S_j = N S$

$\therefore \quad S N = N S$.

UNIQUENESS OF S AND N

HOMOGENEOUS LINEAR SYSTEM

Theorem 9: Let T be a linear operator on a vector space E which is complex if T has any non real eigen values. Then there is only one way of expressing T as $S+N$ where S is diagonalizable, N is nilpotent and $SN = NS$.

Proof: Let E be a vector space and T an operator on E. If E is real vector space then we assume that all the eigen values of T are real.

Let $\alpha_1...\alpha_k$ be the distinct eigen values of T of multiplicities $r_1...r_k$ respectively.

Let E_j be the generalized eigen space of E belonging to eigen value $\alpha_j, k_j \leq k$.

We know that,

$E = E_1 \oplus E_2 \oplus .... \oplus E_k$

Further, $T = T_1 \oplus T_2 \oplus .... \oplus T_k$

Where $T_i = T \mid_{E_j}$ $1 \leq i \leq k$

We have, $S_i = \alpha_i I$ and $N_i = T_i - S_i$

S_i is diagonal and N_i is nilpotent.

If we define S to be $S = \oplus \sum_{i=1}^{k} S_i$ and $N = \oplus \sum_{i=1}^{k} N_i$

Then we have $T = S + N$, S diagonalizable and N nilpotent

Further $SN = NS$

Let S' and N' be operator on E such that S' is diagonalizable, N' is nilpotent.

$T = S' + N'$ and $S'N' = N'S'$

We show that, $S' = S$ and $N' = N$

Consider

$$S'T = S'(S'+N') = S'S' + S'N' = S'S' + N'S'$$
$$= (S'+N')S'$$
$$= TS'$$

Also,

$$N'T = N'(S'+N') = N'S' + NN' = (S'+N')N' = TN'$$

Thus, N' and S' both commute with T.

By definition, $\qquad E_j = \ker(T - \alpha_j I)^{r_j}$

$\therefore \quad S'E_j \subseteq E_j \quad$ and similarly $\quad N'E_j \subseteq E_j \qquad 1 \le j \le k$

This is because if $x \in E_j$ then $(T - \alpha_j I)^{r_j} x = 0$

Now, $(T - \alpha_j I)^j S'x = S'(T - \alpha_j I)^j x = S' \cdot 0 = 0$

$$(\because S' \text{ and } T \text{ commute})$$

$\because \quad S'x \in \ker(T - \alpha_j I)^{r_j} = E_j$

$\therefore \quad S' E_j \subseteq E_j$

Similarly $\quad N'E_j \subseteq E_j$

Thus we consider $\quad S' \Big|_{E_j}$ and $N' \Big|_{E_j}$ as operator on E_j.

As $\quad E = \oplus \sum\limits_{i=1}^{k} E_j \qquad$ we write

$$S' = \oplus \sum_{i=1}^{k} S' \Big|_{E_j} \quad \text{and} \quad N' = \oplus \sum_{i=1}^{k} N' \Big|_{E_j}$$

Since $\quad S = \oplus \sum\limits_{j=1}^{k} S_j \quad$ and $\quad N = \oplus \sum\limits_{j=1}^{k} N_j$

$\therefore \quad$ The theorem will be proved if we show that

$$S' \Big|_{E_j} = S_j \quad , \quad N' \Big|_{E_j} = N_j \qquad 1 \le j \le k.$$

HOMOGENEOUS LINEAR SYSTEM

Since S' is diagonalizable, it follows that $S' \mid_{E_j}$ is diagonalizable.

$S_j = \alpha_j I$ is diagonal.

It follows that $S' \mid_{E_j} - S_j$ is diagonalizable.

$$T_j = T \mid_{E_j} = (S' + N') \mid_{E_j} = S' \mid_{E_j} + N' \mid_{E_j}$$

But we also have, $T_j = S_j + N_j$

$\therefore \quad S' \mid_{E_j} + N' \mid_{E_j} = S_j + N_j$

$\qquad S' \mid_{E_j} - S_j = N_j - N' \mid_{E_j}$

Consider $\quad N_j - N' \mid_{E_j}$

We first show that, N_j and $N' \mid_{E_j}$ commutes with each other.

N' commutes with $T \cdot N' \mid_{E_j}$ commutes with $T \mid_{E_j}$

i.e. $N' \mid_{E_j}$ also commutes with S_j $\qquad S_j = \alpha_j I$.

$N' \mid_{E_j}$ commutes with $T_j - S_j = N_j$

Since N' is nilpotent, $N' \mid_{E_j}$ is also nilpotent.

N_j is nilpotent.

$\therefore \quad N' \mid_{E_j} - N_j$ is nilpotent.

Thus we have $\quad S_j - S' \mid_{E_j} = N' \mid_{E_j} - N_j$

while the left hand side is diagonalizable and right hand side is nilpotent. Thus we have an operator which is both diagonal as well as nilpotent such an operator must be zero.

$$\therefore \quad S' \Big|_{E_j} = S_j \quad \text{and} \quad N' \Big|_{E_j} = N_j \quad \text{for} \quad 1 \le j \le k$$

$$\therefore \quad S' = S \quad \text{and} \quad N' = N.$$

Theorem 10: Let A be any operator on a real or complex vector space. Let its characteristic polynomial be $p(t) = \sum_{k=0}^{n} a_k t^k$. then $p(A) = 0$, that is

$$\sum_{k=0}^{n} a_k A^k (x) = 0 \quad \text{for all} \quad x \in E.$$

Proof: let $\alpha_1 .... \alpha_k$ be the roots of the characteristic polynomial $p(t)$ of multiplicity $r_1 .... r_k$ respectively. Then we can write

$$p(t) = \alpha \ (t - \alpha_1)^{r_1} (t - \alpha_2)^{r_2} ... (t - \alpha_k)^{r_k} \quad , \quad \alpha = \text{constant}$$

We can write,

$$p(A) = \alpha \ (A - \alpha_1 I)^{r_1} (A - \alpha_2 I)^{r_2} ... (A - \alpha_k I)^{r_k}$$

We know that,

$$E = E_1 \oplus E_2 \oplus .... \oplus E_k \quad \text{Where} \quad E_i = \ker \ (A - \alpha_i I)^{r_i}$$

Consider $p(A)x$ for $x \in E$.

As $E = \oplus \sum_{i=1}^{k} E_j$, $x = x_1 + x_2 .... + x_k$ where $x_i \in E_i$

$$\therefore \quad p(A)x = \sum_{i=1}^{k} p(A)x_i = 1$$

As the order of the factors in the expression for $p(A)$ is not important we

HOMOGENEOUS LINEAR SYSTEM

have,

$$p(A)x_1 = \alpha \ (A - \alpha_1 I)^{r_2} .... (A - \alpha_k I)^{r_k} (A - \alpha_1 I)^{r_1}(x_1) = 0$$

$$\left\{ \because \ x_1 \in \ker \ (A - \alpha_k I)^{r_i} \right\}$$

Similarly $p(A)x_2 = 0$, $p(A)x_3 = 0 .... \ p(A)x_k = 0$

$\therefore \ p(A)x = 0$

Definition: Let T be an operator on R^n and $T_c : C^n \to C^n$ its complexification. If T_C is diagonalizable then we say that T is semi simple.

Theorem 11: For any operator $T \in L(R^n)$ there are unique operators S, N on R^n such that $T = S + N$, $SN = NS$, S is semi simple and N is nilpotent.

Proof: If all the roots of characteristic polynomial of T are real then we know that there exist unique operators S and N such that $T = S + N$ S diagonal N nilpotent, and $SN = NS$.

Thus, the theorem holds in this case.

We assume that one of the roots of the characteristic Polynomial is a complex root.

Thus we go complexificator of the space R^n in C^n and T_C complexification of the operator T.

Since the characteristic polynomial for T and T_C are same.

It follows that T_C and T have same eigen value (including complex roots).

Thus T_C is a complex operator on the complex space C^n. Therefore there exist unique operator S_o and N_o ,

such that $T_C = S_o + N_o$, $S_o N_o = N_o S_o$

S_o is diagonalizable and N_o nilpotent.

As T_C is complexification of operator T we know that $\sigma T_C = T_C \sigma$ where σ gives complex conjugation.

$$T_C = \sigma T_C \sigma^{-1} = \sigma (S_o + N_o) \sigma^{-1}$$

$$= \sigma S_o \sigma^{-1} + \sigma N_o \sigma^{-1} = S_1 + N_1$$

Where $S_1 = \sigma S_o \sigma^{-1}$ and $N_1 = \sigma N_o \sigma^{-1}$.

Consider,

$$S_1 N_1 = \sigma S_o \sigma^{-1} \quad \sigma N_o \sigma^{-1} = \sigma S_o N_o \sigma^{-1} = \sigma N_o S_o \sigma^{-1}$$

$$= \sigma N_o \sigma^{-1} \sigma S_o \sigma^{-1} = N_1 S_1$$

As N_o is nilpotent, $N_o{}^n = 0$ for some +ve integer n .

Consider,

$$N_1{}^n = \left(\sigma N_o \sigma^{-1} \right)^n = \sigma N_o{}^n \sigma^{-1} = 0$$

To show that, S_1 is diagonalizable.

Consider, $S_1 = \sigma S_o \sigma^{-1}$

It is clear that $S_1 = S_o{}^-$.

Considering images of points of C^n under S_1.

So if $Q S_o Q^{-1}$ is diagonal then

$\overline{Q} \overline{S_o} \overline{Q^{-1}}$ is also diagonal,

But $\overline{Q^{-1}} = \left(\overline{Q} \right)^{-1}$

$\therefore \quad \overline{Q} \overline{S_o} \overline{Q^{-1}}$ is diagonal

$\therefore \quad \overline{S_o} = S_1$ is diagonalizable.

Thus $T_C = S_o + N_o = S_1 + N_1$

Using uniqueness, we have

HOMOGENEOUS LINEAR SYSTEM

$S_o = S_1, N_o = N_1$

That is $\quad \sigma S_o \sigma^{-1} = S_o$, $\quad \sigma N_o \sigma^{-1} = N_o$.

That is $\qquad \sigma S_o = S_o \sigma$, $\quad \sigma N_o = N_o \sigma$

Hence there exist operators S and N on R^n such that

$S_o = S_c$ and $\quad N_o = N_c$ that is,

S_o and N_o are complexification of the operators S and N .

$\therefore \quad T_c = S_o + N_o = S_c + N_c = (S + N)_c$

We now show that the mapping $\quad T \rightarrow T_C$ of real operators on R^n to complexification on C^n is 1-1.

Then let T_1 and T_2 be two operators on R^n such that $T_1 C = T_2 C$.

By definition , for any element of C^n of the type $\sum \alpha_i v_i$ where α_i is complex (C) $v_i \in R^n$.

$$\therefore \quad T_{1C}\left(\sum \alpha_i v_i\right) = T_{2C}\left(\sum \alpha_i v_i\right)$$

That is $\qquad \sum \alpha_i T_1(v_i) = \sum \alpha_i T_2(v_i)$

Thus for $\quad a \qquad v \in R^n$ we have,

$T_{1C} v = T_{2C} v$

That is $\quad T_1 v = T_2 v$

$\therefore \quad T_1 = T_2$

$\therefore \quad T_C = (S + N)_c$

$\Rightarrow \quad T = S + N$

As $N_o = N_c$ is nilpotent and as $N_o \Big|_{R^n} = N$.

It follows that N is nilpotent.

S is semi simple as S_c which is same as S_o is diagonal

DYNAMICAL SYSTEM – A SHORT COURSE

we have,

$S_o N_o = N_o S_o$ that is

$S_c N_c = N_c S_c$

$(S N)_c = (N S)_c$ $\therefore$ $(S N - N S)_c = 0$

$\Rightarrow S N - N S = o \Rightarrow S N = N S$

Uniqueness of S and N in $T = S + N$ follows from the uniqueness of S_o

and N_o in the decomposition

$T_c = S_0 + N_o$ and $S_o = (S)_c$, $N_o = (N)_c$.

METHOD OF SOLVING THE EXAMPLE:

If T is an operator, we shall denote by T_0, its matrix in the standard basis system.

We can get the decomposition,

where $T_o = S_o + N_o$

where S_o = semi simple / diagonalizable.

N_o = Nilpotent part.

We first find the eigen values of the operators along with the multiplicities and then the corresponding generalized eigen space.

We then define $S_1, S_2 ...$ whose direct sum is S.

S is diagonal in the basis system.

Consisting of independent generalized eigen vectors belonging to distinct eigen values.

S_o is the transformation of S in the standard basis system.

N_o is then given by $T_o - S_o$.

Ex 4: The given matrix is $\begin{bmatrix} 1 & -1 \\ 1 & 3 \end{bmatrix}$

172

HOMOGENEOUS LINEAR SYSTEM

Here, $T_o = \begin{bmatrix} 1 & -1 \\ 1 & 3 \end{bmatrix}$ $E = R^2$

Solution. We now find the eigen values of the operator along with the multiplicities.

$\therefore$ Eigen values $\Rightarrow \begin{vmatrix} 1-\lambda & -1 \\ 1 & 3-\lambda \end{vmatrix} = 0$

$\Rightarrow (1-\lambda)(3-\lambda)+1 = 0$

$\Rightarrow 3-\lambda-3\lambda+\lambda^2+1 = 0$

$\Rightarrow \lambda^2 - 4\lambda + 4 = 0$

$\Rightarrow (\lambda-2)^2 = 0$

$\Rightarrow \lambda = 2$

Since there is only one eigen value $\lambda = 2$ which have the multiplicity 2, we have

$R^2 = E = E_1$

Then by definition we have

$E_1 = \ker (T_o - 2I)^2$

We choose,

$f_1 \equiv e_1 = (1,0)$ and $f_2 \equiv e_2 = (0,1)$

As two linearly independent generalized eigen vectors belonging to the eigen value $\lambda = 2$.

Here $S = S_1 = 2 \cdot I = 2\begin{bmatrix} 1 & 0 \\ 0 & 1 \end{bmatrix} = \begin{bmatrix} 2 & 0 \\ 0 & 2 \end{bmatrix}$

Obviously as S itself is in the standard basis system we have

$S_o \equiv S = \begin{bmatrix} 2 & 0 \\ 0 & 2 \end{bmatrix}$

$\therefore$ $N_o = T_o - S_o$

$$= \begin{bmatrix} 1 & -1 \\ 1 & 3 \end{bmatrix} - \begin{bmatrix} 2 & 0 \\ 0 & 2 \end{bmatrix} = \begin{bmatrix} -1 & -1 \\ 1 & 1 \end{bmatrix}$$

$$\therefore \quad N_o^2 = \begin{bmatrix} -1 & -1 \\ 1 & 1 \end{bmatrix}\begin{bmatrix} -1 & -1 \\ 1 & 1 \end{bmatrix} = \begin{bmatrix} 0 & 0 \\ 0 & 0 \end{bmatrix}$$

Thus in the basis consisting of generalized eigen vectors

$$f_1 \equiv e_1 = (1,0) \qquad \text{and} \qquad f_2 \equiv e_2 = (0,1)$$

We have the decomposition $\quad T_o = S_o + N_o$

We have with

$$S_o = \begin{bmatrix} 0 & 0 \\ 0 & 2 \end{bmatrix} \quad \text{and} \quad N_o = \begin{bmatrix} -1 & -1 \\ 1 & 1 \end{bmatrix}$$

Further also,

$$e^T = e^{T_o} = e^{S_o + N_o} = e^{S_o} \cdot e^{N_o}$$

Now,

$$e^{S_o} = e^{\begin{bmatrix} 2 & 0 \\ 0 & 2 \end{bmatrix}} = \begin{bmatrix} e^2 & 0 \\ 0 & e^2 \end{bmatrix} \quad \text{and}$$

$$e^{N_o} = I + N_o + \frac{N_o^2}{2!} + \ldots$$

$$= I + N_o$$

$$= \begin{bmatrix} 1 & 0 \\ 0 & 1 \end{bmatrix} + \begin{bmatrix} -1 & -1 \\ 1 & 1 \end{bmatrix}$$

$$= \begin{bmatrix} 0 & -1 \\ 1 & 2 \end{bmatrix}$$

$$\therefore \quad \exp(T) = \exp(T_o) = \begin{bmatrix} e^2 & 0 \\ 0 & e^2 \end{bmatrix}\begin{bmatrix} 0 & -1 \\ 1 & 2 \end{bmatrix}$$

$$= \begin{bmatrix} 0 & -e^2 \\ e^2 & 2e^2 \end{bmatrix}$$

For real $t, e^{tT} = e^{tT_0} = e^{t(S_o + N_0)}$

174

$$= e^{t\,S_0} \cdot e^{t\,N_o}$$

$$= \begin{bmatrix} e^{2t} & 0 \\ 0 & e^{2t} \end{bmatrix} \begin{bmatrix} 1-t & -t \\ t & 1+t \end{bmatrix}$$

$$= e^{2t} \begin{bmatrix} 1-t & -t \\ t & 1+t \end{bmatrix}$$

Ex 5: Let the given matrix be as

$$T_o = \begin{bmatrix} 1 & 1 \\ 0 & 1 \end{bmatrix} \quad \text{Let} \quad E \equiv R^2$$

Now the eigen values is given by

$$\begin{vmatrix} 1-\lambda & 1 \\ 0 & 1-\lambda \end{vmatrix} = 0$$

$$\Rightarrow \quad (1-\lambda)^2 = 0 = 1-\lambda = 0$$

$$\therefore \quad \lambda = 1$$

$\therefore$ There is only one eigen value with multiplicity 2.

Now we have,

$$R^2 = E = E_1$$

Then by definition we have.

$$E_1 = \ker (T_o - I)^2$$

We choose,

$$f_1 \equiv e_1 = (1,0) \quad \text{and} \quad f_2 \equiv e_2 = (0,1)$$

as two linearly independent generalized eigen vector belonging to the eigen value $\lambda = 1$.

Here $\quad S = S_1 = \lambda I = \begin{bmatrix} 1 & 0 \\ 0 & 1 \end{bmatrix}$

Obviously as S itself is in the standard basis system we have

$$S_o \equiv S = \begin{bmatrix} 1 & 0 \\ 0 & 1 \end{bmatrix}$$

$$\therefore \quad N_o = T_o - S_o$$

$$= \begin{bmatrix} 1 & 1 \\ 0 & 1 \end{bmatrix} - \begin{bmatrix} 1 & 0 \\ 0 & 1 \end{bmatrix} = \begin{bmatrix} 0 & 1 \\ 0 & 0 \end{bmatrix}$$

$$\therefore N_o^2 = \begin{bmatrix} 0 & 1 \\ 0 & 0 \end{bmatrix} \begin{bmatrix} 0 & 0 \\ 0 & 0 \end{bmatrix} = \begin{bmatrix} 0 & 0 \\ 0 & 0 \end{bmatrix}$$

Thus in the basis consisting of generalized eigen vectors.

$$f_1 \equiv e_1 = (1,0) \quad f_2 \equiv e_2 = (0,1)$$

We have the decomposition $T_o = S_o + N_o$

We have with

$$S_o = \begin{bmatrix} 1 & 0 \\ 0 & 1 \end{bmatrix} \quad , \quad N_o = \begin{bmatrix} 0 & 1 \\ 0 & 0 \end{bmatrix}$$

Further also,

$$e^T = e^{T_o} = e^{S_o + N_o} = e^{S_o} e^{N_o}$$

Now,

$$e^{S_o} = e^{\begin{bmatrix} 1 & 0 \\ 0 & 1 \end{bmatrix}} = \begin{bmatrix} e & 0 \\ 0 & e \end{bmatrix} \qquad \text{and}$$

$$e^{N_o} = I + N_o + \frac{N_o^2}{2!} + \ldots$$

$$= I + N_o$$

$$= \begin{bmatrix} 1 & 0 \\ 0 & 1 \end{bmatrix} + \begin{bmatrix} 0 & 1 \\ 0 & 0 \end{bmatrix} = \begin{bmatrix} 1 & 1 \\ 0 & 1 \end{bmatrix}$$

$$\therefore \quad e^T = e^{T_o} = e^{S_o} \cdot e^{N_o}$$

$$= \begin{bmatrix} e & o \\ o & e \end{bmatrix} \begin{bmatrix} 1 & 1 \\ 0 & 1 \end{bmatrix}$$

HOMOGENEOUS LINEAR SYSTEM

$$= \begin{bmatrix} e & e \\ 0 & e \end{bmatrix}$$

For real t, $e^{tT} = e^{tT_o} = e^{tS_0} \cdot e^{tN_o}$

$$\therefore \quad e^{tS_o} = \begin{bmatrix} e^t & 0 \\ 0 & e^t \end{bmatrix} \qquad \text{and}$$

$$e^{tN_o} = I + t N_o = \begin{bmatrix} 1 & 0 \\ 0 & 1 \end{bmatrix} + \begin{bmatrix} 0 & t \\ 0 & 0 \end{bmatrix}$$

$$= \begin{bmatrix} 1 & t \\ 0 & 1 \end{bmatrix}$$

$$\therefore \quad e^{tT} = e^{tT_o} = e^{tS_0} \cdot e^{tN_o}$$

$$= \begin{bmatrix} e^t & 0 \\ 0 & e^t \end{bmatrix} \cdot \begin{bmatrix} 1 & t \\ 0 & 1 \end{bmatrix}$$

$$= \begin{bmatrix} e^t & t e^t \\ 0 & e^t \end{bmatrix}$$

$$= e^t \begin{bmatrix} 1 & t \\ 0 & 1 \end{bmatrix}$$

Ex 6: $T_o = \begin{bmatrix} 1 & 1 \\ 0 & -1 \end{bmatrix}$

Now the eigen values are

$$\Rightarrow \begin{vmatrix} 1-\lambda & 1 \\ 0 & -1-\lambda \end{vmatrix} = 0$$

$$\Rightarrow (1-\lambda)(-1-\lambda) = 0$$

$$\Rightarrow -1 - \lambda + \lambda + \lambda^2 = 0$$

$$\Rightarrow \lambda^2 - 1 = 0$$

$$\Rightarrow \lambda = \pm 1$$

There are two distinct eigen values +1 , -1 of multiplicity 1.

177

$\therefore\ E = R^2 = E_1 \oplus E_2$

where, $\quad E_1 = \ker\ (T_o - I)\ \ \left\{\begin{matrix} for\ \lambda = 1 \\ \lambda = -1 \end{matrix}\right\}$

$\qquad E_2 = \ker\ (T_o + I)$

Thus E_1 and E_2 are the eigen spaces belonging to the eigen values $\lambda = +1$ and $\lambda = -1$ respectively.

We next find E_1 and E_2.

Then by definition,

$E_1 = \{(\alpha, \beta) / \alpha, \beta \in R\ \}\qquad$ such that

$E_1 = T_o - I$

$\qquad = \begin{bmatrix} 1 & 1 \\ 0 & -1 \end{bmatrix} - \begin{bmatrix} 1 & 0 \\ 0 & 1 \end{bmatrix} = \begin{bmatrix} 0 & 1 \\ 0 & -2 \end{bmatrix}\begin{bmatrix} \alpha \\ \beta \end{bmatrix} = \begin{bmatrix} 0 \\ 0 \end{bmatrix}$

$\Rightarrow\ \begin{bmatrix} \beta \\ -2\beta \end{bmatrix} = \begin{bmatrix} 0 \\ 0 \end{bmatrix}$ i.e. $\beta = 0$

$\therefore\ E_1 = \{(t, 0),\, t \in R\}$

will choose $\quad f_1 = (1, 0)$

$\therefore\ E_2 = T_o + I = \begin{bmatrix} 1 & 1 \\ 0 & -1 \end{bmatrix} + \begin{bmatrix} 1 & 0 \\ 0 & 1 \end{bmatrix} = \begin{bmatrix} 2 & 1 \\ 0 & 0 \end{bmatrix}$

$\therefore\quad \begin{bmatrix} 2 & 1 \\ 0 & 0 \end{bmatrix}\begin{bmatrix} \alpha \\ \beta \end{bmatrix} = \begin{bmatrix} 0 \\ 0 \end{bmatrix}$

$\Rightarrow\ 2\alpha + \beta = 0$

$\Rightarrow\ \beta = -2\alpha$

$\therefore\ E_2 = \{(t, -2t), t \in R\}\ ,\quad f_2 = (1, -2)$

Next to write,

$S_1 = [1]\ ,\quad S_2 = [-1]$

HOMOGENEOUS LINEAR SYSTEM

$\therefore S = S_1 \oplus S_2 = \begin{bmatrix} 1 & 0 \\ 0 & -1 \end{bmatrix}$ is the representation of the matrix so in the basis

system (f_1, f_2)

We know that

$S = Q S_o Q^{-1}$

where $Q = (P^T)^{-1}$, $P = \begin{pmatrix} P_{11} & P_{12} \\ P_{21} & P_{22} \end{pmatrix}$

and $f_1 = P_{11}e_1 + P_{12}e_2$

$f_2 = P_{21}e_1 + P_{22}e_2$

As, $f_1 = (1, 0)$, $P_{11} = 1$, $P_{12} = 0$

$f_2 = (1, -2)$, $P_{21} = 1$, $P_{22} = -2$

$\therefore P = \begin{bmatrix} 1 & 0 \\ 1 & -2 \end{bmatrix}$

$\therefore P^T = \begin{bmatrix} 1 & 1 \\ 0 & -2 \end{bmatrix}$

$\therefore Q = (P^T)^{-1} = \dfrac{\begin{bmatrix} -2 & -1 \\ 0 & 1 \end{bmatrix}}{-2} = \begin{bmatrix} 1 & \frac{1}{2} \\ 0 & -\frac{1}{2} \end{bmatrix}$

$\therefore S_o = Q^{-1} S Q = P^T S Q$

$= \begin{bmatrix} 1 & 1 \\ 0 & -2 \end{bmatrix} \begin{bmatrix} 1 & 0 \\ 0 & -1 \end{bmatrix} \begin{bmatrix} 1 & \frac{1}{2} \\ 0 & -\frac{1}{2} \end{bmatrix}$

$= \begin{bmatrix} 1 & -1 \\ 0 & 2 \end{bmatrix} \begin{bmatrix} 1 & \frac{1}{2} \\ 0 & -\frac{1}{2} \end{bmatrix} = \begin{bmatrix} 1 & 1 \\ 0 & -1 \end{bmatrix}$

$\therefore N_o = T_o - S_o$

$= \begin{bmatrix} 1 & 1 \\ 0 & -1 \end{bmatrix} - \begin{bmatrix} 1 & 1 \\ 0 & -1 \end{bmatrix} = \begin{bmatrix} 0 & 0 \\ 0 & 0 \end{bmatrix}$

Thus the operator is diagonalizable.

Now, $e^T = e^{T_o} = e^{S_o} = e^{\begin{bmatrix} 1 & 1 \\ 0 & -1 \end{bmatrix}}$

$$= \begin{bmatrix} e & e \\ 0 & -e \end{bmatrix}$$

For real t ,

$$e^{tT} = e^{tT_o} = e^{tS_o}$$

$$\therefore \; e^{tS_o} = \begin{bmatrix} e^t & e^t \\ 0 & -e^t \end{bmatrix} = e^t \begin{bmatrix} 1 & 1 \\ 0 & -1 \end{bmatrix}$$

$$e^{T_o} = e^{S_o} = e^{Q^{-1}SQ} = Q^{-1} e^S Q$$

$$= \begin{bmatrix} 1 & 1 \\ 0 & -2 \end{bmatrix} \begin{bmatrix} e & 0 \\ 0 & -e \end{bmatrix} \begin{bmatrix} 1 & \frac{1}{2} \\ 0 & -\frac{1}{2} \end{bmatrix}$$

$$= \begin{bmatrix} e & -e \\ 0 & 2e \end{bmatrix} \begin{bmatrix} 1 & \frac{1}{2} \\ 0 & -\frac{1}{2} \end{bmatrix}$$

$$= \begin{bmatrix} e & \frac{e}{2} + \frac{e}{2} \\ 0 & -2\frac{e}{2} \end{bmatrix} \qquad = \begin{bmatrix} e & e \\ 0 & -e \end{bmatrix}$$

For real t ,

$$e^t \cdot T_o = e^{\begin{bmatrix} t & t \\ 0 & -t \end{bmatrix}} = \begin{bmatrix} e^t & e^t \\ 0 & -e^t \end{bmatrix}$$

$$= e^t \begin{bmatrix} 1 & 1 \\ 0 & -1 \end{bmatrix}$$

Ex 7: $T_o = \begin{bmatrix} 0 & 1 \\ 1 & 0 \end{bmatrix}$ $\qquad E = R^2$

Eigen values are given by

$$\Rightarrow \quad \begin{vmatrix} -\lambda & 1 \\ 1 & -\lambda \end{vmatrix} = 0 \quad \Rightarrow \quad \lambda^2 - 1 = 0$$

$$\Rightarrow \lambda = \pm 1$$

There are two distinct eigen values +1 and -1 with multiplicities 1.

HOMOGENEOUS LINEAR SYSTEM

$\therefore\ E = R^2 = E_1 \oplus E_2$

where $E_1 = \ker(T_o - I)$ for $\lambda = +1$

       $E_2 = \ker(T_o + I)$ $\lambda = -1$

are the eigen spaces

By definition,

$E_1 = \{(\alpha, \beta),\ \alpha, \phi \in R\}$ such that

$E_1 = T$

$$= \begin{bmatrix} 0 & 1 \\ 1 & 0 \end{bmatrix} - \begin{bmatrix} 1 & 0 \\ 0 & 1 \end{bmatrix} = \begin{bmatrix} -1 & 1 \\ 1 & -1 \end{bmatrix} \begin{bmatrix} \alpha \\ \beta \end{bmatrix} = \begin{bmatrix} 0 \\ 0 \end{bmatrix}$$

$\Rightarrow\ -\alpha + \beta = 0$

$\Rightarrow\ -\alpha + \beta = 0$

$\Rightarrow\ \alpha - \beta = 0$

$\Rightarrow\ \alpha = \beta$

$\therefore\ E_1 = \{(t,\ t), t \in R\}$ We choose $f_1 = (1,\ 1)$

$\therefore E_2 = T_o + I = \begin{bmatrix} 0 & 1 \\ 1 & 0 \end{bmatrix} + \begin{bmatrix} 1 & 0 \\ 0 & 1 \end{bmatrix} = \begin{bmatrix} 1 & 1 \\ 1 & 1 \end{bmatrix} \begin{bmatrix} \alpha \\ \chi \end{bmatrix}$

$\Rightarrow\ \ \alpha + \beta = 0$ i.e. $\alpha = -\beta$

$E_2 = \{(t, -t),\ t \in R\}$ We choose $f_2 = (1, -1)$

$\therefore\ S_1 = [1]$ and $S_2 = [-1]$

$\therefore\ S = S_1 \oplus S_2 = \begin{bmatrix} 1 & 0 \\ 0 & -1 \end{bmatrix}$ is the matrix representation in the basis system

$(f_1,\ f_2)$

We know that,

$S = Q S_o Q^{-1}$

Where $Q = (P^T)^{-1}$ $P = \begin{pmatrix} P_{11} & P_{12} \\ P_{21} & P_{22} \end{pmatrix}$

and $\quad f_1 = P_{11}e_1 + P_{12}e_2$

$\qquad f_2 = P_{21}e_1 + P_{22}e_2$

As $\quad f_1 = (1, 1) \qquad P_{11} = 1, \; P_{12} = 1$

$\qquad f_2 = (1, -1) \qquad P_{21} = 1, \; P_{22} = -1$

$\therefore \; P = \begin{bmatrix} 1 & 1 \\ 1 & -1 \end{bmatrix} \qquad \therefore \; P^T = \begin{bmatrix} 1 & 1 \\ 1 & -1 \end{bmatrix}$

$$Q = (P^T)^{-1} = \dfrac{\begin{bmatrix} -1 & -1 \\ -1 & 1 \end{bmatrix}}{-2} = \begin{bmatrix} \frac{1}{2} & \frac{1}{2} \\ \frac{1}{2} & -\frac{1}{2} \end{bmatrix}$$

$\therefore \; S_o = Q^{-1} \, S \, Q = P^T \, S \, Q$

$$= \begin{bmatrix} 1 & 1 \\ 1 & -1 \end{bmatrix} \begin{bmatrix} 1 & 0 \\ 0 & -1 \end{bmatrix} \begin{bmatrix} \frac{1}{2} & \frac{1}{2} \\ \frac{1}{2} & -\frac{1}{2} \end{bmatrix}$$

$$= \begin{bmatrix} 1 & -1 \\ 1 & 1 \end{bmatrix} \begin{bmatrix} \frac{1}{2} & \frac{1}{2} \\ \frac{1}{2} & -\frac{1}{2} \end{bmatrix} = \begin{bmatrix} 0 & 1 \\ 1 & 0 \end{bmatrix}$$

$\therefore \; N_o = T_o - S_o = \begin{bmatrix} 0 & 1 \\ 1 & 0 \end{bmatrix} - \begin{bmatrix} 0 & 1 \\ 1 & 0 \end{bmatrix}$

$$\therefore \; N_o = \begin{bmatrix} 0 & 0 \\ 0 & 0 \end{bmatrix}$$

Thus the operator is diagonalizable.

Now, $\quad e^T = e^{T_0} = e^{S_0} = e^{\begin{bmatrix} 0 & 1 \\ 1 & 0 \end{bmatrix}} = \begin{bmatrix} 0 & e \\ e & 0 \end{bmatrix}$

$$= e^{Q^{-1} S Q} = Q^{-1} \, e^S \, Q$$

$$= \begin{bmatrix} 1 & 1 \\ 1 & -1 \end{bmatrix} \begin{bmatrix} e & 0 \\ 0 & -e \end{bmatrix} \begin{bmatrix} \frac{1}{2} & \frac{1}{2} \\ \frac{1}{2} & -\frac{1}{2} \end{bmatrix}$$

$$= \begin{bmatrix} e & -e \\ e & e \end{bmatrix} \begin{bmatrix} \frac{1}{2} & \frac{1}{2} \\ \frac{1}{2} & -\frac{1}{2} \end{bmatrix} = \begin{bmatrix} 0 & e \\ e & 0 \end{bmatrix}$$

HOMOGENEOUS LINEAR SYSTEM

For real t

$$e^{tT_o} = e^{\begin{bmatrix} 0 & t \\ t & 0 \end{bmatrix}} = \begin{bmatrix} 0 & e^t \\ e^t & 0 \end{bmatrix}$$

Ex 8: $T_o = \begin{bmatrix} 0 & 1 \\ -4 & 0 \end{bmatrix}$

$\therefore$ The eigen values are

$$\begin{vmatrix} -\lambda & 1 \\ -4 & -\lambda \end{vmatrix} = 0 \Rightarrow \lambda^2 + 4 = 0$$

$$\Rightarrow \lambda = \pm 2i$$

For the eigen value $+2i$, we find the complex eigen space that belongs to this eigen value.

It is given by $\{(z_1 , z_2) \, z_1 , z_2 \in e\}$ such that

$$= \{z_1 , 2iz_1), z_1 \in \mathbb{C}\}$$

We now choose the eigen vector say $(1 , 2i)$ belonging to eigen value $2i$.

Thus we choose the real basis $\{V, U\}$ where $V = (0,2)$ and $U = (1,0)$.

Thus the matrix S is given by

$$S = \begin{bmatrix} 0 & -2 \\ 2 & 0 \end{bmatrix} \qquad \left\{ \because \ a + ib = \begin{bmatrix} a & -b \\ b & a \end{bmatrix} \right\}$$

Thus S is the transformation of the matrix S_o in the basis $\{V, U\}$.

Thus we have,

$$S = QS_o Q^{-1} \Rightarrow S_o = Q^{-1} S Q$$

$$Q = (P^T)^{-1} \quad P = P_{ij} \quad \text{is given by}$$

$$V = P_{11} e_1 + P_{12} e_2$$

$$U = P_{21} e_1 + P_{22} e_2$$

As $V = (0 , 2) \quad U = (1 , 0)$ we have

$$P = \begin{bmatrix} 0 & 2 \\ 1 & 0 \end{bmatrix} \qquad \therefore \ P^T = \begin{bmatrix} 0 & 1 \\ 2 & 0 \end{bmatrix}$$

$$\therefore \quad Q = \left(P^T\right)^{-1} = \frac{\begin{bmatrix} 0 & -1 \\ -2 & 0 \end{bmatrix}}{-2} = \begin{bmatrix} 0 & \frac{1}{2} \\ 1 & 0 \end{bmatrix}$$

$$\therefore \ S_o = Q^{-1} S Q = P^T S Q$$

$$= \begin{bmatrix} 0 & 1 \\ 2 & 0 \end{bmatrix} \begin{bmatrix} 0 & -2 \\ +2 & 0 \end{bmatrix} \begin{bmatrix} 0 & \frac{1}{2} \\ 1 & 0 \end{bmatrix}$$

$$= \begin{bmatrix} 2 & 0 \\ 0 & -4 \end{bmatrix} \begin{bmatrix} 0 & \frac{1}{2} \\ 1 & 0 \end{bmatrix} = \begin{bmatrix} 0 & 1 \\ -4 & 0 \end{bmatrix}$$

$\therefore$ Here $\quad N_o = T_o = S_o = 0$

$\therefore \ S_o$ operator is semi simple.

Now, $\quad e^{T_o} = e^{S_0} = e^{Q^{-1} S Q} = e^{P^T S Q} = P^T e^S Q$

$$= \begin{bmatrix} 0 & 1 \\ 2 & 0 \end{bmatrix} \begin{bmatrix} \cos 2 & -\sin 2 \\ \sin 2 & \cos 2 \end{bmatrix} \begin{bmatrix} 0 & \frac{1}{2} \\ 1 & 0 \end{bmatrix}$$

$$= \begin{bmatrix} \sin 2 & \cos 2 \\ 2\cos 2 & -2\sin 2 \end{bmatrix} \begin{bmatrix} 0 & \frac{1}{2} \\ 1 & 0 \end{bmatrix}$$

$$= \begin{bmatrix} \cos 2 & \frac{1}{2}\sin 2 \\ -2\sin 2 & \cos 2 \end{bmatrix}$$

For $\quad e^{t T_o} = Q^{-1} e^{t S} Q$

$$= \begin{bmatrix} 0 & 1 \\ 2 & 0 \end{bmatrix} \begin{bmatrix} \cos 2t & -\sin 2t \\ \sin 2t & \cos 2t \end{bmatrix} \begin{bmatrix} 0 & \frac{1}{2} \\ 1 & 0 \end{bmatrix}$$

$$= \begin{bmatrix} \sin 2t & \cos 2t \\ 2\cos 2t & -2\sin 2t \end{bmatrix} \begin{bmatrix} 0 & \frac{1}{2} \\ 1 & 0 \end{bmatrix}$$

$$= \begin{bmatrix} \cos 2t & \frac{1}{2}\sin 2t \\ -2\sin 2t & \cos 2t \end{bmatrix}$$

HOMOGENEOUS LINEAR SYSTEM

Ex 9 : $T_o = \begin{bmatrix} -1 & 1 & -2 \\ 0 & -1 & 4 \\ 0 & 0 & 1 \end{bmatrix}$ $\qquad E = R^3$

Now the eigen values are

$\begin{vmatrix} (-1-\lambda) & 1 & -2 \\ 0 & (-1-\lambda) & 4 \\ 0 & 0 & (1-\lambda) \end{vmatrix} = 0$

$\Rightarrow \qquad (-1-\lambda)(-1-\lambda)(1-\lambda) = 0$

$\Rightarrow \qquad (-1-\lambda)^2 (1-\lambda) = 0$

$\Rightarrow \qquad \lambda = -1 \text{ and } \lambda = 1$

The eigen values are 1 of multiplicity 1 and -1 of multiplicity 2.

Thus we have,

$\qquad E = R^3 = E_1 \oplus E_2 \qquad \text{where}$

$\qquad E_1 = \ker (T_o - II)^1 \quad \text{and}$

$\qquad E_2 = \ker (T_o + II)^2$

Now, $\quad E_1 = \{(\alpha, \beta, \gamma), \ \alpha, \beta, \gamma \in R\} \quad$ such that

$\begin{bmatrix} (-1-1) & 1 & -2 \\ 0 & (-1-1) & 4 \\ 0 & 0 & (1-1) \end{bmatrix} \cdot \begin{bmatrix} \alpha \\ \beta \\ \gamma \end{bmatrix} = \begin{bmatrix} 0 \\ 0 \\ 0 \end{bmatrix} \qquad$ that is,

$\begin{bmatrix} -2 & 1 & -2 \\ 0 & -2 & 4 \\ 0 & 0 & 0 \end{bmatrix} \begin{bmatrix} \alpha \\ \beta \\ \gamma \end{bmatrix} = \begin{bmatrix} 0 \\ 0 \\ 0 \end{bmatrix}$

$\therefore \ -2\alpha + \beta - 2\gamma = 0$

$-2\beta + 4\gamma = 0$

$\Rightarrow \ \beta = 2\gamma \text{ and } \quad \alpha = 0$

Thus $\quad E_1 = \{(0, t, \tfrac{1}{2}), t \in R\} \qquad (\beta = t)$

We choose $(0, 1, \tfrac{1}{2})$ as an eigen vector belonging to eigen value +1.

$$\therefore \quad f_1 = \left(0 , 1, \tfrac{1}{2}\right)$$

Now E_2 is given by

$$(T_o + I)^2 \begin{pmatrix} \alpha \\ \beta \\ \gamma \end{pmatrix} = \begin{pmatrix} 0 \\ 0 \\ 0 \end{pmatrix}$$

Thus $E_2 = \{(\alpha , \beta , \gamma), \ \alpha\,\beta\,\gamma \in R\}$ such that

$$\begin{bmatrix} (-1+1) & 1 & -2 \\ 0 & (-1+1) & +4 \\ 0 & 0 & (1+1) \end{bmatrix} \cdot \begin{bmatrix} \alpha \\ \beta \\ \gamma \end{bmatrix} = \begin{bmatrix} 0 \\ 0 \\ 0 \end{bmatrix}$$

$$\Rightarrow \begin{bmatrix} 0 & 1 & -2 \\ 0 & 0 & +4 \\ 0 & 0 & 2 \end{bmatrix}^2 \begin{bmatrix} \alpha \\ \beta \\ \gamma \end{bmatrix} = \begin{bmatrix} 0 \\ 0 \\ 0 \end{bmatrix}$$

$$\Rightarrow \begin{bmatrix} 0 & 1 & -2 \\ 0 & 0 & +4 \\ 0 & 0 & 2 \end{bmatrix} \begin{bmatrix} 0 & 1 & -2 \\ 0 & 0 & -4 \\ 0 & 0 & 2 \end{bmatrix} \begin{bmatrix} \alpha \\ \beta \\ \gamma \end{bmatrix} = \begin{bmatrix} 0 \\ 0 \\ 0 \end{bmatrix}$$

$$\Rightarrow \begin{bmatrix} 0 & 0 & 0 \\ 0 & 0 & 8 \\ 0 & 0 & 4 \end{bmatrix} \begin{bmatrix} \alpha \\ \beta \\ \gamma \end{bmatrix} = \begin{bmatrix} 0 \\ 0 \\ 0 \end{bmatrix}$$

that is $\begin{bmatrix} 0 \\ 8\gamma \\ 4\gamma \end{bmatrix} = \begin{bmatrix} 0 \\ 0 \\ 0 \end{bmatrix}$

$$\therefore \quad E_2 = \{(t , s , 0), \ t , s \in R\}$$

As E_2 is 2-dimensional we choose two independent members from E_2 to get f_2 and f_3

We choose $\quad f_2 = (1 , 0 , 0)$

$$f_3 = (0 , 1 , 0)$$

Thus we have

HOMOGENEOUS LINEAR SYSTEM

$$S_1 = 1[1] = [1]$$

$$S_2 = -1\begin{bmatrix} 1 & 0 \\ 0 & 1 \end{bmatrix} = \begin{bmatrix} -1 & 0 \\ 0 & -1 \end{bmatrix}$$

$$\therefore \quad S = \begin{bmatrix} 1 & 0 & 0 \\ 0 & -1 & 0 \\ 0 & 0 & -1 \end{bmatrix}$$

Thus S is the transformation of the matrix S_o in the basis $\{f_1 \ f_2 \ f_3\}$.

$S = Q S_o Q^{-1}$ that is $S_o = Q^{-1} S Q$

Where $Q = (P^T)^{-1}$

Here, $f_1 = 0e_1 + 1e_2 + \frac{1}{2}e_3$

$\qquad f_2 = 1e_1 + 0e_2 + 0e_3$

$\qquad f_3 = 0e_1 + 1e_2 + 0e_3$

$$\therefore \quad P = \begin{bmatrix} 0 & 1 & \frac{1}{2} \\ 1 & 0 & 0 \\ 0 & 1 & 0 \end{bmatrix} \text{ and } P^T = \begin{bmatrix} 0 & 1 & 0 \\ 1 & 0 & 1 \\ \frac{1}{2} & 0 & 0 \end{bmatrix}$$

$$\therefore \quad Q = (P^T)^{-1} = \dfrac{\begin{bmatrix} 0 & 0 & 1 \\ \frac{1}{2} & 0 & 0 \\ 0 & \frac{1}{2} & -1 \end{bmatrix}}{\frac{1}{2}}$$

$$= \begin{bmatrix} 0 & 0 & 2 \\ 1 & 0 & 0 \\ 0 & 1 & -2 \end{bmatrix}$$

$$\therefore \quad S_o = P^T S Q = \begin{bmatrix} 0 & 1 & 0 \\ 1 & 0 & 1 \\ \frac{1}{2} & 0 & 0 \end{bmatrix}\begin{bmatrix} 1 & 0 & 0 \\ 0 & -1 & 0 \\ 0 & 0 & -1 \end{bmatrix}\begin{bmatrix} 0 & 0 & 2 \\ 1 & 0 & 0 \\ 0 & 1 & -2 \end{bmatrix}$$

$$= \begin{bmatrix} 0 & -1 & 0 \\ 1 & 0 & -1 \\ \frac{1}{2} & 0 & 0 \end{bmatrix}\begin{bmatrix} 0 & 0 & 2 \\ 1 & 0 & 0 \\ 0 & 1 & -2 \end{bmatrix}$$

$$= \begin{bmatrix} -1 & 0 & 0 \\ 0 & -1 & 4 \\ 0 & 0 & 1 \end{bmatrix}$$

Now, $N_o = T_o - S_o = \begin{bmatrix} -1 & 1 & -2 \\ 0 & -1 & 4 \\ 0 & 0 & 1 \end{bmatrix} - \begin{bmatrix} -1 & 0 & 0 \\ 0 & -1 & 4 \\ 0 & 0 & 1 \end{bmatrix}$

$$= \begin{bmatrix} 0 & 1 & -2 \\ 0 & 0 & 0 \\ 0 & 0 & 0 \end{bmatrix}$$

$\therefore \quad N_o = N_o^2 = \begin{bmatrix} 0 & 1 & -2 \\ 0 & 0 & 0 \\ 0 & 0 & 0 \end{bmatrix} \begin{bmatrix} 0 & 1 & -2 \\ 0 & 0 & 0 \\ 0 & 0 & 0 \end{bmatrix}$

$$= \begin{bmatrix} 0 & 0 & 0 \\ 0 & 0 & 0 \\ 0 & 0 & 0 \end{bmatrix}$$

Thus the operator is diagonalizable.

$e^{T_0} = e^{S_0 + N_0} = e^{S_0} \cdot e^{N_0}$

Now $\quad e^{S_o} = e^{\begin{bmatrix} -1 & 0 & 0 \\ 0 & -1 & 4 \\ 0 & 0 & 1 \end{bmatrix}} = \begin{bmatrix} -e & 0 & 0 \\ 0 & -e & e^4 \\ 0 & 0 & e \end{bmatrix}$

$e^{N_o} = I + N_o + \dfrac{N_o^2}{2!} + \dots = I + N_o$

$$= \begin{bmatrix} 1 & 0 & 0 \\ 0 & 1 & 0 \\ 0 & 0 & 1 \end{bmatrix} + \begin{bmatrix} 0 & 1 & -2 \\ 0 & 0 & 0 \\ 0 & 0 & 0 \end{bmatrix} = \begin{bmatrix} 1 & 1 & -2 \\ 0 & 1 & 0 \\ 0 & 0 & 1 \end{bmatrix}$$

$e^T = e^{T_o} = \begin{bmatrix} -e & 0 & 0 \\ 0 & -e & e^4 \\ 0 & 0 & e \end{bmatrix} \begin{bmatrix} 1 & 1 & -2 \\ 0 & 1 & 0 \\ 0 & 0 & 1 \end{bmatrix}$

$$= \begin{bmatrix} -e & -e & 2e \\ 0 & -e & e^4 \\ 0 & 0 & e \end{bmatrix}$$

For real t, $e^{tT_0} = e^{t(S_o+N_o)} = e^{tS_o} e^{tN_o}$

$$= e^{\begin{bmatrix} -t & 0 & 0 \\ 0 & -t & 4t \\ 0 & 0 & t \end{bmatrix}} e^{\begin{bmatrix} 1 & t & -2t \\ 0 & 1 & 0 \\ 0 & 0 & 1 \end{bmatrix}}$$

Ex 10: $\quad T_o = \begin{bmatrix} 0 & 2 & 0 \\ -2 & 0 & 0 \\ 2 & 0 & 6 \end{bmatrix} \quad \in = R^3$

The eigen values are given by

$$\begin{vmatrix} -\lambda & 2 & 0 \\ -2 & -\lambda & 0 \\ 2 & 0 & 6-\lambda \end{vmatrix} = 0$$

$\Rightarrow \quad -\lambda(-\lambda(6-\lambda)-2(-2(6-\lambda))) = 0$

$\Rightarrow (6-\lambda)(\lambda^2+4) = 0 \quad \Rightarrow \lambda = 6, \; \lambda = \pm 2i$

We find the eigen space belonging to real eigen value 6. It is given by $\{(\alpha,\beta,\gamma), \; \alpha,\beta,\gamma \in R\}$.

$$\begin{bmatrix} -6 & 2 & 0 \\ -2 & -6 & 0 \\ 2 & 0 & 0 \end{bmatrix} \begin{bmatrix} \alpha \\ \beta \\ \gamma \end{bmatrix} = \begin{bmatrix} 0 \\ 0 \\ 0 \end{bmatrix}$$

$\Rightarrow \quad -6\alpha + 2\beta = 0$

$\Rightarrow \quad -2\alpha - 6\beta = 0$

$\Rightarrow \quad 2\alpha = 0$

$\Rightarrow \quad \alpha = \beta = 0$

Thus the eigen space belonging to eigen value 6 is $\{(0,0,t), \; t \in R\}$.

We choose a non zero eigen vector in this space as $f_1 = (0, 0, 1)$.

For the eigen value $+2i$, we find the complex eigen space that belongs to this eigen value.

It is given by $\{(z_1, z_2, z_3), z_1 z_2 z_3 \in \mathcal{C}\}$

$$\begin{bmatrix} -2i & 2 & 0 \\ -2 & -2i & 0 \\ 2 & 0 & 6-2i \end{bmatrix} \begin{bmatrix} z_1 \\ z_2 \\ z_3 \end{bmatrix} = \begin{bmatrix} 0 \\ 0 \\ 0 \end{bmatrix}$$

that is, $\quad -2i z_1 + 2 z_2 = 0$

$$-2 z_1 - 2i z_2 = 0$$

$$2 z_1 + (6-2i) z_3 = 0$$

that is $\quad z_2 = i z_1 \quad$ and $\quad z_3 = -\frac{2}{(6-2i)} z_1 = \frac{1}{-3+i} z_1$

$$= \frac{1}{i-3} \left(\frac{i+3}{i+3} \right) z_1 = \frac{i+3}{10} z_i$$

Thus the eigen space belonging to the complex eigen value $+2i$ is

$\left\{ z_1, i z_1, \frac{-(i+3)}{10} z_1 \right\}$

We choose the complex eigen vector as

$\phi = \left(1, i, \frac{-(i+3)}{10} \right) (z_1 = 1)$

$\therefore \; V = \left(0, 1, -\frac{1}{10} \right), \qquad U = \left(1, 0, -\frac{3}{10} \right)$

The matrix of T_o in the basis $\{V, U\}$ is

$$\begin{bmatrix} 0 & -2 \\ +2 & 0 \end{bmatrix} \qquad (a+ib) = 0 + 2i$$

Thus we consider the matrix S.

$$S = \begin{bmatrix} 6 & 0 & 0 \\ 0 & 0 & -2 \\ 0 & 2 & 0 \end{bmatrix}$$

This is the matrix of S_o in the basis $\{f, V, U\}$ or (f_1, f_2, f_3)

where, $\quad f_1 = (0, 0, 1)$

$$f_2 = \left(0, 1, -\frac{1}{10} \right)$$

$$f_3 = \left(1,\, 0,\, -\tfrac{3}{10}\right)$$

$$\therefore\; P = \begin{bmatrix} 0 & 0 & 1 \\ 0 & 1 & -\tfrac{1}{10} \\ 1 & 0 & -\tfrac{3}{10} \end{bmatrix}$$

$$\therefore\; P^T = \begin{bmatrix} 0 & 0 & 1 \\ 0 & 1 & 0 \\ 1 & -\tfrac{1}{10} & -\tfrac{3}{10} \end{bmatrix}$$

$$\therefore\; Q = \left(P^T\right)^{-1} = \frac{\begin{bmatrix} -\tfrac{3}{10} & -\tfrac{1}{10} & 1 \\ 0 & -1 & 0 \\ -1 & 0 & 0 \end{bmatrix}}{-1} = \begin{bmatrix} \tfrac{3}{10} & \tfrac{1}{10} & -1 \\ 0 & 1 & 0 \\ 1 & 0 & 0 \end{bmatrix}$$

$$S_o = P^T S Q = \begin{bmatrix} 0 & 0 & 1 \\ 0 & 1 & 0 \\ 1 & -\tfrac{1}{10} & -\tfrac{3}{10} \end{bmatrix} \begin{bmatrix} 6 & 0 & 0 \\ 0 & 0 & -2 \\ 0 & 2 & 0 \end{bmatrix} \begin{bmatrix} \tfrac{3}{10} & \tfrac{1}{10} & -1 \\ 0 & 1 & 0 \\ 1 & 0 & 0 \end{bmatrix}$$

$$= \begin{bmatrix} 0 & 2 & 0 \\ 0 & 0 & -2 \\ 6 & -\tfrac{6}{10} & \tfrac{2}{10} \end{bmatrix} \begin{bmatrix} \tfrac{3}{10} & \tfrac{1}{10} & -1 \\ 0 & 1 & 0 \\ 1 & 0 & 0 \end{bmatrix}$$

$$S_o = \begin{bmatrix} 0 & 2 & 0 \\ -2 & 0 & 0 \\ 2 & 0 & -6 \end{bmatrix} = T_o$$

$$\therefore\; N_o = 0$$

The operator is semi simple.

$$\therefore\; e^{T_o} = e^{S_o} = e^{Q^{-1}SQ} = Q^{-1} e^S Q$$

As $S = \begin{bmatrix} 6 & 0 & 0 \\ 0 & 0 & -2 \\ 0 & 2 & 0 \end{bmatrix}$ then e^S is given by,

$$e^S = \begin{bmatrix} e^6 & 0 & 0 \\ 0 & 0 & 0 \\ 0 & 0 & 0 \end{bmatrix} \begin{bmatrix} 0 & 0 & 0 \\ 0 & 0 & -2 \\ 0 & 2 & 0 \end{bmatrix}$$

$$= \begin{bmatrix} e^6 & 0 & 0 \\ 0 & \cos 2 & -\sin 2 \\ 0 & \sin 2 & \cos 2 \end{bmatrix}$$

$$\therefore e^{T_o} = \begin{bmatrix} 0 & 0 & 1 \\ 0 & 1 & 0 \\ 1 & -\frac{1}{10} & -\frac{3}{10} \end{bmatrix} \begin{bmatrix} e^6 & 0 & 0 \\ 0 & \cos 2 & -\sin 2 \\ 0 & \sin 2 & \cos 2 \end{bmatrix} \begin{bmatrix} \frac{3}{10} & \frac{1}{10} & -1 \\ 0 & 1 & 0 \\ 1 & 0 & 0 \end{bmatrix}$$

$$= \begin{bmatrix} 0 & \sin 2 & \cos 2 \\ 0 & \cos 2 & -\sin 2 \\ e^6 & & \end{bmatrix} \cdot \begin{bmatrix} \frac{3}{10} & \frac{1}{10} & -1 \\ 0 & 1 & 0 \\ 1 & 0 & 0 \end{bmatrix}$$

$$= \begin{bmatrix} \cos 2 & \sin 2 & 0 \\ -\sin 2 & \cos 2 & 0 \\ & & e^6 \end{bmatrix}$$

Ex 11: $\quad T_o = \begin{bmatrix} 0 & -1 & 0 & 0 \\ 1 & 0 & 0 & 0 \\ 0 & 0 & 0 & -1 \\ 2 & 0 & 1 & 0 \end{bmatrix} \quad E = R^4$

The eigen values are given by,

$$\begin{vmatrix} -\lambda & -1 & 0 & 0 \\ 1 & -\lambda & 0 & 0 \\ 0 & 0 & -\lambda & -1 \\ 2 & 0 & 1 & -\lambda \end{vmatrix} = 0$$

$$\Rightarrow (-\lambda)(\lambda(\lambda^2+1)) + 1(\lambda^2+1) = 0$$

$$\Rightarrow (\lambda^2+1)(+\lambda^2+1) = 0$$

$$\Rightarrow \lambda = \pm i \text{ of multiplicity 2.}$$

We find the generalized eigen space wz the multiplicity 2; belonging to eigen value $+i$.

It is given by $\{(z_1, z_2, z_3, z_4),\ z_1 z_2 z_3 z_4 \in \mathcal{C}\}$

where,

HOMOGENEOUS LINEAR SYSTEM

$$\left(T_o - iI\right)^2 \begin{bmatrix} z_1 \\ z_2 \\ z_3 \\ z_4 \end{bmatrix} = \begin{bmatrix} 0 \\ 0 \\ 0 \\ 0 \end{bmatrix}$$

$$\begin{bmatrix} 0 & -1 & 0 & 0 \\ 1 & 0 & 0 & 0 \\ 0 & 0 & 0 & -1 \\ 2 & 0 & 1 & 0 \end{bmatrix}$$

$-2z_1 + 2iz_2 = 0$

$-2iz_1 - 2z_2 = 0$

$-2z_1 - 2z_3 + 2iz_3 = 0$

$-4iz_1 - 2z_3 - 2iz_3 - 2z_4 = 0$

i.e., $z_1 = iz_2$

$\qquad -z_3 + iz_4 = iz_2$

Since the generalized eigen space is of dimension 2 as a complex vector space.

Therefore we choose two independent generalized eigen vectors in this space.

We select say,

$\qquad \phi_1 = (i, 1, 0, 1)$ and

$\qquad \phi_2 = (i, 1, -i, 0)$

$\therefore V_1 = (1, 0, 0, 0)$ { img and real parts of ϕ_1 }

$U_1 = (0, 1, 0, 1)$

$\therefore V_2 = (1, 0, -1, 0)$ { img and real parts of ϕ_2 }

$U_2 = (0, 1, 0, 0)$

Thus in the basis $\{f_1, f_2, f_3, f_4\}$ with

$\qquad f_1 = V_1 \ , \ f_2 = U_1 \ , \ f_3 = V_2 \ , \ f_4 = U_2$

The matrix S is given by,

$$S = \begin{bmatrix} 0 & -1 & 0 & 0 \\ 1 & 0 & 0 & 0 \\ 0 & 0 & 0 & -1 \\ 0 & 0 & 1 & 0 \end{bmatrix} \left\{ \begin{bmatrix} a & -b & 0 & 0 \\ b & a & 0 & 0 \\ 0 & 0 & a & -b \\ 0 & 0 & b & a \end{bmatrix} \begin{array}{l} \text{where } a = 0 \text{ and } b = +1 \\ \\ \\ \because a + i\, b = +i \end{array} \right\}$$

Thus we get;

$$S_o = Q^{-1} S Q \quad \text{Where } P \text{ is given by}$$

$$P = \begin{bmatrix} 1 & 0 & 0 & 0 \\ 0 & 1 & 0 & 0 \\ 1 & 0 & -1 & 0 \\ 0 & 1 & 0 & 0 \end{bmatrix}$$

$$\therefore \ P^T = \begin{bmatrix} 1 & 0 & 1 & 0 \\ 0 & 1 & 0 & 1 \\ 0 & 0 & -1 & 0 \\ 0 & 0 & 0 & 0 \end{bmatrix}$$

$$\therefore \ \left(P^T \right)^{-1} = \dfrac{\begin{bmatrix} 1 & 0 & 1 & 0 \\ 0 & 0 & 0 & 1 \\ 0 & 0 & -1 & 0 \\ 0 & 1 & 0 & -1 \end{bmatrix}}{1}$$

$$Q = \begin{bmatrix} 1 & 0 & 1 & 0 \\ 0 & 0 & 0 & 1 \\ 0 & 0 & -1 & 0 \\ 0 & 1 & 0 & -1 \end{bmatrix}$$

$$S_o = \begin{bmatrix} 0 & -1 & 0 & 0 \\ 1 & 0 & 0 & 0 \\ 0 & 1 & 0 & -1 \\ 1 & 0 & 1 & 0 \end{bmatrix}$$

$$N_o = T_o - S_o = \begin{bmatrix} 0 & 0 & 0 & 0 \\ 0 & 0 & 0 & 0 \\ 0 & -1 & 0 & 0 \\ 1 & 0 & 0 & 0 \end{bmatrix}$$

HOMOGENEOUS LINEAR SYSTEM

$$\therefore\ e^{tT_o} = e^{tS_o+tN_o} = e^{tS_o}\cdot e^{tN_o}$$

$$\therefore\ e^{tS_o} = Q^{-1}e^{tSQ}$$

As $S=\begin{bmatrix}0&-1&0&0\\1&0&0&0\\0&0&0&-1\\0&0&1&0\end{bmatrix}\quad \therefore\ tS=\begin{bmatrix}0&-t&0&0\\t&0&0&0\\0&0&0&-t\\0&0&t&0\end{bmatrix}$

$$\therefore\quad \exp(tS)=\begin{bmatrix}\cos t&-\sin t&0&0\\\sin t&\cos t&0&0\\0&0&\cos t&-\sin t\\0&0&\sin t&\cos t\end{bmatrix}$$

$e^{tS_o}=Q^{-1}e^{tS}Q$

$$=\begin{bmatrix}1&0&1&0\\0&1&0&1\\0&0&-1&0\\0&0&0&0\end{bmatrix}\begin{bmatrix}\cos t&-\sin t&0&0\\\sin t&\cos t&0&0\\0&0&\cos t&-\sin t\\0&0&\sin t&\cos t\end{bmatrix}\begin{bmatrix}1&0&1&0\\0&0&0&1\\0&0&-1&0\\0&1&0&-1\end{bmatrix}$$

$$=\begin{bmatrix}\cos t&-\sin t&\cos t&-\sin t\\\sin t&\cos t&\sin t&\cos t\\0&0&-\cos t&0\\0&0&0&0\end{bmatrix}\begin{bmatrix}1&0&1&0\\0&0&0&1\\0&0&-1&0\\0&1&0&-1\end{bmatrix}$$

$$=\begin{bmatrix}\cos t&-\sin t&0&0\\\sin t&\cos t&0&0\\0&0&\cos t&0\\0&0&0&0\end{bmatrix}$$

Ex. 12) Show that a matrix $\lfloor a_{ij}\rfloor$ such that $a_{ij}=0$ for $i\le j$ is nilpotent

Solution. Let $A=\lfloor a_{ij}\rfloor$ with $a_{ij}=0$ for $i\le j$.

Consider A^2

$$\left(A^2\right)_{ij}=\sum_{k=1}^{n}a_{ik}a_{kj}=\sum_{k=j+1}^{i-1}a_{ik}a_{kj}$$

For the matrix A, non zero terms are given by a_{ij} for $j\le i-1$.

For the matrix A^2 non zero terms are restricted to $j < i - 2$.

(i.e., (i, j) terms may be non-zero provided $j < i - 2$)

Proceeding with higher powers of A , the region over which terms may be non zero gets more and more restricted.

In particular for (i, j) the term of the matrix A^i may non zero provided $j < i - 1 = 0$ i.e. , no term of the matrix A^i is non zero.

$A^i = (0)$

i.e., A is nilpotent.

Ex. 13) If N is a nilpotent operator on an n-dimensional vector space then $N^n = 0$.

Solution. Since n is nilpotent $\exists$ +ve integer m such that $N^m = 0$.

The problem will be solved if we show that $m \leq n$.

If possible, let us assume that $m > n$.

Consider the spaces $\ker T^i$, $0 < i$

$(0) = \ker T^o \subseteq \ker T \subseteq \ker T^2 \subseteq\ \subseteq \ker T^m = E$ (with dim)

Since we know that if $\ker T^i = \ker T^{i+1}$,

then $\ker T^i = \ker T^j$ for all $j \geq i$.

Since we are assuming that $m > n$,

it therefore follows that,

$(0) = \ker T^o \subsetneq \ker T^1 \subsetneq \ker T^2 \subsetneq ...... \subsetneq \ker T^{m-1} \subsetneq \ker T^m = E$ (with dim n)

Since all the $m + 1$ spaces namely $\ker T^0, \ker T^1, ... \ker T^m$ are distinct and strictly increasing in dimension with dim of $\ker T^m = n$,

It follows that there must be $m + 1$ number of non-negative integers in between 0 and n. This is not possible as $m > n$.

Thus the assumption must be wrong

i.e. $T^i = 0$ for some $i \leq n$.

$\Rightarrow T^n = 0$.

Ex. 14) Find necessary and sufficient conditions on a,b,c,d in order that

the operator $\begin{bmatrix} a & b \\ c & d \end{bmatrix}$ is a

i) diagonalizable ii) Semi simple iii) Nilpotent.

Ex. 15) what values of a, b, c, d make the following operators semi simple

/ nilpotent.

$$\begin{bmatrix} 0 & a \\ -1 & 2 \end{bmatrix} \begin{bmatrix} 1 & -1 \\ 1 & b \end{bmatrix} \begin{bmatrix} 1 & 0 & 0 \\ 2 & 1 & 1 \\ 0 & 0 & c \end{bmatrix} \begin{bmatrix} 0 & d & 0 \\ 1 & 0 & d \\ d & 1 & 0 \end{bmatrix}.$$

NILPOTENT CANONICAL FORM:

For $n \geq 2$, we define an elementary nilpotent block of size n (i.e., an $n \times n$ matrix) as that $n \times n$ matrix all of whose entries are zero excepting the terms immediately below the diagonal with value 1.

Thus $\begin{bmatrix} 0 & 0 \\ 1 & 0 \end{bmatrix} \begin{bmatrix} 0 & 0 & 0 \\ 1 & 0 & 0 \\ 0 & 1 & 0 \end{bmatrix} \begin{bmatrix} 0 & 0 & 0 & 0 \\ 1 & 0 & 0 & 0 \\ 0 & 1 & 0 & 0 \\ 0 & 0 & 1 & 0 \end{bmatrix}$

are example of elementary nilpotent canonical blocks of size 2,3 and 4 respectively.

The matrix $[0]$ is defined to be an elementary nilpotent block of size 1.

Let N be an operator on vector space V with basis $\{e_1, e_2 \ldots e_n\}$ such that

$N(e_1) = e_2$

$N(e_2) = e_3$

$:$

:

$$N(e_{n-2}) = e_{n-1}$$
$$N(e_{n-1}) = e_n$$
$$N(e_n) = 0$$

The matrix of N with respect to the basis $\{e_1, e_2 .... e_n\}$ is

Which is an elementary canonical (nilpotent) block of size n.

Consider the operator N^2.

$$N^2(e_1) = e_3$$
$$N^2(e_2) = e_4$$

:

:

$$N^2(e_{n-2}) = e_n$$
$$N^2(e_{n-1}) = 0$$
$$N^2(e_n) = 0$$

Similarly,

$$N^3(e_1) = e_4$$
$$N^3(e_2) = e_5$$

:

:

$$N^3(e_{n-2}) = 0$$
$$N^3(e_{n-1}) = 0$$
$$N^3(e_n) = 0$$

It follows that $N^n(e_i) = 0 \qquad 1 \le i \le n$

i.e. N is nilpotent.

However $N^K \ne 0$ for $K < n$ as

$$N^K(e_1) = e_{1+K} \ne 0.$$

Thus every elementary nilpotent block of size ' n ' represents a nilpotent operator in $L(R^n)$ whose order of nil potency is n.

However, every nilpotent operator may not be represented by an elementary nilpotent block.

Ex. 16) : Operators given by matrices,

$\begin{bmatrix} 0 & 0 \\ \alpha & 0 \end{bmatrix}, \begin{bmatrix} 0 & 0 & 0 \\ \alpha & 0 & 0 \\ \beta & \gamma & 0 \end{bmatrix}$ are nilpotent but are not given by single elementary

nilpotent block.

CYCLIC SUBSPACE

Let T be an operator on a vector space V. A Subspace W of V is said to be a cyclic subspace of V with respect to the operator T provided.

(1) $TW \subseteq W$ and their exists $x \in W$ such that W is spanned by $x, Tx, T^2x.....$

For any $y \in V$, then the subspace generated by $y, Ty, T^2y, T^3y....$ is a cyclic subspace of V with respect to the operator T.

If a cyclic subspace W of V is spanned by $x, Tx, T^2x.....$

for some $x \in V$, we say that x is a cyclic vector of W.

Let N be a nilpotent operator on V and $x \in V$.

If n is the order of nil potency of the operator N, then certainly

$N^n(x) = 0$ $\quad \forall x \in V$

However for $x \in V$, we may have

$N^k x = 0$ for some $k \leq n$.

The smallest k such that $N^k x = 0$ for given $x \in V$ is known as nil of x with respect to the operator N and is denoted by Nil (x, N)

For example, consider $N = \begin{pmatrix} 0 & 0 \\ \alpha & 0 \end{pmatrix}$ with $\alpha \neq 0$.

So obviously, $N^2 \equiv (0)$

However, $N \begin{pmatrix} 0 \\ y \end{pmatrix} = \begin{pmatrix} 0 \\ 0 \end{pmatrix}$ $y \in R$

But $N \begin{pmatrix} x \\ y \end{pmatrix} \neq \begin{pmatrix} 0 \\ 0 \end{pmatrix}$ whenever $x \neq 0$.

Thus nil $((0,y),N) = 1$

$\quad$ nil $((x,y), x \neq 0, N) = 2$

Thus if $x \neq 0$ in V and N is nilpotent operator with order of nil potency equal to n, then $1 \leq nil\,(x,N) \leq n$.

Further , if $k = nil\,(x\,,\,N)$.

then $N^i(x) \neq 0$ $\qquad 1 \leq i \leq k-1$

If N is a nilpotent operator on V and $x \in V$, then the cyclic subspace of V generated by x is denoted by $Z(x,N)$.

Theorem 12: Let $nil\,(x,N) = n$, then the vectors $N^k x$, $0 \leq k \leq n-1$ form a basis for $Z(x,N)$.

Proof: By the definition , any element of $Z(x,N)$ is spanned by

$x, N(x), N^2(x)...N^n(x), N^{n+1}(x)....$

As $N^n(x)$, $N^{n+1}(x) = 0$

It follows that the cyclic space $Z(x,N)$ is spanned by x, $N(x)..N^{n-1}(x)$.

We now show that the elements $x, N(x), N^2(x)..., N^{n+1}(x)$ are L.I.

If possible, let us assume that $\exists$ exist elements $a_0, a_1 ... a_{n-1}$ in the field (R/C) not all zero such that

$$\sum_{k=0}^{n-1} a_k \cdot N^k(x) = 0$$

Since the coefficients $a_0, a_1 ... a_{n-1}$ are not all zero, we can find smallest suffix ' j ' such that $a_j \neq 0$

i.e. $a_i = 0$ for $0 \leq i \leq j-1$.

Therefore $\displaystyle\sum_{k=0}^{n-1} a_k \, N^k(x) = 0$ becomes $\displaystyle\sum_{k=j}^{n-1} a_k \, N^k(x) = 0$.

Since $0 \leq j \leq n-1$

We can consider the operator N^{n-1-j}

$\therefore \quad N^{n-1-j}\left(\displaystyle\sum_{k=j}^{n-1} a_k \, N^k(x)\right) = 0$

i.e. $\displaystyle\sum_{k=j}^{n-1} a_k \, N^{n-1-j+k}(x) = 0$

i.e. $a_j \, N^{n-1}(x) + a_{j+1} \, N^n(x) + \dots = 0$.

i.e. $a_j \, N^{n-1}(x) = 0 \qquad \left(\because N^i(x) = 0 \ for \ i \geq n\right)$

$\Rightarrow \quad a_j = 0 \ as \ N^{n-1}(x) \neq 0 \qquad \left(\because nil\left(x, \ N\right) = n\right)$

a contradiction as $a_j \neq 0$.

Thus the n elements $x, N(x), N^2(x), N^3(x) \dots N^{n-1}(x)$ spans the cyclic space $Z(x, N)$ and are also linearly independent.

$\{x, N(x) \dots N^{n-1}(x)\}$ is a basis of $Z(x, N)$ where $n = nil\left(x, \ N\right)$

Obviously the cyclic space $Z(x, N)$ is invariant under N.

Thus we can consider the restriction of N to $Z(x, N)$ as an operator on $Z(x, N)$.

Since $Z(x, N)$ has $\{x, N(x) \dots N^{n-1}(x)\}$ as a basis, we can consider the matrix of N with respect to ordered basis. It is clear that the matrix of N is

$$\begin{bmatrix} 0 & 0 & \dots & 0 & 0 \\ 1 & 0 & \dots & 0 & 0 \\ 0 & 1 & & 0 & 0 \\ \cdot & \cdot & & 0 & 0 \\ & & & 1 & 0 \\ 0 & 0 & & 0 & 0 \end{bmatrix}$$ an elementary nilpotent block of size n .

(1) If $\sum_{k=0}^{q} a_k N^k(x) = 0$ then $a_k = 0$ for $k < nil\,(x, N)$.

Solution: If $q < nil\,(x, N) = n$. Then we can define,

$a_{q+1} = 0$, $a_{q+2} = 0 \dots a_{n-1} = 0$ and then the above sum becomes,

$$\sum_{k=0}^{n-1} a_k N^k(x) = 0$$

But $x, N(x) \dots N^{n-1}(x)$ are $l_i I$ so, we must have,

$a_k = 0$, $0 \le k \le n-1$

Let us now assume that $q \ge n$.

Then, $\sum_{k=0}^{q} a_k N^k(x) = 0$ becomes

$a_o x + a_1 N(x) + \dots + a_{n-1}N^{n-1}(x) + a_n N^n(x) + \dots + a_q N^q(x) = 0$.

But $n = nil\,(x, N)$

$\Rightarrow \quad 0 = N^n(x) = N^{n+1}(x) = N^{n+2}(x) \dots$

Thus we have,

$a_o x + a_1 N(x) + \dots + a_{n-1}N^{n-1}(x) = 0$

But $x, N(x) \dots N^{n-1}(x)$ are L.I.

Therefore , $a_0 = a_1 = \dots = a_{n-1} = 0$

i.e. $a_k = 0$ for $0 \le k \le n-1$

Theorem 13: Let $n = nil\,(x, N)$ if $p(t)$ is a polynomial such that

HOMOGENEOUS LINEAR SYSTEM

$p(N)x = 0$ then t^n divides $p(t)$ that is, there is a polynomial $p_1(t)$ such

that $p(t) = t^n p_1(t)$.

Proof: Let us assume that $p(t)$ is a polynomial of degree n and

$p(t) = a_0 + a_1 t + ... + a_q t^q$ for coefficient $a_0, a_1 .... a_q \in F$.

(F in R for real vector space F is C for complex vector space).

$\therefore \ p(N) = a_0 + a_1 N + ... + a_q N^q$

$\therefore \ p(N)x = 0$

$\Rightarrow \ (a_0 I + a_1 N + ... a_q N^q)x = 0$

i.e. $\sum_{k=0}^{q} (a_k N^k)x = 0$ as $n = nil \ (x, N)$

$N^k x = 0 \quad$ for $k \geq n$.

$\therefore \ \sum_{k=0}^{q} (a_k N^k)x = 0$

$\Rightarrow \ \sum_{k=0}^{n-1} (a_k N^k)x = 0$

But the set $N^k x$, $k = 0 .... n-1$ is linearly independent.

$\therefore \ a_k = 0$, $k = 0$ to $n-1$

Thus in the polynomial $p(t) = a_0 + a_1 t + ... + a_q t^q$.

The coefficients $a_0, a_1 .... a_{n-1}$ are all zero

$\therefore \ p(t) = a_n t^n + a_{n+1} t^{n+1} + ... + a_q t^q$

$\quad = t^n (a_n + a_{n+1} t + ... a_q t^{q-n})$

$\quad = t^n \ p_1(t) \quad$ where $p_1(t) = a_n + a_{n+1} + ... + a_q t^{q-1}$

Thus the polynomial $p(t)$ is divisible by $t^i \quad 0 < i < n$.

Let N be a nilpotent operator.

If $x \neq 0$ then $nil\ (x, N) \geq 1$.

If $y = N(x)$ then $Z(y) \subseteq Z(x)$.

Further, if $x \neq 0$ then $Z(y) \subsetneq Z(x)$.

Also $nil\ (y, N) = nil\ (x, N) - 1$

Also for a nilpotent operator N on a vector space V ,

$\ker N \gneq (0)$ i.e. $\ker N$ is non-trivial.

If possible, let $\ker N = 0$ for $N : V \to V$

Since it is 1-1 and onto

$\therefore N$ is invertible

$\therefore$ N^{-1} is an operator on V .

As N is nilpotent, we must have

$N^k = (0)$ for some $k \geq 1$

$\therefore \left(N^{-1}\right)^k \cdot N^k = (0)$

$\left(N^{-1} N\right)^k = 0$

i.e. $I^k = 0$

$I^k = 0$ which is not true for non zero spaces V .

So if N is nilpotent on V then

dim $\ker N \geq 1$

$\therefore$ dim $N(V) < \dim V$

i.e. $N(V) \subsetneq V$

Theorem 14: Let N be a nilpotent operator on a real or complex vector space E . Then E has a basis giving N a matrix of the form

$A = diag\ \{A_1, A_2 ... A_r\}$

HOMOGENEOUS LINEAR SYSTEM

where A_j is an elementary nilpotent block and the size of A_k is non increasing function of k . The matrices $A_1....A_r$ are uniquely determined by the operator N .

Proof: We shall first show that the space E is a direct sum of cyclic spaces.

We prove this by strong induction on dimension of vector space

Since $(0)=(0)$

The statement obviously holds for vector spaces with dim (0).

We assume that the statement holds for all spaces and nilpotent operators on them, with dimension less than that of E .

Consider $N(E)$. As $N(E) \subsetneq E$

$\dim N(E) < \dim E$

Hence by induction hypothesis, the statement holds for $N(E)$.

That is, $N(E) = Z(y_1) \oplus Z(y_2) \oplus .... \oplus Z(y_k)$ for some $y_1, y_2...y_k \in N(E)$

As $y_i \in N(E)$ $\quad 1 \le i \le k$

$\qquad y_i = N(x_i)$ for $x_i \in E$, $1 \le i \le k$

We consider the cyclic spaces,

$Z(x_1) \, Z(x_2)....Z(x_k)$

We show that they are independent space.

For this $\mu_i \, Z(x_i)$ $\quad 1 \le i \le k$ $\qquad$ such that

$\mu_1 + \mu_2 + ... + \mu_k = 0$

We then show that $\mu_i = 0$ $\quad 1 \le i \le k$

Operator N on $\sum_{i=1}^{k} \mu_i = 0$ to get $\sum_{i=1}^{k} N(\mu_i) = 0$

that is, $N(\mu_1) + N(\mu_2) + ... + N(\mu_k) = 0$

205

But $\mu_i \in Z(x_i)$ so $N(\mu_i) \in Z(N(x_i)) = Z(y_i)$

But, $N(E) = \oplus \sum_{i=1}^{k} Z(y_i)$

Implies the spaces $Z(y_1), Z(y_2)...Z(y_k)$ are independent.

So $\sum_{i=1}^{k} N(\mu_i) = 0$ and $N(\mu_i) \in Z(y_i)$

$\Rightarrow$ $N(\mu_i) = 0$, $1 \le i \le k$

As, $\mu_i \in Z(x_i)$ $1 \le i \le k$

$$\mu_i = \sum_{r=0}^{n_{i-1}} a_{i_r} N^r(x_i) \quad \text{where} \quad n_i = nil\ (x_i, N)$$

$$\therefore N(\mu_i) = N \sum_{r=0}^{n_{i-1}} a_{i_r} N^r(x_i)$$

$$= \sum_{r=0}^{n_{i-1}} a_{i_r} N^r(N(x_i))$$

$$= \sum_{r=0}^{n_{i-1}} a_{i_r} N^r(y_i)$$

But $N(\mu_i) = 0$

$\therefore$ $\sum a_{i_r} N^r(y_i) = 0$

The polynomial $a_{i_r} t^r$ is divisible by $t^{n_{i-1}}$ $\left\{ \begin{array}{l} \because\ n_i = nil\ (x_i) \\ \therefore\ n_{i-1} = nil\ (y_i) \end{array} \right\}$

But $y_i \ne 0$ as $Z(y_i)$ $1 \le i \le k$

$\therefore$ $nil\ (y_i) \ge 1$

i.e. $n_{i-1} \ge 1$

i.e. the polynomial $\sum a_{i_r} t^r$ is divisible by t

i.e. we can write,

$\sum a_{i_r} t^r$ as $t \cdot p_1(t)$, for some polynomial $p_1(t)$.

$$\therefore \quad \mu_i = p(N)x_i \quad \left(where \; p(t) \equiv \sum a_{i_r} t^r\right)$$

$$= N\, p_1(N)x_i$$

$$= p_1(N)\, N(x_i)$$

$$= p_1(N)\, y_i$$

$$\in Z(y_i) \qquad 1 \le i \le k$$

But we have,

$$\mu_1 + \mu_2 + \dots + \mu_k = 0$$

with $\mu_i \in Z(y_i)$ and $Z(y_1)..Z(y_k)$ are independent.

$$\therefore \quad \mu_i = 0, \quad 1 \le i \le k$$

Thus the spaces

$Z(x_i)\, Z(x_2)....Z(x_k)$ are independent.

So their sum is a direct sum.

We thus consider $\oplus \sum_{i=1}^{k} Z(x_i)$.

Let $k = \ker N$

Consider $k \cap N(E)$ a subspace of k .

Let L be a subspace of k such that,

$$k = (k \cap N(E)) \oplus L$$

We show that,

$$E = \oplus \sum_{i=1}^{k} Z(x_i) \oplus L$$

For this we show that, the spaces $\oplus \sum_{i=1}^{k} Z(x_i)$ and L are independent

Let us assume that $U \in L \cap \oplus \sum_{i=1}^{k} Z(x_i)$

We show that $\qquad \mu = 0$

As $\quad \mu \in \oplus \sum_{i=1}^{k} Z(x_i)$

We have $\quad \mu = v_1 + \dots v_k \qquad$ where $\quad v_i \in Z(x_i)$

But $\quad \mu \in L \subseteq k = \ker of N$

$\therefore \quad N(\mu) = 0$

$\therefore \quad N\left(\sum_{i=1}^{k} v_i\right) = 0 \quad$ i.e.

$N(v_1) + N(v_2) + \dots + N(v_k) = 0$

But $\quad v_i \in Z(x_i) \Rightarrow N(v_i) \in Z(y_i)$

But $\quad Z(y_1), Z(y_2) \dots Z(y_k)$ are independent.

$\therefore \quad N(v_i) = 0 \qquad 1 \le i \le k$

$\therefore$ By the argument similar to one to show that $\quad \mu_i = 0, \quad 1 \le i \le n,$

we get $\quad v_i = 0 \qquad \therefore \mu = \sum_{i=1}^{k} v_i = 0$

Thus $\mu \in \oplus \sum_{i=1}^{k} Z(x_i) \cap L \Rightarrow \mu = 0$

$\oplus \sum_{i=1}^{k} Z(x_i) \quad$ and L are independent.

Hence the sum $\quad \oplus \sum_{i=1}^{k} Z(x_i) + L$ is a direct sum.

We now show that $E = \oplus \sum_{i=1}^{k} Z(x_i) + L$

Obviously, $\quad \oplus \sum_{i=1}^{k} Z(x_i) + L \subseteq E$

Let $\quad x \in E \quad , \quad N(x) \in N(E)$

and $\quad N(E) = Z(y_1) \oplus Z(y_2) \oplus \dots \oplus Z(y_k)$

$\therefore \quad N(x) = \beta_1 + \beta_2 + ... + \beta_k \quad$ Where $\quad \beta_i \in Z(y_i)$

But $\quad \beta_i \in Z(y_i) = Z(N(x_i))$

We have $\quad \beta_i = N(\alpha_i), \quad$ for some $\quad \alpha_i \in Z(x_i)$.

$\therefore \quad N(x) = N(\alpha_1) + N(\alpha_2) + ... + N(\alpha_k)$

i.e. $\quad N(x - \alpha_1 - \alpha_2 ..... - \alpha_k) = 0$

i.e. $\quad x - \alpha_1 - \alpha_2 ... - \alpha_k \in k$

but $\quad k = (k \cap N(E)) \oplus L$

$\therefore \quad x - \alpha_1 - \alpha_2 .... - \alpha_k = \gamma + \delta$

Where $\quad \gamma \in k \cap N(E) \quad , \quad \delta \in L$

$\therefore \quad \gamma \in N(E) = Z(y_1) \oplus Z(y_2) \oplus .... \oplus Z(y_k)$

i.e., $\gamma = \epsilon_1 + \epsilon_2 + ... + \epsilon_k \quad$ where $\quad \epsilon_i \in Z(y_i) \subseteq Z(x_i)$

thus $\quad x - \alpha_1 - \alpha_2 .... \alpha_k = \epsilon_1 + \epsilon_2 + ... + \epsilon_k + \delta$

i.e. $\quad x = \alpha_1 + \epsilon_1 + \alpha_2 + \epsilon_2 + ... + \alpha_k + \epsilon_k + \delta$

i.e., $\quad x \in \sum_{i=1}^{k} Z(x_i) + L$

$\therefore \quad E = \oplus \sum_{i=1}^{k} Z(x_i) + L$

Thus we have

$E = Z(x_1) \oplus Z(x_2) \oplus .... \oplus Z(x_k) \oplus L$

Consider L . It is a subspace of k .

Let $\{\tau_1, \tau_2, ... \tau_l\}$ be the basis of L .

i.e., L is a direct sum of spaces generated by $\tau_1, \tau_2, ... \tau_l$.

However as $\tau_i \in L \subseteq k$, $N(\tau_i) = 0$, $\quad 1 < i < l$

$\therefore \quad Z(\tau_i)$ is same as the spaces generated by τ_i , $1 \le i \le l$.

$$\therefore \quad L = \oplus \sum_{i=1}^{l} Z(\tau_i)$$

$$E = Z(x_1) \oplus Z(x_2) \oplus ... \oplus Z(x_k) + Z(\tau_1) \oplus Z(\tau_2)...\oplus Z(\tau_l).$$

Union of basis of $Z(x_1)$, $Z(x_2)...Z(\tau_l)$ will form a basis of E.

It is clear that all these spaces $Z(x_1)..Z(\tau_l)$ are invariant under N.

Thus taking restriction of N to these cyclic spaces, matrix of N will by obtain by putting matrices of restriction of N to these cyclic spaces along the diagonal.

But the matrix of restriction of N to any cyclic space is an elementary nilpotent block. Thus the matrix of N with respect to the basis which is the union of basis of the component cyclic spaces is diagonal $\{A_1, A_2....A_r\}$ where $A_1, A_2....A_r$ are elementary nilpotent blocks.

Since the sizes of the elementary nilpotent blocks are given by the nil of the cyclic vectors of the corresponding cyclic spaces and since in a direct sum decomposition, the order of direct summands is immaterial, it follows that by arranging the various cyclic spaces we get the matrix of $N = diag \{A_1, A_2...A_r\}$ such that size of $A_i \geq size of A_{i+1}$ $\quad 1 \leq i \leq r-1$

Thus we can find a basis of E such that the matrix of the operator N with respect to this basis is diag $\{A_1, A_2....A_r\}$ where $A_1, A_2....A_r$ are elementary nilpotent blocks corresponding to the cyclic spaces decomposition of E.

Since an elementary nilpotent block of size n is uniquely given by the matrix

$$\begin{bmatrix} 0 & 0 & 0 & .... & 0 \\ 1 & 0 & 0 & & 0 \\ 0 & 1 & 0 & & 0 \\ : & 0 & 1 & & 0 \\ : & & 0 & & : \\ 0 & & .... & & 1 \end{bmatrix}$$

and the size of an elementary block is determined by the dimension of the cyclic spaces.

Uniqueness of the elementary nilpotent blocks appearing in the matrix of N will follow if we show that the dimension of the cyclic spaces which appear in the decomposition of E.

We consider the decomposition.

$$E = Z(x_1) \oplus Z(x_2) \oplus \dots + Z(x_k) \oplus L$$

Since L is a subspace of k the kernel of N, the contribution to the dimensions from each cyclic subspace decomposition of L will be 1 thus the uniqueness is proved if we show that dim $Z(x_i)$ is uniquely determined by N.

The result obviously holds for dim $E = 0$ we use strong induction to prove the statement N on E.

So let us assume that the statement holds for all space with dimension less than that of E.

$$E = Z(x_1) \oplus Z(x_2) \oplus \dots + Z(x_k) \oplus L$$

$$\Rightarrow \quad N(E) = Z(y_1) \oplus Z(y_2) \oplus \dots + Z(y_k), \text{ where } \quad y_i = N(x_i) \qquad 1 \le i \le k$$

We know that $\dim Z(x_i) = \dim Z(y_i) + 1$

By induction hypothesis, dimensions of $Z(y_1), Z(y_2), \dots Z(y_k)$ are determined by N.

$\therefore$ dimensions of $Z(x_i)$ are determined by N.

The matrix that we get for the nilpotent operator N in the form diag $\{A_1, A_2 \dots A_r\}$ where $A_1, A_2 \dots A_r$ are elementary nilpotent blocks of decreasing size is known as the canonical (nilpotent) form of the operator N.

A canonical form of a matrix is the canonical form of the operator which is given by the matrix.

Theorem 15: Two nilpotent $n \times n$ matrices or two nilpotent operators on the same vector space are similar iff they have the same canonical form.

Proof. Obviously if two matrices have same canonical form then they are similar to a matrix (giving the canonical form) and hence are similar to each other.

Conversely, if two matrices are similar they represent same operator (nilpotent) has a unique canonical form.

Thus these two matrices will have same canonical form.

Theorem 16: The number of blocks (elementary nilpotent blocks) appearing in the canonical form of an operator is equal to dim ker N.

Proof. We consider the direct sum decomposition of the vector space E as

$$E = Z(x_1) + Z(x_2) + ... + Z(x_r)$$

So that $Z(x_i)$ corresponds to the elementary nilpotent block A_i.

As $Z(x_1)Z(x_2)..Z(x_r)$ are cyclic spaces with respect to operator N, we can decompose N as

$$N = N_1 \oplus N_2 \oplus ... \oplus N_r$$

Where $N_i = N \Big|_{Z(x_i)}$

It follows that,

$\ker N = \ker N_1 \oplus \ker N_2 .... \oplus \ker N_r$

$\therefore$ dim ker N = dim ker N_1 + dim ker N_2 + ... + dim ker N_r.

Since each N_i corresponds to an elementary (nilpotent) block and dimension of kernel of elementary nilpotent block is 1.

It follows that,

dim ker $N = 1 + 1 .... r\ time\ = r$

HOMOGENEOUS LINEAR SYSTEM

Let N be a nilpotent operator on space E.

Consider the canonical form of N.

Let δ_i denote the $\dim \ker N^i$ $0 \le i$

Let V_i denote the number of elementary nilpotent blocks of size i which appear in the canonical form of N.

If $n = \dim E$, we just consider $V_1, V_2 .... V_n$.

Theorem 17: $\delta_m = \sum_{1 \le k < m} k v_k + m \sum_{m \le j \le n} v_j .... \; m = 1....n.$

$$v_1 = 2\delta_1 - \delta_2$$

$$v_k = -\delta_{k-1} + 2\delta_k - \delta_{k+1} \qquad 1 < k < n$$

$$v_n = \delta_n - \delta_{n-1}$$

Proof. The total number of blocks in the canonical form of N is given by

$$v_1 + v_2 .... + v_n$$

But we know that the total number of blocks is equal to $\dim \ker N$ i.e. δ_1

$$\therefore \qquad \delta_1 = v_1 + v_2 .... + v_n$$

We note that if N_i is an elementary nilpotent block of size say n then the domain on which N_i is an operator is R^n

Further,

$\dim \ker N_i = 1$

$\dim \ker N_i^2 = 2$

$\dim \ker N_i^3 = 3$

$\quad \vdots$

$\dim \ker N_i^{n-1} = n-1$

$\dim \ker N_i^n = n$

$\dim \ker N_i^{n+1} = n$

and so on.

i.e., $\dim \ker N_i^j = j \qquad 0 \le j \le n$

$\dim \ker N_i^j = n \qquad n \le j$

Thus to the dimension of ker of T, there is a contribution of 1 from each constituent block in the canonical form of T.

Thus we have,

$\delta_1 = \dim \ker T = v_1 + v_2 + \dots + v_n$

For $\dim \ker T^2$, there will be contribution of 1 from each elementary block of sizes and of 2 from each elementary block of size ≥ 2.

$\therefore \quad \delta_2 = v_1 + 2(v_2 + v_3 + \dots + v_n)$

For $\dim \ker T^3$, contribution is 1 from elementary block of sizes 1, contribution is 2 from elementary block of size 2, contribution is 3 from elementary blocks of size ≥ 3.

$\therefore \quad \delta_3 = v_1 + v_2 + 3(v_3 + v_4 + \dots + v_n)$

Repeating this argument we haves,

$\delta_4 = v_1 + 2v_2 + 3v_3 + 4(v_4 + v_5 + \dots + v_n)$

$\therefore \quad \delta_{n-1} = v_1 + 2v_2 + 3v_3 \dots (n-1)(v_{n-1}) + (n-1)v_n$

$\delta_n = v_1 + 2v_2 + 3v_3 \dots + (n-1)v_{n-1} + nv_n$

i.e., $\delta_m = \sum_{1 \le k < m} k v_k + m \sum_{m \le j \le n} v_j \qquad 1 \le m \le n$

It is also clear that,

$v_1 = 2\delta_1 - \delta_2$

Further for $1 < k < n, \quad v_k = 2\delta_k - \delta_{k-1} - \delta_{k+1}$

and $\qquad\qquad\qquad v_n = \delta_n - \delta_{n-1}$

HOMOGENEOUS LINEAR SYSTEM

Since, $T^o \equiv I$,

$\delta_0 = \dim \ker T^o = 0$

Also, $\delta_n = \delta_{n+1} = ... = n$

Thus we can write,

$v_1 = 2\delta_1 - \delta_2 = 2\delta_1 - \delta_2 - \delta_0 = \delta_0 + 2\delta_1 - \delta_2$

$v_n = \delta_n - \delta_{n-1} = \delta_{n+n} - \delta_{n-n}$

$\qquad = \delta_{n+n} - \delta_{n-1} - \delta_{n+1}$

$\qquad = -\delta_{n-1} + 2\delta_n - \delta_{n+1}$

Thus we have for $1 \leq k \leq n$, $v_k = -\delta_{k-1} + 2\delta_k - \delta_{k+1}$

We can consider $\delta_0, \delta_1 .... \delta_n$ to be known quantities use these equations to get values of $v_1, v_2 .... v_n$ with the help of $v_1, v_2 .... v_m$ we can write the canonical form of the operator T.

Verify that each of the following operators is nilpotent and find its canonical form.

a). $\begin{bmatrix} 0 & 0 & 0 \\ 1 & 0 & 0 \\ 0 & 2 & 0 \end{bmatrix}$

Obviously the operator A whose matrix is nilpotent (3 is order of nil potency).

We find $\delta_0, \delta_1, \delta_3$

Consider $\ker A$

$(\alpha, \beta, \gamma) \in R^3$ is in $\ker A$ iff

$$\begin{bmatrix} 0 & 0 & 0 \\ 1 & 0 & 0 \\ 0 & 2 & 0 \end{bmatrix}\begin{bmatrix} \alpha \\ \beta \\ \gamma \end{bmatrix} = \begin{bmatrix} 0 \\ 0 \\ 0 \end{bmatrix} \quad i.e. \quad \begin{bmatrix} 0 \\ \alpha \\ 2\beta \end{bmatrix} = \begin{bmatrix} 0 \\ 0 \\ 0 \end{bmatrix}$$

i.e. $\alpha = \beta = 0$

$\therefore \quad \ker A = \{(0,0,t) \, , \, t \in R\}$

$\therefore \dim \ker A = 1 \qquad \therefore \delta_1 = 1$

Since the operator A is a linear operator on R^3, we need to know v_0, v_1, v_3. Thus v_0, v_1, v_3

Since v_0, v_1, v_3 are non negative integers, we must have,

$v_1 = 0, v_2 = 0, v_3 = 0$

Or $v_1 = 0, v_2 = 1, v_3 = 0$

Or $v_1 = 0, v_2 = 0, v_3 = 1$

First two possibilities are ruled out as in that case the 3×3 matrix does not get filled up. Thus we must have

$0 = v_1 = v_2 \, , \, v_3 = 1.$

Thus the canonical form of A is

$$\begin{bmatrix} 0 & 0 & 0 \\ 1 & 0 & 0 \\ 0 & 1 & 0 \end{bmatrix} \quad \text{(an elementary block of size 3)}$$

[b]. $\begin{bmatrix} 0 & 2 & -2 \\ 0 & 0 & 4 \\ 0 & 0 & 0 \end{bmatrix}$

Obviously the operator A whose matrix is 3×3 and is nilpotent (3 is order of nil potency).

We find v_0, v_1, v_3.

HOMOGENEOUS LINEAR SYSTEM

Consider δ_1, ker A. $\{(\alpha,\beta,\gamma)\in R^3\}$ is in ker A iff

$$\begin{bmatrix} 0 & 2 & -2 \\ 0 & 0 & 4 \\ 0 & 0 & 0 \end{bmatrix}\begin{bmatrix} \alpha \\ \beta \\ \gamma \end{bmatrix}=\begin{bmatrix} 0 \\ 0 \\ 0 \end{bmatrix} \ i.e. \ \begin{bmatrix} 2\alpha-2\beta \\ 4\gamma \\ 0 \end{bmatrix}=\begin{bmatrix} 0 \\ 0 \\ 0 \end{bmatrix}$$

i.e. $\beta=\gamma=0$

$\therefore$ ker $A=\{(t,0,0),\ t\in R\}$

$\therefore$ $\delta_1=1$ $(v_3=1,v_2=v_1=0)$

The canonical form is

$$\begin{bmatrix} 0 & 0 & 0 \\ 1 & 0 & 0 \\ 0 & 1 & 0 \end{bmatrix}$$

[c].
$$\begin{bmatrix} 0 & 0 & 0 & 0 \\ 0 & 0 & 0 & 0 \\ 6 & 7 & 0 & 0 \\ 8 & 9 & 0 & 0 \end{bmatrix}$$

$\therefore$ ker $A=\{(\alpha,\beta,\gamma,\delta)\in R^4\}$

$$\begin{bmatrix} 0 & 0 & 0 & 0 \\ 0 & 0 & 0 & 0 \\ 6 & 7 & 0 & 0 \\ 8 & 9 & 0 & 0 \end{bmatrix}\begin{bmatrix} \alpha \\ \beta \\ \gamma \\ \delta \end{bmatrix}=\begin{bmatrix} 0 \\ 0 \\ 0 \\ 0 \end{bmatrix}$$

i.e. $6\gamma+7\beta=0$

$8\gamma+9\beta=0$

i.e. $\alpha=\beta=0$

Thus ker $A=\{(0,0,t,s),t,s\in R\}$

$\therefore$ $\delta_1=2$

$\therefore$ $v_1+v_2+v_3+v_4=2$

Consider A^2,

$$A^2 = \begin{bmatrix} 0 & 0 & 0 & 0 \\ 0 & 0 & 0 & 0 \\ 0 & 0 & 0 & 0 \\ 0 & 0 & 0 & 0 \end{bmatrix}$$

$$\therefore \quad \delta_2 = \delta_3 = \delta_4 = 4$$

$$\therefore \quad \gamma_1 = 0$$

$$v_2 = 2\delta_2 - \delta_1 - \delta_3 = 8 - 2 - 4 = 2$$

$$v_3 = 2\delta_3 - \delta_2 - \delta_4 = 8 - 4 - 4 = 0$$

$$\therefore \quad v_1 = v_3 = v_4 = 0 \quad and \quad v_2 = 2$$

Thus the canonical form of A is

$$\begin{bmatrix} 0 & 0 & 0 & 0 \\ 1 & 0 & 0 & 0 \\ 0 & 0 & 0 & 0 \\ 0 & 0 & 1 & 0 \end{bmatrix}$$

d).
$$\begin{bmatrix} 0 & 0 & 0 & 0 & 0 \\ 1 & 0 & 0 & 0 & 0 \\ 0 & 0 & 0 & 0 & 0 \\ 0 & 0 & 0 & 0 & 0 \\ 0 & 0 & 2 & 3 & 0 \end{bmatrix}$$

e).
$$\begin{bmatrix} 1 & 1 & 0 & 0 \\ -1 & -1 & 0 & 0 \\ 0 & 1 & 2 & -2 \\ 1 & 0 & 2 & -2 \end{bmatrix}$$

Solution (d) $\Rightarrow \quad (\alpha, \beta, \gamma, \delta, \in) \in R^5$ will in the ker A iff

$$A \begin{bmatrix} \alpha \\ \beta \\ \gamma \\ \delta \\ \in \end{bmatrix} = \begin{bmatrix} 0 \\ 0 \\ 0 \\ 0 \\ 0 \end{bmatrix}$$

$$\begin{bmatrix} 0 & 0 & 0 & 0 & 0 \\ 1 & 0 & 0 & 0 & 0 \\ 0 & 0 & 0 & 0 & 0 \\ 0 & 0 & 0 & 0 & 0 \\ 0 & 0 & 2 & 3 & 0 \end{bmatrix} \begin{bmatrix} \alpha \\ \beta \\ \gamma \\ \delta \\ \in \end{bmatrix} = \begin{bmatrix} 0 \\ 0 \\ 0 \\ 0 \\ 0 \end{bmatrix} \Rightarrow \begin{bmatrix} 0 \\ \alpha \\ 0 \\ 0 \\ 2\alpha + 3\delta \end{bmatrix} = \begin{bmatrix} 0 \\ 0 \\ 0 \\ 0 \\ 0 \end{bmatrix}$$

i.e. $(\alpha, \beta, \gamma, \delta, \in) \in \ker A$

iff $\alpha = 0$ and $2\gamma + 3\delta = 0$ i.e. $\delta = -\frac{2}{3}\gamma$

Then $\ker A = \{0, \beta, \gamma, -\frac{2}{3}\gamma, \in\}$, $\beta, \gamma, \in \in R^5$

$\delta_i = \dim \ker A = 3$

$\therefore$ Now A^2 will be

$$\begin{bmatrix} 0 & 0 & 0 & 0 & 0 \\ 1 & 0 & 0 & 0 & 0 \\ 0 & 0 & 0 & 0 & 0 \\ 0 & 0 & 0 & 0 & 0 \\ 0 & 0 & 2 & 3 & 0 \end{bmatrix} \begin{bmatrix} 0 & 0 & 0 & 0 & 0 \\ 1 & 0 & 0 & 0 & 0 \\ 0 & 0 & 0 & 0 & 0 \\ 0 & 0 & 0 & 0 & 0 \\ 0 & 0 & 2 & 3 & 0 \end{bmatrix} = \begin{bmatrix} 0 & 0 & 0 & 0 & 0 \\ 0 & 0 & 0 & 0 & 0 \\ 0 & 0 & 0 & 0 & 0 \\ 0 & 0 & 0 & 0 & 0 \\ 0 & 0 & 0 & 0 & 0 \end{bmatrix}$$

As $A^2 = 0$, $\ker A^2 = R^5$

$\delta_2 = \dim \ker A^2 = 5$

Now $A^2 = 0 \Rightarrow A^3 = 0 \Rightarrow A^4 = 0 \Rightarrow A^5 = 0$

$\therefore \delta_2 = \delta_3 = \delta_4 = \delta_5 = 5$

$\therefore$ we need to find out v_1, v_2, v_3, v_4, v_5.

$\therefore v_1 = 2\delta_1 - \delta_2 = 6 - 5 = 1$

$v_2 = 2\delta_2 - \delta_1 - \delta_3 = 10 - 3 - 5 = 2$

$v_3 = 2\delta_3 - \delta_2 - \delta_4 = 0 - 0 - 5 = v_4 = v_5$

Thus the canonical form of matrix will consist of two elementary nilpotent blocks of size 2 and one elementary nilpotent block of size 1.

∴ The canonical form is

$$\begin{bmatrix} 0 & 0 & 0 & 0 & 0 \\ 1 & 0 & 0 & 0 & 0 \\ 0 & 0 & 0 & 0 & 0 \\ 0 & 0 & 1 & 0 & 0 \\ 0 & 0 & 0 & 0 & 0 \end{bmatrix}$$

e). $A = \begin{bmatrix} 1 & 1 & 0 & 0 \\ -1 & -1 & 0 & 0 \\ 0 & 1 & 2 & -2 \\ 1 & 0 & 2 & -2 \end{bmatrix} \begin{bmatrix} \alpha \\ \beta \\ \gamma \\ \delta \end{bmatrix} = \begin{bmatrix} \alpha + \beta \\ -\alpha - \beta \\ \beta + 2\gamma - 2\delta \\ \alpha + 2\gamma - 2\delta \end{bmatrix} = \begin{bmatrix} 0 \\ 0 \\ 0 \\ 0 \end{bmatrix}$

$\Rightarrow \alpha + \beta = 0$

$\quad -\alpha - \beta = 0$

$\quad \beta + 2\gamma - 2\delta = 0$

$\quad \alpha + 2\gamma - 2\delta = 0$

$\Rightarrow \alpha = \beta = 0$

$\quad 2\gamma - 2\delta = 0$

$\quad \gamma = \delta$

$\therefore$ ker $A = \{(0,0,\gamma,\gamma), \gamma \in R\}$

$\therefore \delta_1 = \dim \ker A = 1$

As $\therefore \delta_1 = v_1 + v_2 + v_3 + v_4$

we have $\quad 0 = v_1 = v_2 = v_3 \; and \; v_4 = 1$

$\therefore$ canonical form of A

$$\begin{bmatrix} 0 & 0 & 0 & 0 \\ 1 & 0 & 0 & 0 \\ 0 & 1 & 0 & 0 \\ 0 & 0 & 1 & 0 \end{bmatrix}$$

HOMOGENEOUS LINEAR SYSTEM

f) Let N be a matrix in nilpotent canonical form.

Prove that N is similar to

  (a) KN for all non zero $K \in R$

  (b) the transpose of N.

Solution. We assume that A is itself an elementary nilpotent block.

Suppose N is an elementary block of size n.

i.e. N is given by a matrix

$$\begin{bmatrix} 0 & 0 & . . & 0 & 0 \\ 1 & 0 & & 0 & 0 \\ 0 & 1 & & 0 & 0 \\ : & : & & 0 & 0 \\ 0 & 0 & & 1 & 0 \end{bmatrix}$$

i.e. there exist a basis say $\{e_1, e_2 ... e_n\}$ of R^n such that,

$N(e_1) = e_2$

$N(e_2) = e_3$

$\quad :$

$N(e_{n-1}) = e_n$

$N(e_n) = 0$

Now for a given real K, Consider the basis $\{f_1, ... f_n\}$

where $\quad f_i = \dfrac{e_i}{k^{i-1}}$, $\quad 1 \le i \le n.$

$\therefore \quad f_1 = e_1$, $f_2 = \dfrac{e_2}{k}$, $f_3 = \dfrac{e_3}{k^3} ...$

$\qquad f_n = \dfrac{e_n}{k^{n-1}}$

$\therefore \quad N(f_1) = N(e_1) = e_2 = k\, f_2$

$N(f_2) = N\left(\dfrac{e_2}{k}\right) = \dfrac{1}{k} N(e_2) = \dfrac{1}{k} e_3 = k\, f_3$

$$\vdots$$

$$N(f_{n-1}) = k\, f_n$$

$$N(f_n) = 0.$$

$\therefore$ The matrix of N in the basis $\{f_1, f_2 \ldots f_n\}$

is $\begin{bmatrix} 0 & 0 & 0 & \ldots & 0 & 0 \\ k & 0 & 0 & & 0 & 0 \\ 0 & k & 0 & & 0 & 0 \\ 0 & 0 & k & & 0 & 0 \\ 0 & 0 & 0 & & 0 & 0 \\ 0 & 0 & 0 & \ldots & k & 0 \end{bmatrix}$

b) If $\{e_1, e_2 \ldots e_n\}$ is the basis of R^n with respect to which the matrix of N is

$$\begin{bmatrix} 0 & 0 & \ldots & 0 & 0 \\ 1 & 0 & & 0 & 0 \\ 0 & 1 & & 0 & 0 \\ 0 & 0 & & 0 & 0 \\ \vdots & \vdots & & 0 & 0 \\ 0 & 0 & & 1 & 0 \end{bmatrix}$$

then we have

$$N(e_1) = e_2$$
$$N(e_2) = e_3$$
$$\vdots$$
$$N(e_{n-1}) = e_n$$
$$N(e_n) = 0$$

We take consider the ordered set $\{f_1, f_2 \ldots f_n\}$

where $f_1 = e_n = f_2 = e_{n-1}$, $f_3 = e_{n-2} \ldots f_n = e_1$

$\therefore \quad N(f_1) = N(e_n) = 0$

$$N(f_2) = N(e_{n-1}) = e_n = f_1$$

HOMOGENEOUS LINEAR SYSTEM

$$N(f_3) = N(e_{n-2}) = e_{n-1} = f_2$$

$$N(f_n) = N(e_1) = e_2 = f_{n-1}$$

Thus the matrix of N with respect to their basis $\{f_1, f_2 ... f_n\}$ is

$$\begin{bmatrix} 0 & 1 & 0 & 0.. \\ 0 & 0 & 1 & 0. \\ 0 & 0 & 0 & 1. \\ 0 & 0 & 0 & 0 \end{bmatrix}$$

If N is given in the canonical form then the matrix of N in the given basis consist of elementary nilpotent blocks along the diagonal.

Since the operations of multiplication by k or taking transpose are obtained by performing these operations on each constituent block and as each of there operations are similar for elementary nilpotent blocks, the similarity follows for the nilpotent canonical form.

3) Let N be an $n \times n$ nilpotent matrix of rank r. If $N^k = 0$, then show that $R \geq \dfrac{n}{n-r}$.

Solution. Since N is nilpotent we can find a basis of the domain space $E \equiv R^n$ such that the space $E \equiv R^n$. Such that the matrix N can be written in the nilpotent canonical form.

(or N is similar to its nilpotent canonical form)

Let the canonical form given by $\{A_1 .... A_r\}$ are elementary nilpotent blocks.

Since rank of $N = r$

i.e. dim image $N(E) = r$

it follows that dim image $N = n - r$

i.e. $\delta_1 = n - r$

but we know that δ_1 is the total number of elementary nilpotent blocks which constitutes the canonical form of N.

If the size of each elementary nilpotent block constituting the canonical

form of N is $< \dfrac{n}{n-r}$ then all these blocks together along less than n

which is impossible as N is $n \times n$.

Thus the canonical form of N will contain at least

one elementary block of size $\geq \dfrac{n}{n-r}$.

As the order of nil potency of a canonical form is equal to the size of the

largest elementary block constituting the canonical form, it follows that the

order of nil potency $k(N)$ must be $\geq \dfrac{n}{n-r}$.

Ex 17: Two nilpotent operators N, M on a vector space E are similar iff,

dim ker N^i = dim ker M^i , $\quad 0 \leq i \leq n$,

where $n = \dim E$.

Classify the following operators on R^4 by similarity.

$$N_1 = \begin{bmatrix} 0 & 1 & 0 & 0 \\ 0 & 0 & 2 & 0 \\ 0 & 0 & 0 & 3 \\ 0 & 0 & 0 & 0 \end{bmatrix} \qquad N_2 = \begin{bmatrix} 2 & 0 & 0 & 2 \\ 0 & 0 & 0 & 0 \\ 0 & 0 & 0 & 0 \\ -2 & 0 & 0 & -2 \end{bmatrix}$$

$$N_3 = \begin{bmatrix} 0 & 0 & 0 & 0 \\ 4 & 0 & 0 & 0 \\ 0 & 0 & 0 & 4 \\ 0 & 0 & 0 & 0 \end{bmatrix} \qquad N_4 = \begin{bmatrix} 0 & 0 & 0 & 0 \\ 1 & 0 & 0 & 0 \\ 0 & 1 & 0 & 0 \\ -1 & -1 & -1 & 0 \end{bmatrix}$$

$$N_5 = \begin{bmatrix} 0 & 0 & 0 & 100 \\ 0 & 0 & 0 & 0 \\ 0 & 0 & 0 & 0 \\ 0 & 0 & 0 & 0 \end{bmatrix}$$

Solution. $\ker N_1 = \{(\alpha, \beta, \gamma, \delta) \in R^4\}$

$$\begin{bmatrix} 0 & 1 & 0 & 0 \\ 0 & 0 & 2 & 0 \\ 0 & 0 & 0 & 3 \\ 0 & 0 & 0 & 0 \end{bmatrix} \begin{bmatrix} \alpha \\ \beta \\ \gamma \\ \delta \end{bmatrix} = \begin{bmatrix} 0 \\ 0 \\ 0 \\ 0 \end{bmatrix}$$

$\beta = 0$, $\gamma = 0$, $\delta = 0$

$\ker N_1 = \{(t, 0, 0, 0), \ t \in R\}$

$\therefore \dim \ker N_1 = 1$

$$N_1^2 = \begin{bmatrix} 0 & 1 & 0 & 0 \\ 0 & 0 & 2 & 0 \\ 0 & 0 & 3 & 0 \\ 0 & 0 & 0 & 0 \end{bmatrix} \begin{bmatrix} 0 & 1 & 0 & 0 \\ 0 & 0 & 2 & 0 \\ 0 & 0 & 3 & 0 \\ 0 & 0 & 0 & 0 \end{bmatrix}$$

$$\Rightarrow \begin{bmatrix} 0 & 0 & 2 & 0 \\ 0 & 0 & 0 & 6 \\ 0 & 0 & 0 & 0 \\ 0 & 0 & 0 & 0 \end{bmatrix} \begin{bmatrix} \alpha \\ \beta \\ \gamma \\ \delta \end{bmatrix} = \begin{bmatrix} 0 \\ 0 \\ 0 \\ 0 \end{bmatrix}$$

$\therefore \gamma = 0$, $\delta = 0$

$\dim \ker N_1^2 = 2$

$\dim \ker N_1^3 = 3$

To find $\dim \ker N_2$

$$\begin{bmatrix} 2 & 0 & 0 & 2 \\ 0 & 0 & 0 & 0 \\ 0 & 0 & 0 & 0 \\ -2 & 0 & 0 & -2 \end{bmatrix} \begin{bmatrix} \alpha \\ \beta \\ \gamma \\ \delta \end{bmatrix} = \begin{bmatrix} 0 \\ 0 \\ 0 \\ 0 \end{bmatrix}$$

$$2\alpha + 2\delta = 0$$

$$-2\alpha - 2\delta = 0$$

$$\therefore \quad \alpha = -\delta$$

dim ker $N_2 = 3$

$$N_2{}^2 = \begin{bmatrix} 2 & 0 & 0 & 2 \\ 0 & 0 & 0 & 0 \\ 0 & 0 & 0 & 0 \\ -2 & 0 & 0 & -2 \end{bmatrix} \begin{bmatrix} 2 & 0 & 0 & 2 \\ 0 & 0 & 0 & 0 \\ 0 & 0 & 0 & 0 \\ -2 & 0 & 0 & -2 \end{bmatrix} = \begin{bmatrix} 0 & 0 & 0 & 0 \\ 0 & 0 & 0 & 0 \\ 0 & 0 & 0 & 0 \\ 0 & 0 & 0 & 0 \end{bmatrix}$$

$$\therefore \quad \text{dim ker } N_2{}^2 = 4$$

$$\therefore \quad \text{dim ker } N_2{}^3 = 4$$

$$\therefore \quad \text{dim ker } N_2{}^4 = 4.$$

CHAPTER 4

JORDAN AND REAL CANONICAL FORMS

Let λ be a constant (real or complex) for $n \geq 2$, we define a Jordan block belonging to λ or a λ Jordan block as $n \times n$ matrix with entries along the diagonal given by λ.

One's immediately below the diagonal and rest of the entries are 0.

Thus $\begin{bmatrix} \lambda & 0 \\ 1 & \lambda \end{bmatrix}$, $\begin{bmatrix} \lambda & 0 & 0 \\ 1 & \lambda & 0 \\ 0 & 1 & \lambda \end{bmatrix}$, $\begin{bmatrix} \lambda & 0 & 0 & 0 \\ 1 & \lambda & 0 & 0 \\ 0 & 1 & \lambda & 0 \\ 0 & 0 & 1 & \lambda \end{bmatrix}$ and so on.

are Jordan block belonging to λ of size 2,3,4 and so on.

The 1×1 matrix $\begin{bmatrix} \lambda \end{bmatrix}$ is known as Jordan block of size 1 belonging to λ.

Theorem 1: Let $T \in L(E)$ be an operator if E is real we assume that all eigen values of T are real; then E has a basis giving T a matrix in Jordan form, which is made up of different Jordan blocks belonging to different eigen values of T.

Definition: A matrix is said to be in Jordan form if it is expressible as diagonal $\{A_1 .... A_r\}$ where, A_i is a Jordan block belonging to some scalar.

Proof: We first consider the case where T has only one eigen value say λ (of multiplicity $n \equiv \dim E$).

In this case we consider the matrix $S = \lambda I$, then the matrix (operator) N defined as $N = T - S$ is a nilpotent matrix (operator).

Thus we can write

$T = S + N$

As N is nilpotent there exist a basis of E such that the matrix of N with respect to this basis is in the canonical form.

Since $S = \lambda I$, the form of S will remain same in any basis.

DYNAMICAL SYSTEM – A SHORT COURSE

Thus with respect to the basis in which the matrix of N is in the canonical form, the matrix of T will be given of the type diag $\{A_1', A_2'....A_r'\}$

where $A_1', A_2'....A_r'$ are obtained from the respective elementary blocks $\{A_1, A_2....A_r\}$ in the canonical form of N by replacing zeros along the diagonal by λ.

Thus $A_1', A_2'....A_r'$ are all λ Jordan blocks.

$\therefore$ The matrix diag $\{A_1', A_2'....A_r'\}$, of the operator T with respect to the new basis is in the Jordan form.

We next consider the case where the operator T has number of different eigen values say $\lambda_1, \lambda_2....\lambda_k$ of multiplication say $r_1, r_2,...r_k$ respectively.

Thus we can write

$E = E_1 \oplus E_2 \oplus....\oplus E_k$

where E_i is generalized eigen space belonging to eigen value and dim $E_i = r_i$

We also know that E_i is invariant under T and $T\big|_{E_i}$ has only one eigen value namely λ_i.

Thus there will exist a basis of E_i with respect to which the matrix of $T\big|_{E_i}$ will have a Jordan form.

As $\quad T = \oplus \sum_{i=1}^{k} T\big|_{E_i}$

If we consider the union of the basis of $E_1, E_2....E_k$ with respect to which

$T \mid_{E_i}$ has a Jordan form.

We get a basis of E with respect to which the matrix of T will have Jordan blocks/ forms along its diagonal. This is thus a Jordan form of T.

If an operator T has eigen values $\lambda_1, \lambda_2....\lambda_r$ of different multiplicities (we assume $\lambda_1,....\lambda_r$ all real if is real space), the Jordan form of T will consist of Jordan blocks belonging to $\lambda_1, \lambda_2....\lambda_r$ of different sizes and the diagonal of this form will have the eigen values $\lambda_1,....\lambda_r$ repeated according to their multiplicities.

If $\lambda \neq 0$, then the elementary λ Jordan block is such that the kernel of the λ block as well as its powers is trivial {i.e. (0)}.

Thus there is no contribution to the kernel of operator T form a λ Jordan block with $\lambda \neq 0$.

Hence if we consider the operator $T - \lambda_1$ where λ_1 is one of the eigen values then the Jordan form of $T - \lambda_1$ will consist of elementary nilpotent blocks (from λ_1 Jordan blocks and Jordan blocks belonging to $\lambda_2 - \lambda_1$, $\lambda_3 - \lambda_1....\lambda_r - \lambda_1$).

Since each elementary nilpotent block contributes 1 to the dimension of kernel it follows that [$\dim \ker(T - \lambda_1)$ = number of λ_1 Jordan blocks in the Jordan form of T].

Further if $\delta_i(\lambda_j)$ denotes $\dim \ker(T - \lambda_j)^i$ and $V_i(\lambda_j)$ denotes the number of elementary Jordan blocks of size i belonging to λ_j then we have,

$V_i(\lambda_j) = 2 \delta_i(\lambda_j) - \delta_{i-1}(\lambda_j) - \delta_{i+1}(\lambda_j)$, $1 \leq i \leq n_j$ where n_j is the multiplicity of the eigen value λ_j.

DYNAMICAL SYSTEM – A SHORT COURSE

Theorem 2: Let $T : E \to E$ be an operator on real vector space, then E has a basis giving T a matrix composed of Jordan blocks belonging to real eigen values and blocks of the type

$$\begin{bmatrix} a & -b \\ b & a \end{bmatrix}, \begin{bmatrix} a & -b & 0 & 0 \\ b & a & 0 & 0 \\ 0 & 0 & a & -b \\ 0 & 0 & b & a \end{bmatrix}, \begin{bmatrix} a & -b & 0 & 0 \\ b & a & 0 & 0 \\ 1 & 0 & a & -b \\ 0 & 1 & b & a \end{bmatrix}$$

and so on corresponding to complex eigen pair $a \pm ib$. This form is known as the real form of operator T.

Proof: If all the eigen values of the operator T are real then by the previous theorem.

We can get a basis of the space E with respect to which the matrix of T is in the Jordan form.

So we consider the case where T has some non real eigen values. In this case we consider the complexification of the space to E_c and consider the complexified operator T_c on E_c.

All the eigen values of T are the eigen values of T_c. Thus we can find the basis of E_c with respect to which the matrix of T_c is in the Jordan form which will have complex entries.

If λ is a real eigen value of T (and T_c), we know that there is a decomposition of E_c where one of the direct summands is a generalized eigen space belonging to eigen value λ.

It is a complexification of a real vector space which will be a direct summands of the vector space E. Thus the matrix of T_c restricted to the generalized eigen space belonging to real eigen value λ will be same as the matrix of T restricted to the above mentioned direct summands of E.

JORDAN AND REAL CANONICAL FORMS

Thus it will contribute appropriate number of Jordan blocks to the matrix of T (in appropriate basis).

We know that for real operator complex roots occur in pair and as such multiplicity of the complex root $a+ib$ is same as that of $a-ib$.

Thus if the eigen pair $a+ib$ and $a-ib$ of multiplicity ' r ', then a generalized complex eigen space of dimension r will be a direct summand of E_C.

If $\phi_1, \phi_2, ... \phi_r$ form a basis of this complex space, then considering

$$\phi_j = u_j + iv_j \qquad 1 \le j \le r$$

we get a real vector subspace of E of dimension $2r$ with basis $\{v_1, u_1, v_2, u_2 .... v_i, u_i\}$.

The matrix of T restricted to this space will be a block of the type

$$[D], \begin{bmatrix} D & 0 \\ 1 & 0 \end{bmatrix}, \begin{bmatrix} D & 0 & 0 \\ 1 & D & 0 \\ 0 & 1 & D \end{bmatrix} \quad \text{and so on}$$

where $D = \begin{bmatrix} a & -b \\ b & a \end{bmatrix}$, $I = \begin{bmatrix} 1 & 0 \\ 0 & 1 \end{bmatrix}$.

Thus if T is a operator on a real vector space E, then there always exist a basis of E, the matrix of T with respect to which is a real form.

If all the eigen value of T are real, then the real form of T is the Jordan form.

If the operator T is nilpotent (i.e., all eigen values are zero) then the real form of T is nothing but its (nilpotent) canonical form.

Theorem 3: If $\lambda_1, \lambda_2 .... \lambda_n$ are the eigen values (with multiplicities) of an operator T then

(a) Trace of $T = \lambda_1 + \lambda_2 + ... + \lambda_n$

DYNAMICAL SYSTEM – A SHORT COURSE

(b) Det $T = \lambda_1, \lambda_2 .... \lambda_n$.

Follows because there exists a basis with respect to which E has a real form and all eigen values occur in the diagonal.

1) Find the Jordan form of the following operator on C^2.

a) $\begin{bmatrix} 0 & 1 \\ -1 & 0 \end{bmatrix}$ b) $\begin{bmatrix} i & -1 \\ 1 & i \end{bmatrix}$ c) $\begin{bmatrix} 1+i & 2 \\ 0 & 1+i \end{bmatrix}$

Solution: a) The eigen values are given by

$$\begin{vmatrix} -\lambda & 1 \\ -1 & -\lambda \end{vmatrix} = 0 \implies \lambda^2 + 1 = 0$$

$$\implies \lambda = \pm i$$

The Jordan form of the operator is a 2×2 matrix which will be formed by Jordan block for $+i$ and for $-i$.

It is obvious that, the Jordan blocks belonging to $+i$ and $-i$ must be both of size 1.

$\therefore$ The Jordan form of the operator is $\begin{bmatrix} i & 0 \\ 0 & -i \end{bmatrix}$

b) The eigen values are given by

$$\begin{vmatrix} i-\lambda & -1 \\ 1 & i-\lambda \end{vmatrix} = 0$$

$$\implies (i-\lambda)^2 + 1 = 0$$

$$\implies -1 + \lambda^2 - 2i\lambda + 1 = 0$$

$$\implies \lambda(\lambda - 2i) = 0$$

$\therefore$ The eigen values are $\lambda = 0$, $\lambda = 2i$

$\therefore$ Jordan form $\begin{bmatrix} 0 & 0 \\ 0 & 2i \end{bmatrix}$

c) $\begin{bmatrix} 1+i & 2 \\ 0 & 1+i \end{bmatrix}$

The eigen values are given by

$$\begin{vmatrix} 1+i-\lambda & 2 \\ 0 & 1+i-\lambda \end{vmatrix} = 0$$

$\Rightarrow \quad ((1+i)-\lambda)^2 - 2 = 0$

$\Rightarrow \quad \lambda = 1+i, \ 1+i$

Thus an eigen value of multiplicity 2.

The Jordan form of the operator may contain just a single Jordan block of size 2 or 2 Jordan blocks each of size 1.

To resolve this we need to find $\delta_1 (1+i)$

$$= \dim \ker (T-(1+i)I)$$

Kernel of $\quad T-(1+i)I = \{(z_1,z_2), \ z_1,z_2 \in \mathbb{C}\}$ such that

$$\begin{bmatrix} (1+i)-(1+i) & 2 \\ 0 & (1+i)-(1+i) \end{bmatrix} \begin{bmatrix} z_1 \\ z_2 \end{bmatrix} = \begin{bmatrix} 0 \\ 0 \end{bmatrix}$$

i.e. $\quad 2z_2 = 0 \ \Rightarrow z_2 = 0$

$\therefore \ \ker (T-(1+i)I) = \{(z_1,0), \ z_1 \in \mathbb{C}\}$

$\therefore \ \ \delta_1 = 1$

Thus the Jordan form of the operator will consist of a single Jordan block of size 2.

Thus the Jordan form is $\begin{bmatrix} 1+i & 0 \\ 1 & 1+i \end{bmatrix}$

Find Jordan/real form for the operator T on real vector space.

a) $\begin{bmatrix} 1 & 1 \\ 0 & 1 \end{bmatrix}$ 　　　 b) $\begin{bmatrix} 1 & 1 \\ 0 & -1 \end{bmatrix}$ 　　　 c) $\begin{bmatrix} 0 & 1 \\ 1 & 0 \end{bmatrix}$

d) $\begin{bmatrix} 0 & 2 & 0 \\ -2 & 0 & 0 \\ 2 & 0 & 6 \end{bmatrix}$ 　 e) $\begin{bmatrix} 0 & 0 & 8 \\ 0 & 0 & 4 \\ 0 & 1 & -2 \end{bmatrix}$ 　 f) $\begin{bmatrix} 1 & 1 & 1 & 1 \\ 2 & 2 & 2 & 2 \\ 3 & 3 & 3 & 3 \\ 4 & 4 & 4 & 4 \end{bmatrix}$

g) $\begin{bmatrix} 0 & 0 & 0 & -8 \\ 1 & 0 & 0 & 16 \\ 0 & 1 & 0 & -14 \\ 0 & 0 & 1 & 6 \end{bmatrix}$

Solution: Consider $T = \begin{bmatrix} 0 & 0 & 0 & -8 \\ 1 & 0 & 0 & 16 \\ 0 & 1 & 0 & -14 \\ 0 & 0 & 1 & 6 \end{bmatrix}$

Eigen values are given by

$$\begin{vmatrix} -\lambda & 0 & 0 & -8 \\ 1 & -\lambda & 0 & 16 \\ 0 & 1 & -\lambda & -14 \\ 0 & 0 & 1 & 6-\lambda \end{vmatrix} = 0$$

$$\Rightarrow -\lambda \begin{vmatrix} -\lambda & 0 & 16 \\ 1 & -\lambda & -14 \\ 0 & 1 & 6-\lambda \end{vmatrix} - 0 + 0 + 8 \begin{vmatrix} 1 & -\lambda & 0 \\ 0 & 1 & -\lambda \\ 0 & 0 & 1 \end{vmatrix}$$

$$\Rightarrow -\lambda \times \left[-\lambda \begin{vmatrix} -\lambda & -14 \\ 1 & 6-\lambda \end{vmatrix} + 16 \begin{vmatrix} 1 & -\lambda \\ 0 & 1 \end{vmatrix} \right] + 8 \left[1 \begin{vmatrix} 1 & -\lambda \\ 0 & 1 \end{vmatrix} + \lambda \begin{vmatrix} 0 & -\lambda \\ 0 & 1 \end{vmatrix} \right] = 0$$

$$\Rightarrow -\lambda \left[-\lambda \{ -\lambda(6-\lambda) + 14 \} + 16 \right] + 8[1+0] = 0$$

$$\Rightarrow -\lambda \left[-\lambda \{ (-6\lambda + \lambda^2) + 14 \} + 16 \right] + 8 = 0$$

$$\Rightarrow -\lambda \left[6\lambda^2 - \lambda^3 - 14\lambda + 16 \right] + 8 = 0$$

$$\Rightarrow -6\lambda^3 + \lambda^4 - 14\lambda - 16\lambda + 8 = 0$$

JORDAN AND REAL CANONICAL FORMS

$$\Rightarrow \lambda^4 - 6\lambda^3 + 14\lambda^2 - 16\lambda + 8 = 0$$

The eigen values are 2, 2, $1+i$, $1-i$.

Since the operator T on the real space has a pair of complex eigen values, it will have a real form which will consist of Jordan block corresponding to real eigen value 2 of multiplicity 2, and a real block corresponding to the pair $1 \pm i$.

As the complex pair $1 \pm i$ has multiplicity 1, the corresponding block is $\begin{bmatrix} 1 & -1 \\ 1 & 1 \end{bmatrix}$ (with $a = 1, b = 1$).

We now find the Jordan block corresponding to eigen value 2.

Thus we find $\delta_1(2)$ i.e. $\dim \ker (T - 2I)$

$\ker (T - 2I) = \{(\alpha, \beta, \gamma, \delta), \alpha, \beta, \gamma, \delta \in R\}$ such that

$$\begin{bmatrix} -2 & 0 & 0 & -8 \\ 1 & -2 & 0 & 16 \\ 0 & 1 & -2 & -14 \\ 0 & 0 & 1 & 4 \end{bmatrix} \begin{bmatrix} \alpha \\ \beta \\ \gamma \\ \delta \end{bmatrix} = \begin{bmatrix} 0 \\ 0 \\ 0 \\ 0 \end{bmatrix}$$

$\Rightarrow -2\alpha - 8\delta = 0$

$\alpha - 2\beta + 16\delta = 0$

$\beta - 2\gamma, -14\delta = 0$

$\gamma + 4\delta = 0$ i.e. $\alpha = \gamma = -4\delta$

$-4\delta - 2\beta + 16\delta = 0$

i.e. $-2\beta + 12\delta = 0$

$\beta = 68$

$\therefore \ker (T - 2I) = \{(-4\delta, 6\delta, -4\delta, \delta); \delta \in R\}$

$\therefore \dim \ker (T - 2I) = 1 = \delta_1(2)$

DYNAMICAL SYSTEM – A SHORT COURSE

Thus the real form of T will consist of one Jordan block of size 2 i.e.

$$\begin{bmatrix} 2 & 0 \\ 1 & 2 \end{bmatrix}$$

Hence, the real form of T is

$$\begin{bmatrix} 2 & 0 & 0 & 0 \\ 1 & 2 & 0 & 0 \\ 0 & 0 & 1 & -1 \\ 0 & 0 & 1 & 1 \end{bmatrix}$$

a) $\begin{bmatrix} 1 & 1 \\ 0 & 1 \end{bmatrix}$

Eigen values are given by $\begin{vmatrix} 1-\lambda & 1 \\ 0 & 1-\lambda \end{vmatrix} = 0$

$\Rightarrow (1-\lambda)^2 = 0$

$\Rightarrow 1 - 2\lambda + \lambda^2 = 0$

$\Rightarrow \lambda = 1, 1$

b) $\begin{bmatrix} 1 & 1 \\ 0 & -1 \end{bmatrix}$

Eigen values are $1, -1$.

$\therefore$ The Jordan form is $\begin{bmatrix} 1 & 0 \\ 0 & -1 \end{bmatrix}$

d) $\begin{bmatrix} 0 & 2 & 0 \\ -2 & 0 & 0 \\ 2 & 0 & 6 \end{bmatrix}$

Eigen values are given by $\begin{vmatrix} 0-\lambda & 2 & 0 \\ -2 & 0-\lambda & 0 \\ 2 & 0 & 6-\lambda \end{vmatrix} = 0$

$\Rightarrow -\lambda(-\lambda(6-\lambda)) - 2(-2(6-\lambda)) = 0$

$\Rightarrow (6-\lambda)[\lambda^2 + 4] = 0$

$\Rightarrow \quad \lambda = 6 , \pm 2i$

$\therefore$ The Jordan block is given by $\begin{bmatrix} 6 & 0 & 0 \\ 0 & 0 & -2 \\ 0 & 0 & 0 \end{bmatrix}$

e) $\begin{bmatrix} 0 & 0 & 8 \\ 0 & 0 & 4 \\ 0 & 1 & -2 \end{bmatrix}$

Eigen values are

$\Rightarrow -\lambda\left(\left(-\lambda \times -2\right) - 4\right) = 0$

$\Rightarrow -\lambda\left(2\lambda - 4\right) = -2\lambda^2 + 4\lambda = 0$

$\Rightarrow \lambda = 0 , \quad \lambda^2 + 2\lambda - 4 = 0$

$\Rightarrow \lambda = \dfrac{-2 \pm \sqrt{4+16}}{2}$

$\qquad = \dfrac{-2 \pm 2\sqrt{5}}{2} = -1 \pm \sqrt{5}$

$\therefore$ The Jordan block is

$$\begin{bmatrix} 0 & 0 & 0 \\ 0 & -1+\sqrt{5} & 0 \\ 0 & 0 & -1-\sqrt{5} \end{bmatrix}$$

f) $\begin{bmatrix} 1 & 1 & 1 & 1 \\ 2 & 2 & 2 & 2 \\ 3 & 3 & 3 & 3 \\ 4 & 4 & 4 & 4 \end{bmatrix}$

Eigen values are given by

$$\begin{vmatrix} 1-\lambda & 1 & 1 & 1 \\ 2 & 2-\lambda & 2 & 2 \\ 3 & 3 & 3-\lambda & 3 \\ 4 & 4 & 4 & 4-\lambda \end{vmatrix} = 0$$

$$\Rightarrow (1-\lambda)\begin{vmatrix} 2-\lambda & 2 & 2 \\ 3 & 3-\lambda & 3 \\ 4 & 4 & 4-\lambda \end{vmatrix} -1\begin{vmatrix} 2 & 2 & 2 \\ 3 & 3-\lambda & 3 \\ 4 & 4 & 4-\lambda \end{vmatrix}$$

$$+1\begin{vmatrix} 2 & 2-\lambda & 2 \\ 3 & 3 & 3 \\ 4 & 4 & 4-\lambda \end{vmatrix} -1\begin{vmatrix} 2 & 2-\lambda & 2 \\ 3 & 3 & 3-\lambda \\ 4 & 4 & 4 \end{vmatrix} = 0$$

$$\Rightarrow (1-\lambda)\{(2-\lambda)[(3-\lambda)(4-\lambda)-12]-2[3(4-\lambda)-12]+2[12-4(3-\lambda)]\}$$

$$-\{2[(3-\lambda)(4-\lambda)-12]-2[3(4-\lambda)-12]+2[12-4(3-\lambda)]\}$$

$$+\{2[3(4-\lambda)-12]-(2-\lambda)[3(4-\lambda)-12]+2(12-12)\}$$

$$-\{2[12-4(3-\lambda)]-(2-\lambda)[12-4(3-\lambda)]+2(12-12)\}=0$$

$$\Rightarrow (1-\lambda)\{(2-\lambda)[(3-\lambda)(4-\lambda)-12]-2[12-3\lambda-12]+2[12-12+4\lambda]\}$$

$$-\{2[12-3\lambda-4\lambda+\lambda^2-12]-2[12-3\lambda-12]+2[12-12+3\lambda]\}$$

$$+\{2[12-3\lambda-12]\}=0$$

Ex 1: $\begin{bmatrix} 0 & 0 & 14 \\ 1 & 0 & -17 \\ 0 & 1 & 7 \end{bmatrix}$

Eigen values are $\begin{vmatrix} -\lambda & 0 & 14 \\ 1 & -\lambda & -17 \\ 0 & 1 & 7-\lambda \end{vmatrix} = 0$

$$\Rightarrow -\lambda\left[-\lambda(7-\lambda)+17\right]+14[1]=0$$

$$\Rightarrow -\lambda\left[-7\lambda+\lambda^2+17+14\right]=0$$

$$\Rightarrow 7\lambda^2-\lambda^3-17\lambda+14=0$$

$$\Rightarrow \lambda^3-7\lambda^2+17\lambda-14=0$$

$$\begin{array}{r|rrr} 2 & 1 & -7 & 17 & -14 \\ & & 2 & -10 & 14 \\ \hline & 1 & -5 & 7 & 0 \end{array}$$

$$\lambda^2 - 5\lambda + 7 = 0$$

$$\frac{-b \pm \sqrt{b^2 - 4ac}}{2a} = \frac{5 \pm \sqrt{25 - 28}}{2} = \frac{5 \pm \sqrt{-3}}{2}$$

$$= \frac{5 \pm \sqrt{3}\,i}{2}$$

Jordan block is
$$\begin{bmatrix} 2 & 0 & 0 \\ 0 & \frac{5}{2} & -\frac{\sqrt{3}}{2} \\ 0 & \frac{\sqrt{3}}{2} & \frac{5}{2} \end{bmatrix}$$

APPLICATION OF LINEAR ALGEBRA TO DIFFERENTIAL EQUATION

Consider a system of Differential Equation given by

$$x' = A x$$

where A is a λ-Jordan block.

i.e. $A = \begin{bmatrix} \lambda & & & 0 \\ 1 & \lambda & & \\ & 0 & \lambda & \\ & & 1 & \lambda \end{bmatrix}$

we can write, $A = \lambda I + N$

where I is the $n \times n$ identity /unit matrix and

$N = \begin{bmatrix} 0 & & & 0 \\ 1 & 0 & & \\ & 0 & 0 & \\ & & 1 & 0 \end{bmatrix}$ an elementary nilpotent block of size n.

We know that any solution of this system of equation is given by

$$x(t) = e^{tA} \cdot x_0 \quad \text{where} \quad x_0 \in R^n.$$

Consider, $e^{tA} = e^{t\lambda I + tN}$

As $t\lambda I$ commutes with tN, it follows that,

$$e^{tA} = e^{t\lambda} \cdot e^{tN}$$

Since, N is an elementary nilpotent block of size n. we have $N^n = 0$.

$$\exp(tN) = I + tN + \frac{t^2 N^2}{2!} + \dots + \frac{t^{n-1} N^{n-1}}{(n-1)!}$$

Obviously,

$$tN = \begin{bmatrix} 0 & & & 0 \\ 1 & 0 & & \\ & 0 & 0 & \\ & & 1 & 0 \end{bmatrix}$$

As N is an elementary nilpotent block of size n.

We can find a basis say $e_1 \dots e_n$ of R^n such that

$$
\begin{array}{lll}
N(e_1) = e_2 & N^2(e_1) = e_3 & N^3(e_1) = e_4 \\
N(e_2) = e_3 & N^2(e_2) = e_4 & N^3(e_2) = e_5 \\
\vdots & \vdots & \vdots \\
N(e_{n-1}) = e_n & N^2(e_{n-1}) = 0 & N^3(e_{n-2}) = 0 \\
N(e_n) = 0 & N^2(e_n) = 0 & N^3(e_{n-1}) = 0 \\
& & N^3(e_n) = 0
\end{array}
$$

and so on

It then follows that

$$Exp(tN) = \begin{bmatrix} 1 & 0 & \dots & 0 \\ t & 1 & & 0 \\ t^2/2! & t & & 0 \\ \vdots & \vdots & & \\ t^{n-1}/(n-1)! & t^{n-2}/(n-2)! & & 1 \end{bmatrix}$$

It thus follows that,

$$x(t) = e^{tA} \cdot x_0 = e^{\lambda t} \begin{bmatrix} 1 & 0 & \cdots & 0 \\ t & 1 & & 0 \\ t^2/2! & t & & 0 \\ \vdots & \vdots & & \\ t^{n-1}/(n-1)! & t^{n-2}/(n-2)! & & 1 \end{bmatrix} x_0$$

So, if we write

$$x(t) = \begin{pmatrix} x_1(t) \\ x_2(t) \\ \vdots \\ x_n(t) \end{pmatrix} \quad \text{and} \quad x_0 = \begin{pmatrix} c_1 \\ c_2 \\ \vdots \\ c_n \end{pmatrix} \subset R^n$$

Then we have

$$\begin{pmatrix} x_1(t) \\ x_2(t) \\ \vdots \\ x_n(t) \end{pmatrix} = e^{\lambda t} \begin{bmatrix} 1 & 0 & \cdots & 0 \\ t & 1 & & 0 \\ t^2/2! & t & & 0 \\ \vdots & \vdots & & \\ t^{n-1}/(n-1)! & t^{n-2}/(n-2)! & & 1 \end{bmatrix} \begin{bmatrix} c_1 \\ c_2 \\ \vdots \\ c_n \end{bmatrix}$$

i.e., $x_1(t) = e^{\lambda t} \cdot c_1$

$$x_2(t) = e^{\lambda t} \cdot (t c_1 + c_2)$$

$$x_3(t) = e^{\lambda t} \cdot \left(\frac{t^2}{2!} c_1 + t c_2 + c_3 \right)$$

$$\vdots$$

$$x_n(t) = e^{\lambda t} \cdot \left(\frac{t^{n-1}}{(n-1)!} c_1 + \frac{t^{n-2}}{(n-2)!} c_2 + \ldots + c_n \right)$$

i.e. $x_j(t) = e^{\lambda t} \sum_{k=0}^{j-1} \frac{t^k}{k!} c_{j-k} \qquad 1 \leq j \leq n.$

We now consider the case where A is a wal λ block $\lambda = a + ib$, $b \neq 0$.

Thus A is given by

$$A = \begin{bmatrix} D & & & 0 \\ 1 & D & & \\ & 0 & D & \\ & & 1 & D \end{bmatrix} \quad \text{where } D = \begin{bmatrix} a & -b \\ b & a \end{bmatrix} \quad \text{and} \quad I = \begin{bmatrix} 1 & 0 \\ 0 & 1 \end{bmatrix}$$

Since each entry in this matrix is a 2×2 matrix.

It follows that $n = 2m$

i.e. the given operator A is a $2m \times 2m$ matrix.

We reindex the variables, $x_1, x_2 ... x_n$ as $x_1 y_1, x_2 y_2 ... x_m y_m$ (i.e., denote old x_2 by y_1, old x_3 by x_2 and so on).

Since we know that the operator $\begin{bmatrix} a & -b \\ b & a \end{bmatrix}$ operating on R^2 is given by complex multiplication by the complex number $a + ib$ to the complex point $x_1 + i y_1$ (for the point (x_1, y_1) of R^2), we denote $x_1 + i y_1$ by $z_1, x_2 + i y_2$, by $z_2 .... x_m + i y_m$ by z_m so the equivalent system is given by

$$\begin{bmatrix} z_1 \\ z_2 \\ z_3 \\ : \\ z_m \end{bmatrix}' = \begin{bmatrix} a+ib & & & 0 \\ 1 & a+ib & & \\ & & a+ib & \\ 0 & & 1 & a+ib \end{bmatrix} \begin{bmatrix} z_1 \\ z_2 \\ z_3 \\ : \\ z_m \end{bmatrix}$$

A system of m number of differential equation with complex variables.

Thus we write the system as

$z' = A_z$ where $A = (a + ib)I + N$, N is an elementary block of size m .

Thus the solution to the system $z' = A_z$ is given by

JORDAN AND REAL CANONICAL FORMS

$$z_j = e^{(a+ib)t} \sum_{k=0}^{j-1} \frac{t^k}{k!} \, e_{j-k} \quad \text{where} \quad \begin{bmatrix} c_1 \\ c_2 \\ \vdots \\ c_k \end{bmatrix} \quad \text{is the initial point in the complex}$$

space C^m.

Since $c_1 \ldots c_m$ are complex numbers,

we can write,

$$c_k = L_k + i M_k \quad , \quad\quad 1 \le k \le m$$

Further using $z_k = x_k + i y_k \quad 1 \le k \le m$, we get

$$x_j + i y_j = e^{at}(\cos bt + i \sin bt) \sum_{k=o}^{j-1} \left(\frac{t^k}{k!} \left(L_{j-k} + i M_{j-k} \right) \right) \quad\quad 1 \le j \le m$$

Equating real and imaginary parts we get

$$x_j = e^{at} \left\{ \sum_{k=0}^{j-1} \frac{t^k}{k!} \left((\cos bt) L_{j-k} - \sin bt \cdot M_{j-k} \right) \right\}$$

$$y_j = e^{at} \left\{ \sum_{k=0}^{j-1} \frac{t^k}{k!} \left((\cos bt) M_{j-k} - (\sin bt) L_{j-k} \right) \right\} \quad\quad 1 \le j \le m$$

Thus if λ is real and A is a Jordan λ-block then

a solution of the system $x' = A_x$

is a linear combination of $1, t, \dfrac{t^2}{2!}, \ldots \dfrac{t^{n-1}}{(n-1)!}$ multiplicity of $e^{\lambda t}$.

It λ is complex given by $a + ib$, $b \ne 0$ and A is a real block corresponding to eigen pair $a \pm ib$, then any solution of system $x' = A_x$ is a linear combination of $t^k \cdot \cos bt$, $t^k \cdot \sin bt$, $0 \le k \le n-1$ multiplied by e^{at}.

DYNAMICAL SYSTEM – A SHORT COURSE

Theorem 4: Let $A \in L(R^n)$ and let $x(t)$ be a solution of $x' = A_x$. Then each coordinate $x_j(t)$ of the solution is a linear combination of the following functions

$$t^k e^{ta} \cos bt \quad , \quad t^l e^{ta} \sin bt$$

where $a + ib$ runs through all the eigen values of A with $b \geq 0$ and k and l run through all the integers 0 to $n-1$.

Moreover, for each $\lambda = a + ib, k$ and l are less than the size of the largest λ-block in the real canonical form of A.

Proof: We know that there exists a base of R^n such that the matrix of A with respect to this basis is in the real form.

Thus if we transform the equation in terms of new basis, we shall have an equivalent system of equation, say $y' = B y$, where B is A is the real form.

As B consists of various block (J / R) along its diagonal, we can decouple appropriate number of equations from the system $y' = B y$ and solve them independently.

For example, if B has two blocks say B_1 and B_2 along its diagonal of size say n_1 and n_2 respectively then the first n_1 equations will have only B_1 as an operator on right hand side, while the next n_2 equations will have only B_2 as an operator on the right hand side for each block constituting B, the solution is a linear combination of $t^k e^{at}$ $\quad 0 \leq k \leq 1$ less than the size of the block, provided 'a' is a real eigen values, and $t^k e^{at} \cos bt$, $t^k e^{at} \sin bt$, $\quad 0 \leq k \leq 1$ less than size of the block, for the complex eigen values $a + ib$, $b > 0$.

It thus follows that, each component of the solution of the system $X' = A X$

JORDAN AND REAL CANONICAL FORMS

is a linear combination of functions of the type

$t^k e^{at} \cos bt$, $t^l e^{at} \sin bt$.

where $a + ib$, $b \geq 0$ are various eigen values of the operator A. It is also clear that the power k and l of t in the solutions will not exceed the maximum size of any block constituting the real form of the operator A.

Theorem 5: Suppose every eigen value of $A \in L(R^n)$ has negative real part.

Then, $\lim_{t \to \infty} x(t) = 0$, for every solution $x(t)$ of $x' = A_x$.

Proof: We know that any solution of the system $X' = AX$ is a linear combination of the function of the type

$t^k e^{at} \cos bt$, $t^l e^{at} \sin bt$, where k and l are non negative, finite integers, $a \pm ib$, $b > 0$ are eigen values of the operator A.

Since, $a < 0$ (for every eigen value)

$t^k e^{at} \cos bt \to 0$ and $t^l e^{at} \sin bt \to 0$ as $t \to \infty$.

It thus follows that, $x(t) \to 0$ as $t \to \infty$ for every solution $x(t)$ of $X' = AX$.

Theorem 6: If every solution of $X' = AX$ tends to 0 as $t \to \infty$ then every eigen values of A has negative real part.

Proof: If possible, let us assume that $a \pm ib$, is an eigen value of the operator A (there may be other eigen values also) such that $a \geq 0$.

We can choose appropriate basis and constants to get

$x_1(t) = e^{at} \cos bt$

$y_1(t) = e^{at} \sin bt$

$0 = x_2(t) = y_2(t) = x_3(t) = y_3(t) = ...$

245

DYNAMICAL SYSTEM – A SHORT COURSE

Since $a \geq 0$, obviously $x_1(t) \to 0$,

$y_1(t) \to 0$ as $t \to \infty$.

Thus, we can constant a solution of the system, $X' = AX$

which does not tend to 0 as $t \to \infty$. This is a contradiction.

Thus , in this case every eigen value must have a negative real part.

Theorem 7: If every eigen value of $A \in L(R^n)$ has positive real part, then it has unbounded solution.

Proof. Since $a > 0$ for every eigen value $a + ib$, of A the terms $t^k e^{at} \cos bt$, $t^l e^{at} \sin bt$ becomes unbounded as $t \to \infty$.

Then $|x(t)| \to t$ as $t \to \infty$.

Theorem 8: If $A \in L(R^n)$ then the coordinates of every solution to $X' = AX$ are infinitely differentiable functions.

Proof. Since $t^k e^{at} \cos bt$, $t^l e^{at} \sin bt$, $k \geq 0$, $l \geq 0$ are infinitely differentiable function,

$\therefore$ Any linear combination of them is also infinitely differentiable (function). Hence every component of the solution of the system $X' = AX$ is infinitely differentiable.

Theorem 9: i] A linear combination of solution of nth order homogeneous linear differentiable equation $S^{(n)} + a_1 S^{(n-1)} + ... + a_{n-1} S' + a_n S = 0$ is again a solution.

ii] The derivative of its solution is again its solution.

Proof: **i]** It is obvious that, if f_1 and f_2 are solution of $S^{(n)} + 0_1 S^{(n-1)} + ... + a_{n-1} S' + a_n S = 0$, then

$c_1 f_1 + c_2 f_2$ is again a solution for arbitrary constants c_1 and c_2.

ii] Let f be a solution of the DE

246

$$S^{(n)} + a_1 S^{(n-1)} + ... + a_{n-1} S' + a_n S = 0$$

It follows that f is n times differentiable.

We show that f is infinitely differentiable

$$S^{(n)} + a_1 S^{(n-1)} + ... + a_n S = 0$$

is equivalent to the system of equation

$$x_1' = x_2$$

$$x_2' = x_3$$

$$\vdots$$

$$x_{n-1}' = x_n$$

$$x_n' = -a_1 x_{n-1} \,,\, -a_2 x_{n-2} ...... - a_n x_1$$

Obtained, where we denote S by x_1 , S' by x_2 ...$S^{(n-1)}$ by x_n with

$$A = \begin{bmatrix} 0 & 1 & 0 & . & . & . & 0 \\ 0 & 0 & 1 & . & . & . & 0 \\ \vdots & & & & & & \vdots \\ 0 & & & & & & 1 \\ -a_n & -a_{n-1} & . & . & . & & -a_1 \end{bmatrix}$$

We know that each component of the solution of the system $X' = AX$ is infinitely differentiable.

$\therefore\ x_1(t)$ is infinitely differential

i.e. $S(t)$ is infinitely differential.

i.e. every solution of the differential equation

$S^{(n)} + a_1 S^{(n-1)} + ... + a_n S = 0$ is infinitely differentiable.

Thus the solution, we have

$$f^{(n)} + a_1 f^{(n-1)} + \ldots a_{n-1} f' + a_n f = 0$$

Differentiate this equation to get

$$f^{(n+1)} + a_1 f^{(n)} + \ldots a_{n-1} f'' + a_n f' = 0$$

i.e. $\quad (f')^{(n)} + a_1 (f')^{(n-1)} + \ldots + a_{n-1}(f') + a_n(f') = 0$

i.e. f' is a solution of the given n^{th} order homogeneous linear differential equation.

Consider the nth order homogeneous linear differentiable equation

$$S^{(n)} + a_1 S^{(n-1)} + \ldots + a_{n-1} S' + a_n S = 0, \quad \text{where} \quad a_1\ a_2\ \ldots a_n \in R$$

The corresponding equivalent system of first order differentiable equation is given by

$$x_1' = x_2$$

$$x_2' = x_3$$

$$\vdots$$

$$x_{n-1}' = x_n$$

$$x_n^1 = -a_n x_1\,,\, -a_{n-1} x_2 \ldots\ldots - a_1 x_n$$

i.e. $\quad x' = A x$

i.e. $\quad \begin{bmatrix} 0 & 1 & 0 \ldots\ldots & 0 \\ 0 & 0 & 1 & 0 \\ \vdots & & & \\ 0 & & & 1 \\ -a_n & -a_{n-1} \ldots & & -a_1 \end{bmatrix}$

Consider,

$$\begin{bmatrix} 0 & 0 \ldots & 0 & -a_n \\ 1 & 0 & & -a_{n-1} \\ 0 & 1 & & \\ \vdots & \vdots & & \\ 0 & 0 & \ldots & 1 & -a_1 \end{bmatrix}$$

The characteristic polynomial for the system $x' = AX$ is $\lambda^{(n)} + a_1\lambda^{(n-1)} + \ldots + a_{n-1}\lambda + a_n = 0$.

The matrix A^T

$$\begin{bmatrix} 0 & 0 \ldots & 0 & -a_n \\ 1 & 0 & & -a_{n-1} \\ 0 & 1 & & \\ \vdots & \vdots & & \\ 0 & 0 & \ldots & 1 & -a_1 \end{bmatrix}$$

is known as the companion matrix of the polynomial $\lambda^{(n)} + a_1\lambda^{(n-1)} + \ldots + a_n = 0$.

Theorem 10: Let $\lambda \in c$ be a real or complex eigen value of a companion matrix

$$A = \begin{bmatrix} 0 & 0 \ldots & 0 & -a_n \\ 1 & 0 & & -a_{n-1} \\ 0 & 1 & & \\ \vdots & \vdots & & \\ 0 & 0 & \ldots & 1 & -a_1 \end{bmatrix}$$

then the real canonical form of A has only one λ-block.

Proof: Since λ is an eigen value of the matrix A, the real form of A will have a Jordan block/real block corresponding to eigen value λ.

The number of elementary λ blocks which will constitute this block is given by dim ker $(A - \lambda I)$.

Thus the result will hold if we show that $\dim \ker (A - \lambda I) = 1$.

As λ is an eigen value of A, By definition of eigen value it follows that the eigen space belonging to λ is a non-trivial subspace.

$\therefore \dim \ker (A - \lambda I) \geq 1$.

Consider $A - \lambda I$. It is given by

$$
\begin{bmatrix}
-\lambda & 0 & & 0 & -a_n \\
1 & -\lambda & & & -a_{n-1} \\
0 & 1 & & & \\
& & & & \\
& & & -\lambda & -a_2 \\
0 & & 0 & 1 & -a_1 - \lambda
\end{bmatrix}
$$

Let B denote the $n \times (n-1)$ matrix obtained by considering the first $n-1$ columns of $A - \lambda I$, that is,

$$
B = \begin{bmatrix}
-\lambda & 0 & \cdots & 0 \\
1 & -\lambda & & \cdot \\
0 & 1 & & \cdot \\
\vdots & & & \\
& & & -\lambda \\
0 & & 0 & 1
\end{bmatrix}
$$

As any $S \times T$ matrix represents a linear map from R^T to R^S, be the matrix B will represent a linear map from R^{n-1} to R^n.

It further follows that rank of B is $n-1$

i.e. dim image of R^{n-1} under $B = n-1$.

If we identify any point $(\alpha_1 \dots \alpha_{n-1})$ of R^{n-1} with $(\alpha_1, \alpha_2 \dots \alpha_{n-1}, 0)$ of R^n, it is clear that

$$(A - \lambda I) \begin{bmatrix} \alpha_1 \\ \vdots \\ \alpha_{n-1} \\ 0 \end{bmatrix} = B \begin{bmatrix} \alpha_1 \\ \vdots \\ \vdots \\ \alpha_{n-1} \end{bmatrix}$$

Thus B can be considered as a restriction of $A - \lambda I$ to R^{n-1}.

$\therefore$ Image of R^{n-1} under B must be a subspace of image of R^n under $A - \lambda I$.

$\therefore$ Dim img of R^{n-1} under $B \leq$ dim img of R^n under A.

$\therefore$ Dim of img of R^n under $(A - \lambda I) \geq n - 1$.

But $A - \lambda I : R^n \to R^n$

$\therefore$ Dim img of R^n under $(A - \lambda I) +$ dim ker $(A - \lambda I) = \dim of R^n = n$

$\therefore$ dim ker $(A - \lambda I) \leq 1$.

But we already have

dim ker $(A - \lambda I) \geq 1$

$\therefore$ dim ker $(A - \lambda I) = 1$

Definition: A basis of solutions of the nth order homogeneous linear differential equation

$$S^{(n)} + a_1 S^{(n-1)} + \dots + a_{n-1} S' + a_n S = 0$$

is a set of solution S_1 , S_2 ...S_n such that every solution is expressible as a linear combination of $S_1 .... S_n$ in one and only one way.

Theorem 11: The following n functions form a basis solutions for the nth order homogeneous differential equation

$$S^{(n)} + a_1 S^{(n-1)} + \dots + a_{n-1} S' + a_n S = 0$$

(a) the function $t^k e^{t\lambda}$ where λ runs through the distinct real eigen roots of the characteristic polynomial $\lambda^n + a_1 \lambda^{n-1} + ... + a_n$

and k is a non negative integer in the range $0 \le k <$ multiplicity of λ together with

(b) the functions $t^k e^{at} \cos bt$, $t^k e^{at} \sin bt$

where $a + ib$ runs through the complex roots of the polynomial

$\lambda^n + a_1 \lambda^{n-1} + ... + a_n$

Having $b > 0$ and k is a non negative int in the range $0 \le k <$ multiplicity of $a + ib$.

Proof: We first show that each of the functions $t^k e^{t\lambda}$

and $t^k e^{at} \cos bt$, $t^k e^{at} \sin bt$ is a solution of the given n^{th} order differential equation DE.

Consider a collection of all infinitely differentiable real valued functions. It is a real vector space.

On this vector space, we consider a map D which gives differentiation of a given function. D is a linear operator on this space.

Further if λ is any scalar then $\lambda : f \to \lambda f$ is again an operator on this space.

Thus $(b - \lambda)$ is also an operator.

(If λ is complex number, we can consider the functions to be complex valued functions)

So $(D - \lambda)^m$ is also a linear operator on this space for $m \ge 0$.

Consider the characteristic polynomial namely

$t^n + a_1 t^{n-1} + ... + a_{n-1} t + a_n$

If λ is a root of this polynomial of multiplicity say $m (m \ge 1)$ then $(t - \lambda)^m$

is a divisor of this polynomial

i.e. we can write

$p(t)=t^n+a_1 t^{n-1}+...+a_n \equiv q(t)(t-\lambda)^m$, for some other polynomial $q(t)$.

If λ is complex we can consider $p(t)$ to be factored into complex factors.

Thus we can write

$p(D)=q(D)(D-\lambda)^m$

If we show that $(D-\lambda)^m (t^k e^{\lambda t})=0$, $0 \le k \le m$ then it follows that

$p(D)(t^k e^{\lambda t})=0$, $0 \le k \le m-1$

i.e. $t^k e^{\lambda t}$ is a solution of the given differential equation.

We just need to show that $(D-\lambda)^{k+1}(t^k e^{\lambda t})=0$

We show that this holds by induction.

For $k=0$, we show

$(D-\lambda)^m (e^{\lambda t})= D(e^{\lambda t})-\lambda e^{\lambda t}$

$\qquad = \lambda e^{\lambda t}-\lambda e^{\lambda t}=0$

i.e., the formula holds for $k=0$.

Let us assume that it holds for k

i.e., we assume that

$(D-\lambda)(t^{k-1} e^{\lambda t})=0$

So consider

$(D-\lambda)^{k+1}(t^k e^{\lambda t}) =(D-\lambda)^k \cdot (D-\lambda)(t^k e^{\lambda t})$

$\qquad =(D-\lambda)^k [k t^{k-1} e^{\lambda t}+\lambda t^k e^{\lambda t}-\lambda t^k e^{\lambda t}]$

$\qquad =(D-\lambda)^k (k t^{k-1} e^{\lambda t})$

$\qquad =k(D-\lambda)^k (t^{k-1} e^{\lambda t})=0$

$$\therefore (D-\lambda)^k \left(t^{k-1} e^{\lambda t}\right)=0 \Rightarrow (D-\lambda)^{k+1}\left(t^k e^{\lambda t}\right)=0$$

Thus by induction we have

$$(D-\lambda)^k \left(t^k e^{\lambda t}\right)=0 \ , \qquad k=0,1....$$

It thus follows that if λ is a root of multiplicity m of the characteristic poly., then

$$p(D)\{t^k e^{\lambda t}\}=0 \qquad 0\le k \le m-1$$

If λ is a real root, it follows that $t^k e^{\lambda t}$ $\quad 0\le k \le m-1$ are solutions of the given differential equation.

If λ is complex say $\lambda = a+ib$, $b \ne 0$ then

$$e^{\lambda t} = e^{at}\left(\cos bt + i\sin bt\right)$$

Since $p(D)$ is an operator with real coefficient

$$p(D)\{t^k e^{at}\left(\cos bt + i\sin bt\right)\}=0$$
$$\Rightarrow p(D)\left\{t^k e^{at}\cos bt\right\}=0$$
$$p(D)\left\{t^k e^{at}\sin bt\right\}=0$$

Thus if $a+ib$ is a complex root of multiplicity m, then

$t^k e^{at}\cos bt$, $t^k e^{at}\sin bt$, $0\le k \le m-1$ are real valued solution of the given DE.

Thus each of the functions mentioned in the statement is a solution.

Since the DE is of order n, the degree of the characteristic poly. is n and as such some of the multiplicities of the roots of the polynomial must be n.

If λ is a real toot of multiplicity m, then $t^k e^{\lambda t}$ $\quad 0\le k \le m-1$ are m number of solutions.

JORDAN AND REAL CANONICAL FORMS

If $a+ib$ is a root of multiplicity m then

$t^k e^{at} \cos bt$, $t^k e^{at} \sin bt$, $0 \le k \le m-1$ are $2m$ number of functions.

But we note that if $a+ib$ is a root of multiplicity m then $a-ib$ is also a root of multiplicity m .

Thus the pair $a+ib$, $a-ib$ contributes $2m$ to the sum of multiplicities and the number of functions corresponding to this pair is also $2m$.

Thus the number of functions mentioned in the statement is equal to the sum of the multiplicities i.e. n .

We note that the DE

$S^{(n)} + a_1 S^{(n-1)} + ... + a_n S = 0$

is equivalent to the system of first order differential equation namely

$x' = Ax$

with $A = \begin{bmatrix} 0 & 1 & 0 & 0 \\ 0 & 0 & 1 & \\ & & & \\ 0 & & & 1 \\ -a_n & -a_{n-1} & & -a_1 \end{bmatrix}$

and the characteristic polynomial

$t^n + a_1 t^{n-1} + ... + a_1 = 0$

is also the character. Poly. of the operator A .

Thus roots of $t^n + a_1 t^{n-1} + ... + a_n = 0$ are eigen values of the operator A .

We know that any solution of the system $x' = Ax$ is a linear combinations of the function mentioned in the statement.

Thus the n functions given in the statement span the space of solutions of

255

the given DE.

We denote the above n functions by $f_1, f_2, \ldots f_n$.

We now show that every solution of the given differential equation is uniquely expressed as a linear combination of $f_1, f_2, \ldots f_n$.

For any n-tuple $(\alpha_1, \alpha_2 \ldots \alpha_n) \in R^n$, we define,

$S_\alpha = \alpha_1 f_1 + \alpha_2 f_2 + \ldots + \alpha_n f_n$.

It follows that S_α is a solution of given DE.

We define a function $\phi: R^n \to R^n$.

$\phi((\alpha_1, \alpha_2 \ldots \alpha_n)) = \left(S_\alpha(0) \; S_\alpha'(0) \ldots S_\alpha^{n-1}(0) \right)$

We show that ϕ is linear.

Let $\alpha, \beta \in R^n$

let $\alpha = (\alpha_1, \alpha_2 \ldots \alpha_n)$ and $\beta = (\beta_1, \beta_2 \ldots \beta_n)$

$\therefore S_\alpha = \alpha_1 f_1 + \alpha_2 f_2 + \ldots + \alpha_n f_n$

$\beta_\alpha = \beta_1 f_1 + \beta_2 f_2 + \ldots + \beta_n f_n$

$S_{\alpha+\beta} = (\alpha_1 + \beta_1) f_1 + (\alpha_2 + \beta_2) f_2 + \ldots + (\alpha_n + \beta_n) f_n$

$\qquad = S_\alpha + S_\beta$

$\therefore S_{\alpha+\beta}(0) = S_\alpha(0) + S_\beta(0)$

Similarly,

$S'_{\alpha+\beta}(0) = S'_\alpha(0) + S'_\beta(0)$ and so on.

$\phi(\alpha + \beta) = \left(S_{\alpha+\beta}(0), S'_{\alpha+\beta}(0) \ldots, S^{n-1}_{\alpha+\beta}(0) \right)$

$\qquad = \left(S_\alpha(0) + S_\beta(0), S'_\alpha(0) + S'_\beta(0) + \ldots + S^{n-1}_\alpha(0) + S^{n-1}_\beta(0) \right)$

$\qquad = \left(S_\alpha(0), S'_\alpha(0), \; , S_\alpha^{n-1}(0) \right) + \left(S_\beta(0), S'_\beta(0), \; , S_\beta^{n-1}(0) \right)$

$$= \phi(\alpha) + \phi(\beta)$$

Similarly, for any real r,

$$\phi(r\alpha) = r\phi(\alpha)$$

Thus $\phi \in L(R^n)$.

We show that ϕ is onto.

Let $(\gamma_1, \gamma_2, \ldots \gamma_n) \in R^n$.

We show that there exist $\alpha \in R^n$ such that

$$\phi(\alpha) = (\gamma_1, \gamma_2, \ldots \gamma_n)$$

As any solution of the given DE

$$S^{(n)} + a_1 S^{(n-1)} + \ldots + a_n = 0$$

is a component of the solution of DE, $X' = AX$ and for any $X_0 \in R^n$

corresponding solution is given by $X(t) = e^{tA} x_0$.

It follows that taking $x_0 = (\gamma_1, \gamma_2, \ldots \gamma_n)$ there will exist a solution $x(t)$

given by $x(t) = e^{tA} x_0$ such that

$x_1(0) = x_0(1) = \gamma_1$, $x_2(0) = x_0(2) = \gamma_2 \ldots x_n(0) = x_0(n) = \gamma_n$

However we know that the solution S' of the given n^{th} order DE

$S^{(n)} + a_1 S^{(n-1)} + \ldots + a_n = 0$ such that

$x_1 \equiv S_1$

$x_2 \equiv s'$

$x_3 \equiv s''$

$\vdots$

$x_n \equiv s^{(n-1)}$

Thus there will exist a solution S of the DE such that

$$S(0) = \gamma_1$$

$$S^1(0) = \gamma_2$$

$$\vdots$$

$$S^{(n-1)}(0) = \gamma_n$$

As S is one solution we must have

$$S = \alpha_1 f_1 + \alpha_2 f_2 + \dots + \alpha_n f_n$$

It thus follows that,

$$\phi((\alpha_1 \dots \alpha_n)) = (\gamma_1 \dots \gamma_n)$$

$\therefore$ ϕ is onto.

$\therefore$ ker $\phi = \{0\}$.

Let f be a solution of DE and

Let $f = \alpha_1 f_1 + \dots + \alpha_n f_n$

$\qquad = \beta_1 f_1 + \dots + \beta_n f_n$

It follows that

$$\phi(\alpha_1 \dots \alpha_n) = (\beta_1 \dots \beta_n)$$

$$= (f(0), f'(0) \dots f^{n-1}(0))$$

But ϕ is 1-1

$\therefore$ $(\alpha_1 \dots \alpha_n) = (\beta_1 \dots \beta_n)$

$\therefore$ $f_1 \dots f_n$ is basis of solution.

Ex 2. Find a solution S to DE

$$S^{(4)} + 4S^{(3)} + 5S^{(2)} + 4S^1 + 4S = 0$$

such that $S(0) = 0$, $S'(0) = -1$, $S''(0) = -4$, $S'''(0) = 14$

Solution. The characteristic polynomial of DE is

$$t^4 + 4t^3 + 5t^2 + 4t + t = 0$$

JORDAN AND REAL CANONICAL FORMS

The roots are $-2,-2,+i,-i$

As -2 is a real root of multiplicity 2, it will give two basic solutions namely $t^0 e^{-2t}$, $t'e^{-2t}$.

From the complex pair $0 \pm i$, the two basic solution namely $\cos t$, $\sin t$ (multiplicity is 1) $a=0, b=1$

Thus the four basic solutions are $e^{-2t}, te^{-2t}, \cos t, \sin t$. Hence we can write

$$s(t) = c_1 e^{-2t} + c_2 e^{-2t} + c_3 \cos t + c_4 \sin t$$

The constants c_1, c_2, c_3, c_4 can be found out using the four initial conditions:

$$s(0) = 0 , \ s'(0) = -1 , \ s''(0) = -4 , \ s'''(0) = 14$$

$\therefore \ s(0) = 0 \quad \Rightarrow \quad c_1 + c_3 = 0$

$s'(t) = -2c_1 e^{-2t} + c_2 e^{-2t} - 2c_2 te^{-2t} - c_3 \sin t + c_4 \cos t$

$\therefore \ s'(0) = -1 \quad \Rightarrow \quad -2c_1 + c_2 + c_4 = -1$

and $s''(t) = +4c_1 e^{-2t} - 2c_2 e^{-2t} - 2c_2 e^{-2t} + 4c_2 te^{-2t} - c_3 \cos t - c_4 \sin t$

$\therefore \ s''(0) = -4 \Rightarrow 4c_1 - 4c_2 - c_3 = -4$

And $s'''(t) = -8c_1 e^{-2t} + 4c_2 e^{-2t} + 4c_2 e^{-2t} - 8c_2 te^{-2t}$

$\therefore \ s'''(0) = -14 \Rightarrow -8c_1 + 12c_2 - c_4 = 14$

$c_1 = c_3 = 0 , \ c_2 = 1 , \ c_4 = -2$

Thus the solution is $te^{-2t} - 2\sin t$.

Ex 3. Find a map $s : R \rightarrow R$ such that

$s^{(3)} - s^{(2)} + 4s' - 4s = 0,$

$s(0) = 1 , \ s'(0) = -1 , \ s''(0) = 1$

Solution. The charact poly is

$$t^3 - t^2 + 4t - 4 = 0$$

The roots, $t^2(t-1) + 4(t-1) = 0$

$$\Rightarrow (t-1)(t^2 + 4) = 0$$

$$\Rightarrow t = 1 \quad and \quad t = \pm 2i$$

As the basic solution are e^t , $\cos 2t$, $\sin 2t$

A general solution is $c_1 e^t + c_2 \cos 2t + c_3 \sin 2t$

Thus we can write

$$s(t) = c_1 e^t + c_2 \cos 2t + c_3 \sin 2t$$

But $\quad s(0) = 1$

$$\therefore \quad c_1 + c_2 = 1$$

$$s'(t) = c_1 e^t - 2c_2 \sin 2t + 2c_3 \cos 2t$$

$$\therefore \quad s'(0) = -1 \quad gives \quad c_1 + 2c_3 = -1$$

$$s''(t) = c_1 e^t - 4c_2 \cos 2t - 4c_3 \sin 2t$$

$$s''(0) = 1 \quad gives \quad c_1 - 4c_2 = 1.$$

$$\therefore \quad c_2 = 0 , \ c_1 = 1 , \ c_3 = 1$$

Thus the required solution is $e^t - \sin 2t$

Ex 4. Consider the differential equation

$$s^{(4)} + 4s^{(3)} + 5s^{(2)} + 4s' + 4s = 0,$$

Find out for which initial conditions $s(0)$, $s'(0)$, $s''(0)$, $s'''(0)$

there is a solution $s(t)$ such that

(a) $s(t)$ is periodic

(b) $\lim\limits_{t \to \infty} s(t) = 0$

(c) $\lim\limits_{t \to \infty} |s(t)| = \infty$

(d) $|s(t)|$ bdd for $t \geq 0$

(e) $|s(t)|$ is bdd for all $t \in R$.

Solution. The general expression for the solution of the differential equation is

$$s(t) = c_1 e^{-2t} + c_2 t e^{-2t} - c_3 \cos t + c_4 \sin t$$

$s(t)$ is periodic provided $c_1 = c_2 = 0$.

We have,

$$s(0) = c_1 + c_3$$
$$s'(0) = -2c_1 + c_2 + c_4$$
$$s''(0) = 4c_1 - 4c_2 - c_3$$
$$s'''(0) = -8c_1 + 12c_2 - c_4$$

Use this to get c_1 , c_2, c_3 , c_4 in terms of $s(0)$, $s'(0)$, $s''(0)$, $s'''(0)$.

Ex 5: Find all the periodic solutions to

$$s^{(4)} + 2s^{(2)} + s = 0$$

Solution. The char. Poly. is

$$t^4 + 2t^2 + 1 = 0$$

i.e. $(t^2 + 1)^2 = 0$

$\Rightarrow t = \pm i$ (double root)

$\therefore$ The basis of solutions is

Here $a = 0$, $b = 1$ and multiplicity 2.

$t^0 \cos t$, $t \cos t$, $t^0 \sin t$, $t \sin t$

261

Thus any general solution DE is

$$c_1 \cos t + c_2 t \cos t + c_3 \sin t + c_4 t \sin t$$

It will be periodic if $c_2 = c_4 = 0$.

Thus any periodic solution is $c_1 \cos t + c_3 \sin t$

Ex 6: What is the smallest integer $n > 0$ for which the differential equation

$$s^{(n)} + a_1 s^{(n-1)} + \ldots a_n s' + a_n s = 0$$

having among its solutions the functions

$$\sin 2t , \ 4t^2 e^{2t} , \ -e^{-t} \ ?$$

Solution: If $0 \pm 2i$ are roots of the characteristic polynomial of multiplicity at least 1, then $\cos 2t$ and $\sin 2t$ are solution.

Similarly, if 2 is root of multiplicity of at least 3 then $t^2 e^{2t}$ will also be a solution.

Further, if -1 is a root of multiplicity at least one, then e^{-t} is a solution.

Thus we require,

$t^2 + 4, (t-2)^3 , (t+1)$ to be factor of the char. Poly.

Thus the char. Poly. of minimum degree should be,

$$\left(t^2 + 4\right)(t-2)^3 (t+1)$$

$$= \left(t^2 + 4\right)\left(t^3 - 6t^2 + 12t - 8\right)(t+1)$$

$$= \left(t^2 + 4\right)\left(t^4 - 6t^3 + 12t^2 - 8t + t^3 - 6t^2 + 12t - 8\right)$$

$$= \left(t^2 + 4\right)\left(t^4 - 5t^3 + 6t^2 + 4t - 8\right)$$

$$= t^6 - 5t^5 + 6t^4 + 4t^3 - 8t^2 + 4t^4 - 20t^3 + 24t^2 + 16t - 32$$

$$= t^6 - 5t^5 + 10t^4 - 16t^3 + 16t^2 + 16t - 32$$

So the differential equation is,

JORDAN AND REAL CANONICAL FORMS

$$s^{(6)} - 5s^{(5)} + 10s^{(5)} - 16s^{(3)} + 16s^{(2)} + 16s^1 - 32(s) = 0$$

Theorem 13: Let A be an operator on a real vector space E and suppose $\alpha < R_e \lambda < \beta$ for every eigen value λ of A. Then E has a basis such that in the corresponding inner product and norm

$\alpha|x|^2 \le \langle Ax, x \rangle \le \beta|x|^2$, for all $x \in E$.

Proof: we first consider the case where A is an operator with distinct eigen values (each eigen value is of multiplicity 1).

Let the eigen value be

$\lambda_1 \lambda_2 ... \lambda_r \qquad a_1 \pm ib_i$, $a_2 \pm ib_2 + ... as \pm ib_s$

Obviously, if the domain space $E \equiv R^n$ then

$r + 2s = n$

In this case we can write

$E = E_1 \oplus E_2 \oplus ... \oplus E_r \oplus F_1 \oplus F_2 \oplus ... \oplus F_s$

where $E_1 ... E_r$ are one dimensional eigen spaces belonging to eigen values $\lambda_1 \lambda_2 ... \lambda_r$ and $F_1 ... F_r$ are two dimensional real vector spaces such that

F_i has a basis $\{f_i, g_i\}$ with respect to which the matrix of the operator

A is $\begin{bmatrix} a_i & -b_i \\ b_i & a_i \end{bmatrix}$

Let $e_1, e_2 ... e_r$ be each chosen from the spaces $E_1, E_2 ... E_r$ (each non zero). Then we know that the set $\{e_1, e_2 ... e_r\}$

e_r, f_1 , $g_1, f_2, g_2, ... f_s, g_s$

form a basis of the space E and the matrix of A with respectively to this basis is

$$\begin{bmatrix} \lambda_1 & & & & & & & & & \\ & \lambda_2 & & & & & & 0 & & \\ & & \vdots & & & & & & & \\ & & & \lambda_r & & & & & & \\ & & & & a_1-b_1 & & & & & \\ & & & & b_1 & a_1 & & & & \\ & & & & & & a_2-b_2 & & & \\ & & & & & & b_2 & a_2 & & \\ & & 0 & & & & & & & \\ & & & & & & & \vdots & & \\ & & & & & & & & a_s-b_s & \\ & & & & & & & & b_s & a_s \end{bmatrix}$$

We define inner product on E with respect to this basis in such a way that

$\langle e_i , e_i \rangle = 1 \qquad 1 \le i \le r$

$\langle f_i , f_i \rangle = 1 \qquad 1 \le i \le s$

$\langle g_i , g_i \rangle = 1 \qquad 1 \le i \le s$

$\langle e_i , e_j \rangle = 0 \qquad 1 \le i, \ j \le r, i \ne j.$

$\langle f_i , f_j \rangle = 0 \qquad 1 \le i, \ j \le s, i \ne j.$

$\langle g_i , g_j \rangle = 0 \qquad 1 \le i, \ j \le s, i \ne j.$

$\langle e_i , f_j \rangle = 0 \qquad \langle e_i , g_i \rangle = 0$

$\qquad \langle f_i , g_i \rangle = 0$

i.e., the inner product maps the basis into an orthonormal basis.

We are given that

$\alpha < \lambda_i < \beta \qquad 1 < i < r$

$\alpha < a_i < \beta \qquad 1 < i < s$

Consider $\langle Ae_i , e_i \rangle$ as e_i an eigen vector belonging to eigen value λ_i,

we have,

$Ae_i = \lambda_i e_i$.

$\therefore \langle Ae_i , e_i \rangle = \langle \lambda_i e_i , e_i \rangle = \lambda_i$

JORDAN AND REAL CANONICAL FORMS

$$\therefore \langle Ae_i , e_j \rangle = 0 , \quad i \neq j$$

Consider, $\langle Af_i , f_i \rangle$

We note that the matrix of A with respect to the basis $\{f_i, g_i\}$ of f_i is

$$\begin{bmatrix} a_i & -b_i \\ b_i & a_i \end{bmatrix}$$

i.e. $A f_i = a_i f_i + b_i g_i$

$$\therefore A g_i = -b_i f_i + a_i g_i$$

$$\langle Af_i , f_i \rangle = \langle a_i f_i + b_i g_i , f_i \rangle$$

$$= a_i \langle f_i, f_i \rangle + b_i \langle g_i , f_i \rangle$$

$$= a_i$$

$$\alpha < \langle Af_i , f_i \rangle < \beta$$

Similarly,

$$\langle Ag_i , g_i \rangle = \langle -b_i f_i + a_i g_i , g_i \rangle$$

$$= a_i$$

Let $x \in E$.

Using the basis $\{e_1, e_2, \ldots e_r, f_1, g_1, f_2, g_2, \ldots f_s, g_s\}$

We can write

$$x = \alpha_1 e_1 + \alpha_2 e_2 + \ldots + \alpha_r e_r + \beta_1 f_1 + \gamma_1 g_1 + \ldots \beta_s f_s + \gamma_s g_s$$

$$\therefore Ax = \alpha_1 e_1 + \alpha_2 e_2 + \ldots + \alpha_r Ae_r + \beta_1 Af_1 + \gamma_1 Ag_1 + \ldots \beta_s Af_s + \gamma_s Ag_s.$$

$$\therefore \langle Ax, x \rangle = \langle \alpha_1 Ae_1 + \alpha_2 Ae_2 + \ldots + \alpha_r Ae_r + \ldots + \gamma_s Ag_s, \alpha_1 e_1, \alpha_2 e_2 \ldots \alpha_s e_s \rangle.$$

Since, $\langle e_i, e_j \rangle = 0 \quad i \neq j$

$$\langle Ae_i, f_j \rangle = 0$$

$$\langle Ae_i, g_j \rangle = 0$$

$$\langle Af_i, f_j \rangle = 0 \qquad i \neq j$$

$$\langle Af_i, g_j \rangle = 0$$

We have,

$$\langle Ax, x \rangle = \lambda_1 \alpha_1^{\,2} + \lambda_2 \alpha_2^{\,2} + \dots + \lambda_r \alpha_r^{\,2} + \beta_1^{\,2} a_1 + \gamma_1^{\,2} a_1 + \beta_2^{\,2} a_2 + \gamma_2^{\,2} a_2 + \dots + \beta_s^{\,2} a_s + \gamma_s^{\,2} a_s .$$

Since, $\begin{aligned} \alpha < \lambda_i < \beta \qquad & 1 \leq i \leq r \\ \alpha < a_i < \beta \qquad & 1 \leq i \leq s \end{aligned}$

$$\therefore \quad \alpha \left\{ \alpha_1^{\,2} + \alpha_2^{\,2} + \dots + \alpha_r^{\,2} + \beta_1^{\,2} + \gamma_1^{\,2} + \dots + \beta_s^{\,2} + \gamma_s^{\,2} \right\}$$

$$\leq \langle Ax, x \rangle \leq \beta \left\{ \alpha_1^{\,2} + \alpha_2^{\,2} + \dots + \beta_s^{\,2} + \gamma_s^{\,2} \right\}$$

But $\quad \alpha_1^{\,2} + \alpha_2^{\,2} + \dots + \gamma_s^{\,2} = \langle x, x \rangle = |x|^2$

with respect to this basis.

$$\therefore \quad \alpha |x|^2 \leq A \langle x, x \rangle \leq \beta |x|^2$$

We consider the general case.

Let β be a basis of the domain space E such that the matrix of A with respect to this basis in the real form.

i.e. $\quad A = dig \left\{ A_1, A_2, \dots A_r \right\}$ are Jordan blocks belonging to real eigen values or real blocks corresponding to complex eigen values.

The theorem will hold, if it holds for each of the blocks $A_1, A_2, \dots A_r$ respectively. Thus we assume that A is Jordan block namely

$$A = \begin{bmatrix} \alpha & & & & 0 \\ 1 & \alpha & & & \\ 0 & : & : & & \\ & & : & \alpha & \\ & & & 1 & \alpha \end{bmatrix}$$

where α is some real eigen value of the operator A.

We can write $A = S + N$

Where $\qquad S = \alpha I$, $\quad N = \begin{bmatrix} 0 & 0 & & 0 & 0 \\ 1 & 0 & & & \\ 0 & 1 & & & \\ & & & & \\ 0 & 0 & & 1 & 0 \end{bmatrix}$

If $e_1 .... e_n$ is the basis then we have

$N(e_1) = e_2$

$N(e_2) = e_3$

$\qquad \vdots$

$\qquad \vdots$

$N(e_{n-1}) = e_n$

$N(e_n) = 0$

Let $\in > 0$ be arbitrary.

Consider a new basis $\{\overline{e_1}, \overline{e_2}, ... \overline{e_n}\}$

where $\quad \overline{e_1} = e_1$, $\overline{e_2} = \dfrac{e_2}{\in}$, $\overline{e_3} = \dfrac{e_3}{\in^2}$ .... $\overline{e_n} = \dfrac{e_n}{\in^{n-1}}$

Obviously,

$N(\overline{e_1}) = N(e_1) = e_2 = \in \overline{e_2}$

$\therefore N(\overline{e_2}) = N\dfrac{e_2}{\in} =$

$\qquad\qquad = \dfrac{1}{\in} N(e_2) = \dfrac{1}{\in} e_3 = \dfrac{\in^2}{\in} \overline{e_3} = \in \overline{e_3}$ ....

$N(\overline{e_{n-1}}) = \in \overline{e_n}$

$N(\overline{e_n}) = 0$.

Also,

$$S\left(\overline{e_1}\right) = S(e_1) = \alpha\, e_1 = \alpha \overline{e_1}$$

$$S\left(\overline{e_2}\right) = S\left(\frac{e_2}{\in}\right) = \frac{1}{\in} S(e_2) = \frac{\alpha}{\in} e_2 = \alpha \overline{e_2}$$

$$\vdots$$

$$S\left(\overline{e_n}\right) = \alpha\, \overline{e_n}$$

Thus the matrix of A with respect to new basis is

$$
\begin{bmatrix}
\alpha & & & & 0 \\
\in & \alpha & & & \\
& & \vdots & & \\
0 & & & \alpha & \\
& & & \in & \alpha
\end{bmatrix}
$$

$$\therefore \quad A\left(\overline{e_1}\right) = \alpha'\overline{e_1} + \in \overline{e_2}$$

$$A\left(\overline{e_2}\right) = \alpha'\overline{e_2} + \in \overline{e_3}$$

$$\vdots$$

$$A\left(\overline{e_{n-1}}\right) = \alpha'\overline{e_{n-1}} + \in \overline{e_n}$$

$$A\left(\overline{e_n}\right) = \alpha'\overline{e_n}$$

Thus for any $x \in E$, we can write

$$x = x_1 \overline{e_1} + x_2 \overline{e_2} + \ldots + x_n \overline{e_n}$$

$$\therefore \quad Ax = \sum_{i=1}^{n} x_i\, A\overline{e_i} = x_1\left(\alpha'\overline{e_1} + \in \overline{e_2}\right) + x_2\left(\alpha'\overline{e_2} + \in \overline{e_3}\right) + \ldots x_{n-1}\left(\alpha'\overline{e_{n-1}} + \in \overline{e_n}\right) + x_n\, \alpha'\overline{e_n}\,.$$

$$\therefore \quad \langle Ax, x \rangle = \left\langle \begin{array}{l} x_1\left(\alpha'\overline{e_1} + \in \overline{e_2}\right) + x_2\left(\alpha'\overline{e_2} + \in \overline{e_3}\right)\ldots \\ + x_{n-1}\left(\alpha'\overline{e_{n-1}} + \in \overline{e_n}\right) + x_n\, \alpha'\overline{e_n},\ x_1\overline{e_1} + x_2\overline{e_2} + \ldots x_n\overline{e_n} \end{array} \right\rangle$$

(Here α' is eigen value)

$$= \alpha' x_1^2 + \in x_1 x_2 + \alpha' x_2^2 + \in x_2 x_3 + \ldots + \alpha' x_{n-1}^2 + \in x_{n-1} x_n + \alpha' x_n^2\,.$$

$$= \alpha'\left(x_1^2 + x_2^2 + \ldots + x_n^2\right) \quad as \ \in \to 0.$$

$$= \alpha'\, |x|^2$$

But we have $\alpha < \alpha' < \beta$

$$\alpha|x|^2 \leq \alpha'|x|^2 \leq \beta|x|^2$$

$$\alpha|x|^2 \leq \langle Ax, x \rangle \leq \beta|x|^2$$

If A is a real block,

$$\begin{bmatrix} D & & & 0 \\ I & D & & \\ & & : & \\ 0 & & D & \\ & & I & D \end{bmatrix} \quad \text{where} \quad D = \begin{bmatrix} a & -b \\ b & a \end{bmatrix}$$

$$I = \begin{bmatrix} 1 & 0 \\ 0 & 1 \end{bmatrix}$$

We can complexity and use the previous argument.

Thus $\alpha|x|^2 \leq \langle Ax, x \rangle \leq \beta|x|^2$, proves the theorem.

Definition: Consider the differential equation $X' = Ax$, where $A \in L(R^n)$. The origin O is said to be a sink of the differential equation if all the eigen values of A have negative real part. We say that the linear flow e^{tA} is a contraction.

Theorem 14: Let A be an operator on a vector space E. The following statements are equivalent:

a). the origin is a sink for the dynamical system $X' = AX$.

b). for any norm in E there are constants $k > 0$, $b > 0$ such that

$$\left| e^{tA} x \right| \leq k \cdot e^{-tb} |x| \qquad \text{for all} \quad t \geq 0, \ x \in E.$$

c). There exists $b > 0$ and a basis β of E whose corresponding norm satisfies

$$\left| e^{tA} x \right|_\beta \leq e^{-tb} |x|_\beta \qquad \text{for all} \quad t \geq 0, \ x \in E.$$

Solution: we show that $(c) \Rightarrow (b) \Rightarrow (a) \Rightarrow (c)$.

$$(c) \Rightarrow (b)$$

Since $L(R^n)$ is a finite dimensional vector space, all norms on $L(R^n)$ are equivalent.

Thus if $|\cdot|_\beta$ and $|\cdot|$ are the norms with respect to basis β and any other norm then they are equivalent.

Thus we can find positive constants A and B such that for any x,

$$A\,|x|_\beta \le |x| \le B|x|_\beta$$

So consider,

$$\left|e^{tA}\,x\right|_\beta \le B\,e^{-tb}|x|_\beta \le B\cdot e^{-tb}|x|_\beta$$

$$\le B\cdot e^{-tb}\cdot\frac{1}{A}|x|$$

$$=\frac{B}{A}e^{-tb}|x|.$$

Thus with $k = \dfrac{B}{A}$ we have,

$$\left|e^{tA}\,x\right| \le k\,e^{-tb}|x|$$

$$(b) \Rightarrow (a).$$

We know that any solution of DE

$X' = A\,x$ is given by

$x(t) = e^{tA}\cdot x(0)$ where x_0 is the initial value of $x(t)$ and $x_0 \in R^n$.

$$\therefore |x(t)| = \left|e^{tA}\cdot x_0\right| \le k\,e^{-tb}|x_0|$$

$$\to 0 \quad as \quad t \to t\infty .$$

Thus every solution of the DE

$X' = A\,x$ tends to zero as $t \to \infty$.

$\therefore$ This origin is a sink of the given differential equation.

$$(a) \Rightarrow (c).$$

We are given that origin is a sink of the differential equation $X' = Ax$.

By the definition, all eigen values of the operator A have negative real part.

We can therefore find, such that $\alpha < 0$, $\beta < 0$ *such that* $\alpha < \operatorname{Re}\lambda < \beta$, for any eigen value λ of A.

By theorem, there exist a basis say β of the space E such that with respect to the inner product and norm with respect to β

$$\alpha |x|_\beta^2 < \langle Ax, x \rangle_\beta < \beta |x|_\beta^2$$

i.e., $\quad \alpha |x|^2 < \langle Ax, x \rangle < \beta |x|^2$

let $x(t)$ be the solution of the system of DE $X' = AX$.

Then,

$$|x(t)| = \sqrt{\langle x(t), x(t) \rangle}$$

$$\therefore \frac{d}{dt}|x(t)| = \frac{1}{2\sqrt{\langle x(t), x(t) \rangle}}$$

$$= \frac{\langle x'(t), x(t) \rangle}{\sqrt{\langle x(t), x(t) \rangle}}$$

i.e. $\quad \dfrac{d}{dt}|x(t)| = \dfrac{\langle x'(t), x(t) \rangle}{|x(t)|}$

$$= \frac{\langle Ax(t), x(t) \rangle}{|x(t)|} \quad \{\because X'(t) = AX(t)\}$$

$$\therefore \frac{\frac{d}{dt}|x(t)|}{|x(t)|} = \frac{\langle Ax(t), x(t) \rangle}{|x(t)|^2}$$

But $\quad \alpha |x(t)|^2 \le \langle Ax(t), x(t) \rangle \le \beta |x(t)|^2$

$\Rightarrow \quad \alpha \le \dfrac{\langle Ax(t), x(t) \rangle}{|x(t)|^2} \le \beta$

$\alpha \le \dfrac{\frac{d}{dt}|x(t)|}{|x(t)|} \le \beta \qquad i.e.$

$\alpha \le \dfrac{d}{dt} \log|x(t)| \le \beta$

Integrating with respect to t from 0 to t , we get

$\alpha t \le \log|x(t)| - \log|x(0)| \le \beta t$

$\alpha t + \log|x(0)| \le \log|x(t)| \le \beta t + \log|x(0)|$

i.e. $\quad \log\left(e^{\alpha t}|x(0)|\right) \le \log|x(t)| \le \log\left(e^{\beta t}|x(0)|\right)$

$e^{\alpha t}|x(0)| \le |x(t)| \le e^{\beta t}|x(0)| \quad$ true for all t .

Thus we have, $\quad |x(t)| \le e^{\beta t}|x(0)|$

We define $b > 0 \quad$ such that $\quad \beta = -b$.

Thus there exist a basis β of space E , and a +ve real b such that for all $t \ge 0$, we have

$|x(t)|_\beta \le e^{-bt}|x(0)|_\beta$.

Definition: We say that origin is a source of the system of differential equation $X' = AX \quad , \quad A \in L(E)$.

If all the Eigen values of A have +ve real part.

JORDAN AND REAL CANONICAL FORMS

In this case we say that the flow e^{tA} is an expansion.

Theorem 15: If $A \in L(E)$, the following are Equivalent

a)   the origin is source.

b)   For any norm on E , there are constants $L > 0$, $a > 0$ such that

$$\left| e^{tA} x \right| \geq L e^{ta} |x| \qquad \text{for all } t \geq 0, \ x \in E.$$

c)   There exists $a > 0$, , and a basis β of E whose corresponding norm satisfies

$$\left| e^{tA} x \right|_\beta \geq e^{ta} |x|_\beta \text{ for all } t > 0, \ x \in E.$$

Proof: $\qquad\qquad (c) \Rightarrow (b)$

Assume +ve constants A and B such that

$A|\cdot|_\beta \leq |\cdot| \leq B|\cdot|_\beta$ (follows from equivalence of norm).

So consider,

$$\left| e^{tA} x \right| \geq A \left| e^{tA} x \right|_\beta \geq A e^{ta} |x|_\beta$$

$$\geq \frac{A}{B} e^{ta} |x|$$

$$\therefore \ \left| e^{tA} x \right| \geq L \ e^{ta} \ |x|.$$

$$(b) \Rightarrow (a)$$

b) implies that every solution whose initial value is non zero becomes unbounded as $t \to +\infty$.

This will be true only if all eigen values of A will have +ve real part (because if an eigen value $\lambda = u \pm iv$ is such that $u \leq 0$ then a solution $\left(e^{ut} \cos vt \ and \ e^{ut} \sin vt \right)$ remains founded).

Therefore the origin is a source.

$$(a) \Rightarrow (c)$$

Since origin is source, it follows that real part of each eigen values is positive.

Therefore we can choose $\alpha > 0,\ \beta > 0$ such that

$\alpha \le \operatorname{Re}\lambda \le \beta$ for every eigen value λ of A.

So $\exists$ a basis of E such that

$$\alpha|x|^2 \le \langle Ax\ ,\ x \rangle \le \beta|x|^2 \qquad \forall\ x \in E.$$

It follows that,

$$e^{\alpha t}\left|x(0)\right| \le \left|x(t)\right| \le e^{\beta t}\left|x(0)\right|$$

We have, $\left|x(t)\right| \ge e^{\alpha t}\left|x(0)\right|$ $\qquad \forall t > 0$

We define $\alpha = a$. Thus we have a basis of E and $a > 0$ such that

$$\left|x(t)\right| \ge e^{at}\left|x(0)\right|$$

$$\left|e^{ta}\,x(0)\right| \ge e^{at}\left|x(0)\right|$$

i.e. $\left|e^{ta}\,x\right| \ge e^{at}\left|x\right|.$

HYPERBOLIC FLOW

We say that the flow e^{+A} is hyperbolic if all the eigen values of the operator A have non zero real part.

Obviously every contraction as well as expansion is a hyperbolic flow. If the operator is such that some eigen values of A have +ve real part and remaining eigen value have –ve real part. Then a flow e^{+A} is neither contraction nor expansion but

a hyperbolic flow.

Theorem 16: Let e^{tA} be a hyperbolic linear flow $\left(i.e.,\ A \in L(E)\right)$ then E

has a direct sum decomposition $E = E^S \oplus E^U$ invariant under A , such that the induced flow on E^S is contraction and induced flow on E^U is expansion. Further this decomposition is unique.

Proof: We can write the operator A in its real form consisting of Jordan/Real blocks corresponding to different eigen values of A.

We can also write E as direct sum of generalized eigen spaces belonging to differential eigen values of A.

We reindex the eigen values of A in such a way that all eigen values with negative real part come first followed by the eigen values with positive real part.

We reindex the corresponding eigen spaces also.

Let E^s be the direct sum of generalized eigen spaces belonging eigen values will negative real parts and E^U be the direct sum of generalized eigen spaces belonging eigen values with +ve real parts.

Thus we have $E = E^S \oplus E^U$

Since every generalized eigen space is invariant under A, it follows that E^S and E^U are invariant under A.

Thus $A \Big|_{E^U}$ is an operator say A_4 and $A \Big|_{E^S}$ is an operator say A_S .

We then have the decomposition $A = A_S + A_U$

Since $A = A_S$ on E^S , it follow that

$$e^{tA} \Big|_{E_S} = e^{tA_S}$$

Similarly,

$$e^{tA} \Big|_{E^U} = e^{tA_U}$$

DYNAMICAL SYSTEM – A SHORT COURSE

Since all the eigen values of A_S are –ve and all the eigen values of A_u are +ve.

$e^{tA} \Big|_{E^S} = e^{tA_S}$ is contraction while

$e^{tA} \Big|_{E^U} = e^{tA_U}$ is expansion.

To prove uniqueness of the decomposition, let us assume that

$$E = F^s \oplus F^u$$

where F^s and F^u are subspaces of E such that the flow induced by e^{tA} on F^s is contraction and the flow induced by e^{tA} on F^u is expansion.

Let $\quad v \in F^s$

$$v \in E \qquad \left(\because E = F^s \oplus F^u \right)$$

But $\quad E = E^s \oplus E^u$

$\therefore \quad v = x + y \qquad$ For some $x \in E^s \quad$ and $\; y \in E^u$

As $\; v \in F^s$, by the definition

$\left| e^{tA} v \right| \to 0 \;$ as $\; t \to \infty$

i.e., $\quad e^{tA} v \to 0 \quad$ as $\; t \to \infty$.

Using $\quad v = x + y$, we get

$$e^{tA} v = e^{tA} x + e^{tA} y$$

As F^s and F^U are invariant subspaces,

$$e^{tA} x \in F^s \; , \; e^{tA} y \in F^u$$

$\therefore \; e^{tA} x \to 0 \qquad$ as well as $\quad e^{tA} y \to 0 \;$ as $\; t \to \infty$.

but we know that as $\quad y \in F^u$, the induced flow is expansion.

JORDAN AND REAL CANONICAL FORMS

We can find $a > 0$ such that

$$\left| e^{tA} y \right| \geq \left| e^{ta} y \right| = e^{ta} \left| y \right|$$

As $a > 0$, e^{ta} becomes unbounded.

So thus if $y \neq 0$ then the expression $e^{ta} \left| y \right|$ becomes unbounded.

But we require, $e^{tA} y \to 0$ *as* $t \to \infty$

As $v = x + y$ and $y = 0$

We have $v = x \in E^s$

$\therefore F^s \subseteq E^s$

By similar argument we can show that

$E^s \subseteq F^s$

i.e., $E^s = F^s$

Similarly, $E^U = F^U$. Hence the decomposition is unique.

GENERIC PROPERTY

(The property which is true on substantially big space is kid Generic property)

Let E be a normal linear space. We can consider E to be a metric space.

A set $V \subseteq E$ is dense in E iff $\overline{V} = E$

i.e., every non-empty open set in E has a non-empty intersection with V.

Theorem 16: Let $X_1, X_2, \ldots X_m$ be dense open sets in a non linear space E. Then $X = X_1 \cap X_2 \cap \ldots \cap X_m$ is also dense and open.

Proof: As $X_1, X_2, \ldots X_m$ are all open.

$X = X_1 \cap X_2 \cap \ldots \cap X_m$ is also open.

$\therefore X$ is open.

DYNAMICAL SYSTEM – A SHORT COURSE

Let V be a non empty open set in E.

We show that $V \cap X \neq \phi$.

Now as X_1 is dense, $V \cap X_1 \neq \phi$.

Thus $V \cap X_1$ is non empty open set.

And X_2 is dense, $V \cap X_1 \cap X_2 = \phi$ is non empty.

As $V \cap X_1 \cap X_2$ is non empty open set and X_3 is dense.

$V \cap X_1 \cap X_2 \cap X_3$ is non empty.

Repeating this we get.

$V \cap X = V \cap X_1 \cap X_2 \cap ... \cap X_m$ is non empty.

A GENERIC PROPERTY ON $L(R^n)$

A property of operator $L(R^n)$ which holds on a subset of $L(R^n)$ is said to be generic property if the set on which it holds contain a dense and open set.

Let $T \in L(R^n)$ and let β be a basis of R^n.

Let $A = \left| a_{ij} \right|$ be the matrix of the operator T with respect to the basis β.

We define $\|T\|$ with respect to the basis β as

$$\|T\|_{\beta \max} = \max \left\{ \left| a_{ij} \right|, \ 1 < i < j \leq n \right\}.$$

Theorem 17: The set δ_1 of operators on R^n that have n distinct eigen values is dense is dense and open in $L(R^n)$.

Proof. We first show that the set δ_1 is dense.

For this we show that if A is any operator in $L(R^n)$ and $\epsilon > 0$, then ϵ-nbd of A contains an operator all of whose eigen values are distinct.

278

JORDAN AND REAL CANONICAL FORMS

Let $\lambda_1, \lambda_2, ... \lambda_r$, $a_1 \pm i b_1$, $a_2 \pm i b_2 ... a_s \pm i b_s$ be the eigen values of operator A , repeated according to multiplicity.

Let β be a basis of R^n such that A can be written in the real form say

$$A = \begin{bmatrix} \lambda_1 & & & & & & & \\ 0 & \lambda_2 & & & & & 0 & \\ & 1 & : & & & & & \\ & & & \lambda_r & & & & \\ & & & 1 & & & & \\ & & & & D_1 & & & \\ & & & & 1 & D_2 & & \\ & & & & & & : & \\ & 0 & & & & & & D_s \end{bmatrix}$$

where $D_1 = \begin{bmatrix} a_1 & -b_1 \\ b_1 & a_1 \end{bmatrix}$, $I = \begin{bmatrix} 1 & 0 \\ 0 & 1 \end{bmatrix}$

We consider β – max norm on $L(R^n)$.

We can write $A = S + N$ where

$$S = \begin{bmatrix} \lambda_1 & & 0 \\ & : & \\ 0 & D_1 & \\ & & D_s \end{bmatrix}, \quad N = \begin{bmatrix} 0 & & & & & \\ 1 & 0 & & & & \\ & 1 & 0 & & & \\ & & & : & & \\ & 0 & & 0 & & \\ & & & 1 & 0 & \\ & & & & 1 & 0 \end{bmatrix}$$

where, $1 = \begin{bmatrix} 1 & 0 \\ 0 & 1 \end{bmatrix}$ and $0 = \begin{bmatrix} 0 & 0 \\ 0 & 0 \end{bmatrix}$

For $\in$, we define $\lambda_1', \lambda_2', ... \lambda_r', a_1', a_1' .... a_s'$ such that, they are all distinct and $\left| \lambda_i - \lambda_i' \right| < \in$ and $\left| a_i - a_i' \right| < \in$

(This is possible because every nbd. contain infinity many points)

Let $S' =$ $\begin{bmatrix} \lambda_1' & & & & & \\ & \lambda_2' & & & 0 & \\ & & : & & & \\ 0 & & \lambda_r' & & & \\ & & & D_1' & & \\ & & & & : & \\ & & & & & D_s' \end{bmatrix}$ where, $D_1' = \begin{bmatrix} a_1' & -b_1' \\ b_1' & a_1' \end{bmatrix}$

$A' = S' + N$.

Thus $A' = \begin{bmatrix} \lambda_1' & & & & & \\ 1 & \lambda_2' & & & 0 & \\ & & : & & & \\ 0 & & \lambda_r' & & & \\ & & & D_1' & & \\ & & & I & : & \\ & & & & D_s' & \\ & & & & I & \end{bmatrix}$

Thus A' is an operator on R^n whose eigen values are $\lambda_1', \lambda_2', \dots \lambda_r'$, $a_1' \pm i b_1$, $a_2' \pm i b_2 \dots a_s' \pm i b_s$ all of which are distinct.

Further, it is clear that,

$\left\| A - A' \right\|_{\beta-\max} < \in$

Thus $\in$ nbd of A contains A' (of the δ_1)

$\therefore$ δ_1 is dense in $L(R^n)$.

We next show that δ_1 is open.

If δ_1 is not open, then we can find a member say $'f'$ in δ_1 and operators $\{f_n\}$ each from the complement of δ_1 such that $f_n \to f$ as $n \to \infty$.

Since $f_1, f_2 \dots '$ are not in δ_1 ; each of them has at least one eigen value whose multiplicity at least 2.

Let $'\lambda_j'$ be an eigen value of f_j such that, the multiplicity of $\lambda_j \geq 2$.

Consider the generalized eigen space belonging to the eigen value λ_j of f_j.

Its dimension will be greater or equal to 2.

Since the sequence $\{f_i\}$ is converging to f , it is a bounded sequence.

It therefore follows that the collection of all eigen values of the operators $f_1, f_2 \cdots$ will be bounded. Thus the set $\{\lambda_1 \ \lambda_2 \ \ldots\}$ is a bounded set.

Hence there will exist a subsequence of $\{\lambda_1 \ldots\}$ will be convergent. We denote the subsequence by $\{\lambda_1 \ldots\}$ thus let $\lambda = \lim\limits_{k \to \infty} \lambda_k$.

We first consider the case where all eigen values $\lambda_1 \ \lambda_2 \ldots$ (choose for $f_1, f_2 \cdots$ with multiplicity greater or equal to 2) are real. Thus λ is also real.

As multiplicity of λ_1 is ≥ 2 we can choose two eigen vectors say x_i and y_i linearly independent belonging to eigen value λ_1.

We choose them in such a way that

$|x_i| = 1$, $|y_i| = 1$ and $\langle x_i \ y_i \rangle$

If required we replace x_i by $\dfrac{x_i}{|x_i|}$ and y_i by $\dfrac{y_i - \langle x_i \ y_i \rangle x_i}{|y_i - \langle x_i \ y_i \rangle x_i|}$.

Since the seq $\{x_i\}$ and $\{y_i\}$ is a bdd seq. we can find subsequence of $\{x_i\}$ and $\{y_i\}$ such that $\lim x_i = x$ *and* $\lim y_i = y$.

Since norm and inner product are continuous , it follows that

$|x| = |y| = 1$, $\langle x, y \rangle = 0$ is x and y are linearly independent.

Since x_i, y_i are eigen vector belonging to eigen value λ_i for the operator f_{i_1} we have

$$f_i x_i = \lambda_i x_i,$$
$$f_i y_i = \lambda_i y_i,$$

In the limit, we get

$$f_x = \lambda_x, \quad f_y = \lambda_y$$

$\therefore$ x *and* y are eigen vectors belonging to eigen value λ of f

As x *and* y are linearly independent dim of eigen space of $\lambda \geq 2$.

$\therefore$ Multiplicity of λ is greater or equal to 2.

This is a contradiction as each eigen value of f is of multiplicity 1.

In case some of the eigen values $\lambda_1 \lambda_2 ...$ are complex.

We consider complexification of the space and of the operators. Eigen values will remain same.

We product on C^n defined as for

$$\alpha \in C^n, \ \alpha = (z_1 ... z_n)$$

$$\beta \in C^n, \ \beta = (w_1 .... w_n)$$

$$\langle \alpha \ \beta \rangle = z_1 \overline{w_1} + z_2 \overline{w_2} + ... + z_n \overline{w_n}$$

We thus get an eigen value of f of multiplicity ≥ 2 , a contradiction.

Thus δ_1 is open and dense in $L(R^n)$.

Theorem 18: Semi simplicity is a Generic property in $L(R^n)$.

Proof. Since any operator having all eigen values distinct is semi simple .

The set δ_1 is contained in the collection of all semi simple operators on R^n

JORDAN AND REAL CANONICAL FORMS

Thus semi simplicity is a Generic property we however note that the set of all semi simple operators is not open.

For example: consider the $\begin{bmatrix} 1 & 0 \\ 0 & 1 \end{bmatrix}$.

Since it is already in the diagonal form it is semi simple.

For any $\in \neq 0$,

Consider the operator given by the matrix $\begin{bmatrix} 1 & \in \\ 0 & 1 \end{bmatrix}$.

It has the decomposition $\begin{bmatrix} 1 & 0 \\ 0 & 1 \end{bmatrix} + \begin{bmatrix} 1 & \in \\ 0 & 1 \end{bmatrix}$

And as such it is not semi simple.

Also the distance between $\begin{bmatrix} 1 & 0 \\ 0 & 1 \end{bmatrix}$ *and* $\begin{bmatrix} 1 & \in \\ 0 & 1 \end{bmatrix}$ can be made arbitrarily small by talking $|\in|$ sufficiently small.

Theorem 19: The set $\delta_2 = \{T \in L(R^n) / e^{tT}$ is a hyperbolic flow$\}$ is open and dense in $L(R^n)$.

Proof. We show that δ_2 is dense. Let $A \in L(R^n)$ with eigen values say $\lambda_1, \lambda_2, \dots \lambda_r$, $a_1 \pm i b_1$, $a_2 \pm i b_2 \dots a_s \pm i b_s$ repeated according to multiplicity.

Let β be a basis of R^n with respect to which the matrix of A is in the real form say

$$A = \begin{bmatrix} \lambda_1 & & & & & & \\ 1 & \lambda_2 & & & & 0 & \\ & & \ddots & & & & \\ 0 & & & \lambda_r & & & \\ & & & & D_1 & & \\ & & & & I & : & \\ & & & & & & D_s \\ & & & & & & I \end{bmatrix}$$

283

where $D_i = \begin{bmatrix} a_i & -b_i \\ b_i & a_i \end{bmatrix}$ and $1 = \begin{bmatrix} 1 & 0 \\ 0 & 1 \end{bmatrix}$ $1 \le i \le s$

We can write, $A = S + N$

We choose $\beta -$ max norm on $L(R^n)$.

Let $\in > 0$ be chosen arbitrarily.

We choose $\lambda_1', \lambda_2', \ldots \lambda_r', a_1', a_1' \ldots a_s'$ in such a way that

$\left| \lambda_i - \lambda_i' \right| < \in, 1 \le i \le r$

$\left| a_i - a_i' \right| < \in, 1 \le i \le s$

Further we ensure that,

$\lambda_1', \lambda_2', \ldots \lambda_r', a_1', a_1' \ldots a_s'$ are all non zero.

Let $S' = \begin{bmatrix} \lambda_1' & & & & & & \\ 1 & \lambda_2' & & & & 0 & \\ & & : & & & & \\ & & & \lambda_r' & & & \\ & & & & D_1' & & \\ & & & & I & : & \\ & 0 & & & & D_s' & \\ & & & & & & I \end{bmatrix}$

Where $D_1' = \begin{bmatrix} a_1' & -b_1 \\ b_1 & a_1' \end{bmatrix}$

Let $A' = S' + N$

A' is a real form of an operator real parts of all of whose eigen values are non zero.

$\therefore e^{tA}$ gives an hyperbolic flow.

JORDAN AND REAL CANONICAL FORMS

$\therefore \ A' \in \delta_2$

$\therefore \ \delta_2$ is dense in $L(R^n)$.

We show that δ_2 is open, assume that it is not open.

Let $f \in \delta_2$ such that, $f = \lim_{j \to \infty} f_j$ such that

e^{f_j} Is not an hyperbolic flow.

Thus each f_j will have at least one purely imaginary eigen value.

We denote such an eigen value of f_j by λ_j .

The sequence $\{f_j\}$ is bdd.

$\therefore \ \{\lambda_j\}$ is also a bdd sequence.

Let $\lambda = \lim \lambda_j$ (considering an appropriate subsequence).

Further, let x_j be an eigen vector belonging to eigen value λ_j for the operator f_j .

We can choose x_j such that $|x_j = 1|$.

Let $\quad x = \lim_{j \to \infty} x_j$

$\therefore \ |x| = 1$

Thus we have, $f_j x_j = \lambda_j x_j$

In the $\lim$ as $j \to \infty$, $f x = \lambda x$

As $\lambda = \lim \lambda_j$ with $\text{Re } \lambda_j = 0$,

It follows that $\text{Re } \lambda = 0$.

Thus f has an eigen value with zero real part.

This is a contradiction as e^{tf} is a hyperbolic flow. i.e. $\left(f \in \delta_2\right)$

So δ_2 is open.

$\therefore$ δ_2 is open and dense in $L(R^n)$.

CHAPTER 5

DYNAMICAL SYSTEM AND VECTOR FIELD

Consider a system of differential equation

$$\frac{dx}{dt} = Ax,$$

where $A \in L(E)$.

We know that, $\forall x_o \in E$, we can find a solution $x(t)$ of the given system. It is explicitly given by

$$x(t) = e^{tA} \cdot x_o.$$

Obviously, $x(t)$ depends on t as well as on the initial value x_o.

We can thus denote it by

$\phi(t, x_o), t \in R, x_o \in E, x(t)$ or

$\phi(t, x)$ is thus a function from $R \times E \to E$

(Here $\phi(o, x) = x$).

It is clear that, at time $t = o$,

$x(t) = x(o). \quad (\because e^{tA} x_o = x_o)$

Thus the map $\phi : R \times E \to E$ is such that

$$\phi(o, x) = x, \forall x \in E.$$

We can also denote $\phi(t, x)$ as $\phi_t(x)$.

Thus, $\phi_o(x) = x, \forall x \in E$.

Consider $\phi_s(\phi_t(x))$.

That is we first find $\phi_t(x)$ and then get $\phi_s(\phi_t(x))$.

For this, consider

$$x(t) = e^{tA} \cdot x_o .$$

Obviously, $x(t) = \phi(t, x_o)$.

We consider a solution of

$$x' = Ax$$

which initially is at $x(t)$.

If we denote it by   then

$$y(s) = e^{sA} \cdot x(t)$$

Thus, $y(s) = \phi_s(x(t))$

$$\phi_s(\phi_t(x_o))$$

But $x(t) = e^{tA} \cdot x_o .$

$$\therefore y(s) = e^{sA} \cdot e^{tA} \cdot x_o$$
$$= e^{(s+t)A} \cdot x_o$$
$$= \phi_{s+t}(x_o)$$

$$\therefore \phi_s(\phi_t(x_o)) = \phi_{s+t}(x_o), \forall x_o \in E$$

$$\therefore \phi_s \, \phi_t = \phi_{s+t}$$

We generalised these properties of $x(t)$ i.e. $\phi_t(x)$ to define a dynamical system.

Definition: A dynamical system is a way of describing the passage in time of all points of a given space.

A dynamical system is a $C^1 - map \; R \times \delta \to \delta$, where δ is an open set of an Euclidean space and writing $\phi(t, x)$ as $\phi_t(x)$, the map $\phi_t : \delta \to \delta$

has the properties:

(i). $\phi_o : \delta \to \delta$ is an identity operator.

(ii). The composition $\phi_s \circ \phi_t = \phi_{s+t},\ \forall s, t \in R$.

Thus the system of equations

$$x' = Ax,$$

where $A \in L(E)$ gives rise to a dynamical system

$$x(t) = \phi(t, x) \to (i) \text{ depends on } R \times E.$$

Suppose ϕ is a dynamical system.

As ϕ is C', we can define

$$f(x) = \frac{d}{dt} \phi(t, x)|_{t=o}$$

We thus get a differential equation

$$f(x) = \frac{d}{dt}(\phi(t, x))|_{t=o} = \frac{d}{dt} x(t)|_{t=o}$$

i.e. $x' = f(x)$

A system of different equations of the type

$$x' = f(x)$$

where, R.H.S. does not depends on time explicitly is known as an **autonomous system**.

Instead if we have a system of differential equation as

$$x' = f(t, x)$$

It is known as **non-autonomous system**.

Theorem 1:

1). Let E denote a normed linear space, W will be an open subset of E.

If $f:W \to E$ is a continuous map, by a solution of the differential equation $x' = f(x)$

we mean a function $u:J \to W$ defined on some interval $J \subseteq R$ such that

$$u'(t) = f(u,(t)), \forall t \in J$$

Here J can be an interval of real numbers like $(a,b)\,[a,b],\,(a,\infty),\,(-\infty,b),$ etc.

Theorem 2: Let $W \subseteq E$ be an open subset of a normed linear vector space, $f:W \to E$ a C^1 – map and $x_o \in W$. Then there is some $a > o$ and a unique solution

$x:(-a,a) \to W$ of the differential equation

$x' = f(x)$,

satisfying the initial condition $x(o) = x_o$.

Definition: A function $f:W \to E$, W is an open subset of the normed linear space E is said to be **Lipschitz** on W there exists a constant k such that

$$|f(y) - f(x)| \le k\,|x - y|$$

where, k is known as Lipschitz constant.

A function F is said to be **Locally Lipschitz** if for each point of W has a nbd in which the restriction of F is Lipschitz.

Theorem 3: Let $f:W \to E$ be $C'.\eta$ then F is locally Lipschitz. Two solution curves of $x' = f(x)$ cannot cross.

Similarly, any solution curve of $x' = f(x)$ cannot cross itself.

Theorem 4: Let $W \subseteq E$ be an open and $f:W \to E$ has Lipschitz constant k. Let $y(t)$, $z(t)$ be solutions to $x' = f(x)$ on the closed interval $[t_o, t_1]$, then for all $t \in [t_o, t_1]$,

$$| y(t) - z(t) | \leq | y(t_o) - z(t_o) | (\exp(k(t - t_o)))$$

This thm. will be used in next results.

Theorem 5: Let a $C^1 - map\ F : W \to E$ be given. Suppose two solutions $u(t)$, $v(t)$ of $x' = f(x)$ are defined on the same open interval J containing t_0 and satisfy $u(t_o) = v(t_o)$. Then $u(t) = v(t)$, $\forall t \in J$.

Theorem 6: Let $f(x)$ be C'. Let $y(t)$ be a solution to $x' = f(x)$ defined on the closed interval $[t_o, t_1]$ with $y(t_o) = y_o$. Then there is a nbd $U \subseteq E$ of y_o and a constant k such that

if $z_o \in U$ then there is a unique solution $z(t)$ also defined on $[t_o, t_1]$ with $z(t_o) = z_o$ and z satisfies

$$| y(t) - z(t) | \leq | y_o - z_o | \exp | k(t_1 - t_o) |, \quad \forall t \in [t_o, t_1].$$

Theorem 7: If $F : W \to E$ is locally Lipschitz and $A \subseteq W$ is a compact set then $F|_A$ is Lipschitz.

FLOW OF A DIFFERENTIAL EQUATION

Consider an equation $x' = f(x)$ defined by a C' - function F on W, $W \subseteq E$ is open. For each $Y \in E$, there is a unique solution $\phi(t)$ with $\phi(o) = y$ defined an a maximal open interval $J(Y) \subseteq R$.

It is obvious that $o \in J(Y)$.

Further if $t \in J(Y)$ then $[t, o] \subseteq J(Y)$ (for +ve)

If ϕ is a solution of the diff. equation $x' = f(x)$ with $\phi(o) = y$ then we denote it by $\phi(t, y)$, for $t \in J(Y)$ by making its dependence on t & y explicit.

We also use the notation $\phi_t(y)$ for $\phi(t,\,y)$.

Let $\Omega \subseteq R \times W$ be the following set

$\Omega = \{(t,y) \in R \times W \mid t \in J(Y)\}$

We note that in Ω, the interval over which the first component varies depends on the 2nd app component.

Thus for $(t,y) \in \Omega$, $\phi(t,y)$ is well defined.

The map $(t,y) \rightarrow \phi(t,y)$ is thus function $\phi : \Omega \rightarrow w, \phi$ is called as the **flow** of the differential equation $x' = f(x)$.

Theorem 8: The map ϕ has the following property $\phi_{s+t}(x) = \phi_s\big(\phi_t(x)\big)$

Proof: We shall denote by $J(x)$ the maximal open interval on which the solution $\phi(t,x)$ is defined.

The thm states that, if for some s, t, $\phi_s\big(\phi_t(x)\big)$ is defined, thus so is $\phi_{s+t}(x)$ and the two are equal.

Similarly, if $\phi_{s+t}(x)$ exists then so does $\phi_s\big(\phi_t(x)\big)$ and they are equal.

Let us assume that $\phi_s\big(\phi_t(x)\big)$ exists

i.e. $t \in J(x)$ and $s \in J\big(\phi_t(x)\big)$.

$\phi_{s+t}(x)$ will be defined provided $s + t \in J(x)$

Thus given $t \in J(x), S \in J\big(\phi_t(x)\big)$.

We show that $s + t \in J(x)$.

Let $(\alpha, \beta) \equiv J(x)$.

As $0 \in J(x)$, we have $\alpha > 0, \beta > 0$.

Also, as $t \in (\alpha, \beta), [0,t] \subseteq (\alpha, \beta)$.

DYNAMICAL SYSTEM AND VECTOR FIELD

We first consider the case where $t > o, s > o$.

Thus in order to show that $s + t \in (\alpha, \beta)$, we just need to show that $s + t < \beta \ (\because \alpha < t < s + t)$.

We now define a function ψ on $(\alpha, s + t]$ as

Imp $\rightarrow \psi(r) = \begin{cases} \phi_r(x), \alpha < r \le t \\ \phi_{r-t}(\phi_t(x)), t \le r \le s + t \end{cases}$.

Since $J(x) \equiv (\alpha, \beta)$ and $t \in J(x)$

$\therefore (\alpha, t) \subseteq J(x), \forall r, \alpha < r \le t$.

$\therefore \phi_r(x)$ is well defined

Thus $\psi(r)$ is well defined in (α, β) and is given by $\phi(r, x)$.

For $t \le r \le s + t, o \le r - t \le s$,

As $S \in J(\phi_t(x)), [o, s] \subseteq J(\phi_t(x))$

$\phi_{r-t}(\phi_t(x))$ is well defined for $o \le r - t \le s$

i.e. $\phi_{r-t}(\phi_t(x))$ is well defined for $t \le r \le s + t$.

At t, both the values are same

Thus $\psi(r)$ is well defined in $t \le r \le s + t$

As $\alpha < o$ and $t > o, a \in (\alpha, t)$.

$\therefore \psi(o) = \phi_o(x) = x$

$\psi(r)$ is thus a solution of the diff. equation $x' = f(x)$ and satisfy the initial value $\psi(o) = x$.

It is defined on $(\alpha, s + t]$.

Since $\phi_t(x)$ is also a solution of the Differential equation

$x' = f(x)$ such that $\phi(o, x) = x$

By the uniqueness then, we must have $\phi(t,x)$ same as ψ.

Since (α,β) is the maximal open interval over which $\phi(t,x)$ is defined.

We must have $(\alpha,s+t] \subseteq (\alpha,\beta)$

Further as $s+t \in (\alpha,\beta)$, we consider

$$\phi_{s+t} \cdot (x) = \psi(s+t) = \phi_s(\phi_t(x)).$$

Consider the case where both s and t are negative.

Since $\alpha < t < \beta$, we have $s+t < t < \beta$.

We know that $s+t > \alpha$.

We define a function ψ an $[s+t,\beta]$ as

$$\psi(r) = \begin{cases} \phi_r(x), t \leq r < \beta \\ \phi_{r-t}(\phi_t(x)), s+t \leq r \leq t \end{cases}$$

Since s is negative, $[o,s] \subseteq J(\phi_t(x))$ and thus the function ψ well defined in the interval $[s+t,\beta]$.

It is thus a solution to the differential equation $x' = Ax$ and $\psi(o) = \phi_o(x) = x$ $\{\because \beta > o \,\& \, t < o$.

Thus ψ is a solution to the D.E. $x' = f(x)$ defined in $[s+t,\beta]$ and $\psi(o) = x$.

As $\phi_t(x)$ is also a solution to the diff. equation $x' = f(x)$, with initial value x.

By uniqueness, ψ is same as $\phi(t,x)$.

As $J(x) = (\alpha,\beta)$ it follows that $s+t \in (\alpha,\beta)$

Thus, $\phi_{s+t}(x)$ is well defined.

But $\phi_{s+t}(x) = \psi(s+t) = \phi_s(\phi_t(x))$

DYNAMICAL SYSTEM AND VECTOR FIELD

We now assume that $\phi_{s+t}(x)$ is defined

i.e. $s+t \in J(x)$ and show that $\phi_s(\phi_t(x))$ is well defined and

$\phi_{s+t}(x) = \phi_s(\phi_t(x))$

We now define a function ψ as

$\psi(x) = \phi_{s+t}(x)$, for $o \leq r \leq s$

$\therefore \{s + t \in J(x)$

Thus, $\psi(o) = \phi_t(x)$

Thus ψ is a solution of the diff. equation $X' = f(x)$ with initial value $\phi_t(x)$

$\psi(s) = \phi_s(\phi_t(x))$

But from the definition of ψ, we have

$\psi(s) = \phi_{s+t}(x)$

$\quad = \phi_{s+t}(x) = \phi_s(\phi_t(x))$

Theorem 9: Let Ω is an open set in R×W and $\phi : \Omega \rightarrow W$ is a continuous map.

Proof: Let $(t_o, x_o) \in \Omega$, it means that $t_o \in J(x_o)$.

$\therefore [o, t_o] \subseteq J(x_o)$

Since $J(x_o)$ is an open interval, o, t_0 are interior pts.

of $J(x_o)$, so we can find an $\in > o$ s.t. $[-\in, t_o + \in] \subseteq J(x_o)$.

We can find a nbd of x_o say U in W s.t.

$\forall Y \in U, \phi_t(y)$ is defined U on $[-\in, t_o + \in]$

That is , $[-\in, t_o + \in] \subseteq J(y)$, for every $Y \in U$.

$\therefore [-\in, t_o + \in] \times U \subseteq \Omega$

Thus the nbd $[-\epsilon, t_o + \epsilon] \times U$ of (t_o, x_o) lies entirely in Ω.

$\therefore \Omega$ is an open set

We now show that $\phi : \Omega \to W$ as $(t, x) \to \phi(t, x)$ is continuous.

We show that ϕ is continuous at every pt. of Ω

So let $(t_o, x_o) \in \Omega$, we choose $\epsilon > 0$ such that $[-\epsilon, t_o + \epsilon] \subseteq J(x_o)$

We choose U a nbd of x_o small such that $\overline{U} \subseteq W$ and for every $y \in \overline{U}$,

$[-\epsilon, t_o + \epsilon] \subseteq J(y)$

We thus get a compact set $[-\epsilon, t_o + \epsilon] \times \overline{U}$ in Ω

Since $F \in C'$, f is a locally Lipschitz F will have a Lipschitz constant w.r.t. the variable x in the compact set $[-\epsilon, t_o + \epsilon] \times \overline{U}$, F will be bounded.

As $x_o \in U$, we can find a $\delta > 0$ s.t. $\forall x_1$ with $|x_o - x_1| < \delta$, we have $x_1 \in U$.

We choose δ small enough so that $\delta < \epsilon$.

Consider the point (t_1, x_1), where $|t_1 - t_o| < \delta$ and $|t_1 - x_o| < \delta$.

As $|t_1 - t_o| < \delta$, $t_1 \in [-\epsilon, t_o + \epsilon]$

Since $x_1 \in U$, $[-\epsilon, t_o + \epsilon] \subseteq J(x_1)$.

As $t_1 \in [-\epsilon, t_o + \epsilon]$, $\phi(t_1, x_1)$ is well defined.

So, consider

$|\phi(t_1, x_1) - \phi(t_o, x_o)| = |\phi(t_1, x_1) - \phi(t_1, x_o) + \phi(t_1, x_o) - \phi(t_o, x_o)|$

$\leq |\phi(t_1, x_1) - \phi(t_1, x_o)| + |\phi(t_1, x_o) - \phi(t_o, x_o)|$

Since $\phi(t_1, x_o)$ and $\phi(t_o, x_o)$ are values of the same solution (initial value x_o, here) times t, and to and as a solution is continuous.

It follows that,

$|\phi(t_1, x_o) - \phi(t_o, x_o)| \to 0$ as $|t_1, t_o| \to 0$

Similarly, the expression $|\phi(t_1, x_1) - \phi(t_1, x_o)|$ will be dominated by

(constant). $|x_1, x_o| \cdot \exp(const. |x_1, x_o|) \to 0$ as $|x_1, x_o| \to 0$.

Thus as $\delta \to 0, |\phi(x_1, x_o) - \phi(x_0, x_o)| \to 0$.

$\therefore \phi$ is continuous at (t_o, x_o).

As (t_o, x_o) is arbitrary in Ω.

$\therefore \phi$ is continuous in Ω.

NOTE

Consider a pt. $(t, x) \in \Omega$ i.e. $t \in J(x)$

$\therefore$ a nbd of t will lie in J(x).

We can find a nbd say U of x such that the above nbd of t will lie in J(y),

$\forall y \in U$.

In particular, $t \in J(y), \forall y \in U$,

i.e. $\phi(t, y)$ is well defined for all $y \in U$.

Keeping t fixed, we can consider $\phi(t, y)$ as a function of y.

Thus $\phi(t, y) \equiv \phi_t(y)$ can be considered as a function from U to W.

Theorem 10: The map ϕ_t sends U onto an open set V and ϕ_t is defined

on v and sends V onto U. the composition $\phi_t \cdot \phi_t$ is the identity map on U,

the composition $\phi_t \cdot \phi_t$ is the identity on v.

Proof: Let $x \in U$ and let $y = \phi_t(x) \in v$

Thus, we have $t \in J(x)$.

We show that $-t \in J(y)$.

DYNAMICAL SYSTEM – A SHORT COURSE

We consider a function $\psi(s) = \phi_{s-t}(y)$.

Obviously, ψ is well defined at $s = t$ and is given by y.

Since ψ is a solution of the diff. equation $X' = f(x)$ and is defined at $s=t$, it is well defined in [o, st] (or st, o] if s is –ve).

$\therefore \psi(o)$ is defined.

As $\psi(t) = \phi_o(y) = y, \psi$ is well defined at t.

Since ψ has been defined in terms of the flow ϕ, ψ is also a flow. Thus [o, t] lies in the domain of definition of ψ.

Hence ψ (o) is well defined.

Hence $\phi_{-t}(y)$ is well defined for $y \in V$.

Since $\psi(t) = y$, it follows that ψ is also a flow of $\phi_{-t}(y)$ to y in time t.

Thus we can write,

$$\psi(t) = \phi_t\big(\phi_{-t}(y)\big)$$

But $y = \psi(t), y = \phi_t\big(\phi_{-t}(y)\big)$, for $y \in V$

Thus for every $y \in V$, we have

$y = \phi_t\big(\phi_{-t}(y)\big)$ i.e. $\phi_t \cdot \phi_{-t}$ is identity on V.

To show that $\phi_{-t} \cdot \phi_t$ is identity on U.

Consider $\phi_{-t}\big(\phi_t(x)\big)$, for $x \in U$

We now operate ϕ_t on it to get

$$\phi_t\big(\phi_{-t}(\phi_t x)\big) = \big(\phi_t\ \phi_{-t}\big)\big(\phi_t x\big)$$

$$= \phi_t(x) \ \{\therefore\ \phi_t \cdot \phi_{-t} \text{ is identity on v.}$$

Thus, $\phi_t\big(\phi_{-t}(\phi_t x)\big) = \phi_t(x)$

DYNAMICAL SYSTEM AND VECTOR FIELD

By uniqueness of the flow, we must have

$$\phi_{-t}\,\phi_t(x) = x$$

As $x \in U$ is arbitrary, we have $\phi_{-t} \cdot \phi_t$ on U.

We now show that the set $V = \phi_t(U)$ is open.

We know that ϕ_{-t} is defined on v.

Let v* be a maximal subset of W on which ϕ_{-t} is defined and which contains set v. and so v* must be open in W. this is because as W is open and if v* were closed in W, the pts. on the boundary of v* may contains a nbd outside v* but on which ϕ_{-t} is defined.

This will contradicts the maximality of v*.

Thus we can consider ϕ_{-t} as a map from V to W

But ϕ_{-t} is a continuous map.

As U is an open subset of W, it follows that inverse image of U under ϕ_{-t} i.e. $\phi_t(U)$ i.e. V must be open in V*.

As V* is itself open, V must be open.

CHAPTER 6

NON-LINEAR SINK

Consider a differential equation $X' = f(x)$ and let $F : W \to R^n, W \subseteq R^n$ is open. We suppose $F \in C^1$

A paint $\bar{x}$ of W is said to be an equilibrium point of the diff. equation $X' = f(x)$, if $f(\bar{x}) = 0$

If $\bar{x}$ is an equilibrium point then $x(t) = \bar{x}$ is a solution to the given diff. equation. The solution curve to this particular solution is a singleton set $\{\bar{x}\}$.

If $\phi(t, x)$ denotes the flow of the diff. equation $X' = f(x)$ and $f(\bar{x}) = 0$ then

$\phi(t, \bar{x}) = \bar{x}$, for all time t

For this reason the point $\bar{x}$ is also known as stationary pt. or fixed pt. it is also known as zero or singular pt. of the vector field F.

Suppose is apt. of equilibrium for the diff. equation $X' = f(x)$

i.e. $f(0) = 0$.

Since $F \in C^1$, F is differentiable at O, then by the definition of differentiability, we know that

$$\lim_{|x| \to 0} \frac{|F(x) - DF(0)x|}{|x|} = 0$$

Thus, in a small nbd of $x=0$ $f(x)$ can be approximated by the operator $DF(0)x$.

$DF(0)$ is known as the **Linear part** of the function F.

If all the Eigen values of $DF(0)$ have negative real part then we call o as (non-linear) sink of the diff. equation $X' = f(x)$.

Generalising this, if $\bar{x}$ is an equilibrium pt. of the diff. equation $X' = f(x)$

then if all the Eigen values of the operator $DF(\bar{x})$ have negative real part, $\bar{x}$ is known as (non-linear) **sink** of the diff. equation $X' = f(x)$.

Theorem 1: Let $\bar{x} \in w$ be a sink of diff. equation $X' = f(x)$. Suppose every Eigen value of $DF(\bar{x})$ has real part less than $-c$, $c>o$. then there is a nbd $U \subseteq W$ of $\bar{x}$, such that

(a) $\phi_t(x)$ is defined and in U, $\forall x \in U, t > o$

(b) There is a Euclidean norm on R^n such that

$$|\phi_t(x) - \bar{x}| \le e^{-tc} |x - \bar{x}| \forall x \in U, t \ge o$$

(c) For any norm on R^n, there is a constant $B>o$ s.t.

$$|\phi_t(x) - \bar{x}| \le B \ e^{-tc} |x - \bar{x}| \forall x \in U, t \ge o$$

In particular, $\phi_t(x) \to \bar{x}$ as $t \to \infty, \forall x \in U$.

Proof: Without loss of generality, we can replace $\bar{x}$ by o.

We are thus given that o is a sink of the diff. equation $X' = f(x)$.

We shall denote the operator $DF(o)$ by a.

Thus we are given that for all Eigen value λ of A, $R_e \lambda < -c$.

Let $b>o$ be chosen in such a way that

$R_e \lambda < -b < -c$, $\forall$ Eigen value λ of A

We can thus find a basis of R^n such that w. r. t. the norm and inner product in this basis, we have

$$< Ax, x > \le -b |x|^2, \forall x \in R^n.$$

Using the definition of the derivative of F at o, we know that

$$\lim_{|x| \to o} \frac{|F(x) - Ax|}{|x|} = 0 \quad \{F(o) = 0$$

NON-LINEAR SINK

i.e. $\displaystyle \lim_{|x|\to o} \frac{|F(x)-Ax|\cdot|x|}{|x|^2}=0$

since $-b < -c$, we can find $\epsilon>0$ such that $-b < -b+\epsilon < -c$

We choose such an ϵ and for this $\epsilon>0$, there will exist a $\delta>0$ such that

$|x|<\delta \to \dfrac{|F(x)-A\cdot x|\cdot|x|}{|x|^2}<\epsilon$

We choose δ small enough so that $|x|<\delta \Rightarrow x \in W$

Consider $<f(x)\text{-}Ax.\ x>$

We know that,

$|<F(x)-Ax,x>| \leq |F(x)-Ax|\cdot|x|$

$\therefore \dfrac{|<F(x)-Ax,x>|}{|x|^2} \leq \dfrac{|F(x)-Ax|\cdot|x|}{|x|^2}$

For $|x|<\delta$, we have

$\dfrac{|<F(x)-Ax,x>|}{|x|^2}<\epsilon$

i.e. $<F(x),x>-<Ax,x><\epsilon\cdot|x|^2$

i.e. $<F(x),x><\epsilon\cdot|x|^2+<Ax,x>$

$<\epsilon\cdot|x|^2-b|x|^2<-c|x|^2$

Let x(t) be a solution to the given diff. equation

$|x(t)|=\sqrt{<x(t),x(t)>}$

$\dfrac{d}{dt}|x(t)|=\dfrac{<x'(t),x(t)>}{|x(t)|}=\dfrac{<f(x(t)),x(t)>}{|x(t)|}$

Let us choose an initial value x_o s.t. $|x_o|<\delta$

Let U = {x such that $|x|\leq\delta$}

303

Since the initial value is in U, it follows that for same $t_o > 0, t \in [o, t_o]$ will imply that the solution curve $x(t) \in U$

We show that $x(t) \in U, \forall t \geq 0$

Thus, for $t \in [o, t_o]$, we can use the estimate

$$< f(x(t)), x(t) > \leq -c \cdot |x(t)|^2$$

$$\therefore \frac{d}{dt} |x(t)| = \frac{< f(x(t)), x(t) >}{|x(t)|} < \frac{c |x(t)|^2}{|x(t)|} = -c |x(t)|$$

$$\therefore \frac{d}{dt} |x(t)| < -c |x(t)|$$

i.e. $\dfrac{d |x(t)|}{|x(t)|} < -c \, dt$

$$\Rightarrow |x(t)| < e^{-ct} \cdot |x(o)|$$

{But we know that, $W \subseteq E$ be open,

let $F : W \to E$ be a C' - map.

Let y(t) be a solution on a maximal open interval $J = (\alpha, \beta) \subseteq R$ with $\beta < \infty$.

Thus given any compact set $K \subseteq W$, there is some $t \in (\alpha, \beta)$ s.t. $y(t) \notin k$ }

But as $x_o \in U$ and $x(t) \in U, \forall t \in [o, t_o]$ and U is a compact subset of W, it follows that the solution x(t) must be defined $\forall$ time $t \geq o$.

Further $\forall t \geq o, x(t) \in U$

Thus if $x_o \in U$, it follows that $x(t) \in U, \forall t \geq o$

i.e. $x_o \in U \Rightarrow \phi_t(x_o) \in U$

i.e. $x \in U \Rightarrow \phi_t(x) \in U$ and $\phi_t(x)$ is defined, $\forall t \geq o$.

Also, $|x(t)| \leq e^{-tc} |x_o|$

NON-LINEAR SINK

$$\Rightarrow |\phi_t(x_o)| \le e^{-tc} |x_o|$$

Thus the statement (i) and (ii) holds.

Let $|\cdot|_\beta$ be any norm. due to equivalence of norms we can find, +ve constants say S and T s.t.

$$s|\cdot| \le |\cdot|_\beta \le T|\cdot|$$

$$|\phi_t(x) - \bar{x}|_\beta \le T |\phi_t(x) - \bar{x}| \le T e^{-tc} |x - \bar{x}|$$

$$\le T e^{-tc} |x - \bar{x}|_{\beta \cdot \frac{1}{s}}$$

$$\le B e^{-t/c} |x - \bar{x}|_\beta$$

PHYSICAL EXAMPLE OF A NON-LINEAR SINK

Consider a pendulum consisting of a bob of mass say in attached at the end of (weight less) the rod of length say l.

The forces getting on the bob are given by weight (let g be the accl^n due to gravity) and frictional force proportional to velocity.

We describe the bob by the angular displacement θ from the downward vertical measured in anticlockwise direction sign conversions for anticlockwise division θ is +ve for clockwise division θ is –ve.

If $\phi(t)$ is the angular description of the bob at time t, then at time t, its angular velocity is θ. The linear velocity induced by the angular velocity is $l\theta$.

It follows that the frictional force acting on the bob is : $-k\,l\dfrac{dl}{dt}$, where k is constant.

The force due to the weight of the bob tangential to the motion will be –mg sinϑ.

DYNAMICAL SYSTEM – A SHORT COURSE

The accl^n of the bob is: $l\dfrac{d^2\theta}{dt^2}$

Thus, by the 2nd law of motion, we have

$$ml\ddot{\theta} = -kl\dot{\theta} - mg\sin\theta$$

i.e. $\ddot{\theta} = -\dfrac{k}{m}\dot{\theta} - \dfrac{g}{l}\sin\theta$

If we use a new variable say w and denote $\dot{\theta}$ be w then this 2nd order diff. equation is equivalent to the system of first order diff. equation namely

$$\dot{\theta} = w;\ \dot{w} = -\dfrac{g}{l}\sin\theta - \dfrac{k}{m}w$$

It is a non-linear autonomous system with

$$F(\theta, w) = \big(F_1(\theta, w), F_2(\theta, w)\big)$$

with $F_1(\theta, w) = w;\ F_2(\theta, w) = -\dfrac{g}{l}\sin\theta - \dfrac{k}{m}w$

The pts. of equilibria are given by

$\theta = n\pi, n = o, \pm 1, \pm 2, \cdots$ and w=o

To consider the behaviour of the system at pts. of equilibria, we obtain $Df(\theta, w)$.

If f is a C-map from R^n to R^m and $\{e_1, e_2, \cdots e_n\}, \{u_1, u_2, \cdots u_n\}$ are the basis of R^n to R^m respectively then at any pt. $\bar{x}$ of R^n,

$$DF(\bar{x})\,\bar{e}j = \sum_{i=1}^{m} Dj\, Fi(\bar{x})\bar{U}i$$

The matrix of the operator $Df(\theta, w)$ is given by

$$\begin{bmatrix} \dfrac{\partial f_1}{\partial\theta} & \dfrac{\partial f_1}{\partial w} \\ \dfrac{\partial f_2}{\partial\theta} & \dfrac{\partial f_2}{\partial w} \end{bmatrix}$$

NON-LINEAR SINK

$$\therefore Df(\theta, w) = \begin{bmatrix} 0 & 1 \\ \dfrac{-g}{l}\cos\theta & \dfrac{-k}{m} \end{bmatrix}$$

We consider the behaviour of the system at the pts. of equilibrium (o, o)

$$Df(0,0) = \begin{bmatrix} 0 & 1 \\ \dfrac{-g}{l} & \dfrac{-k}{m} \end{bmatrix}$$

The Eigen values are given by

$$\begin{vmatrix} -\lambda & 1 \\ \dfrac{-g}{l} & \dfrac{-k}{m} - \lambda \end{vmatrix} = 0$$

$$-\lambda\left(\frac{-k}{m} - \lambda\right) + \frac{g}{l} = 0$$

$$\Rightarrow \lambda +^2 \frac{k}{m}\lambda + \frac{g}{l} = 0$$

$$\therefore \lambda = \frac{-k/m \ I\sqrt{k^2/m^2 - 4g/l}}{2}$$

It follows that (irrespective whether λ is real or complex) $\operatorname{Re}\lambda < 0$.

Thus this system has a non-linear sink at the equilibrium pt. (0, 0).

We can also consider the behaviour of the system at $(\pi, 0)$.

The matrix of $DF(\pi, 0)$ is: $\begin{bmatrix} 0 & 1 \\ \dfrac{+g}{l} & \dfrac{-k}{m} \end{bmatrix}$

The Eigen values are given by

$$\begin{bmatrix} -\lambda & 1 \\ \dfrac{g}{l} & \dfrac{-k}{m} - \lambda \end{bmatrix} = 0$$

i.e. $\quad \lambda^2 + \dfrac{k}{m}\lambda - \dfrac{g}{l} = 0$

DYNAMICAL SYSTEM – A SHORT COURSE

$$\therefore \lambda = \frac{-k/m \pm \sqrt{k^2/m^2 - 4g/l}}{2}$$

In this case, both the Eigen values are real, one is +ve and other is –ve.

Definition: Suppose $\bar{x} \in W$ is an equilibrium of the diff. equation $X' = F(x)$, where $F: W \to E$ is a $C'-map$ from an open set W of the normed linear space E into E. then $\bar{x}$ is said to be a stable equilibrium if for every nbd U_1 of $\bar{x}$ in U such that every solution $x(t)$ with $x(o)$ in U_1 is defined and in U for all t>o.

Definition: If U_1 can be chosen so that in addition to the properties described in the previous definition $\lim_{t\to\infty} x(t) = \bar{x}$ then $\bar{x}$ is asymptotically **stable**.

Thus a pt. of equilibrium is asymptotically stable if it is stable and solution x(t) approaches $\bar{x}$ as $t \to \infty$.

Definition : An equilibrium $\bar{x}$ that is not stable is called **unstable**.

If all the Eigen values of the operator $DF(\bar{x})$ have –ve real part then the equilibrium pt. $\bar{x}$ is not just stable but asymptotically stable.

If all the Eigen values of the operator $DF(\bar{x})$ are purely imaginary, then we get an example of a stable equilibrium which is not asymptotically stable.

(1) Thus if λ_1, λ_2 are Eigen values of the operator A such that $\lambda_1, \lambda_2 < 0$ then any component of the solution is of the type $c_1 e^{\lambda_1 t} + c_2 e^{\lambda_1 t}$.

Which can be chosen to be sufficiently close to the origin and $\lim_{t\to\infty} c_1 e^{\lambda_1 t} + c_2 e^{\lambda_1 t} \to 0$.

We thus get origin as asymptotically stable equilibrium.

NON-LINEAR SINK

(2) If the Eigen values of A are say $\pm ib, b \neq 0$ then any component of the solution is of the type $c_1 \cos bt + c_2 \sin bt$

Which will always remain bounded, if we choose c_1 and c_2 such that $|G| + |c_2| < \delta$ then for all $t \geq 0, |c_1 \cos bt + c_2 \sin bt| < \delta$,

However, $\lim_{t \to \infty} c_1 \cos bt + c_2 \sin bt \to 0$.

Thus, in this case 0 is a stable Equilibrium which is not asymptotically stable.

If the Eigen value of A are say λ_1, λ_2 with $\lambda_2 < 0 < \lambda_1$ then even if $|c_1| + |c_2| < \delta$,

$|c_1 e^{\lambda_1 t} + c_2 e^{\lambda_1 t}| \nless \delta$, after some time t provides $c_1 \neq 0$.

Thus O is an unstable equilibrium in this case.

Theorem 2: Let $W \subseteq E$ be open and $F : W \to E$ continuously differentiable. Suppose $F(\bar{x}) = 0$ and $\bar{x}$ is a stable equilibrium pt. of the equation $X' = F(x)$. Then no Eigen value of $DF(\bar{x})$ has +ve real part.

Proof: We assume that at least one Eigen value of $DF(\bar{x})$ has a +ve real root.

Without loss of generality, we can assume that $\bar{x} \equiv 0$ (We may replace F by $f(x - \bar{x})$)

Thus we are assuming that $DF(0)$ has at least one Eigen value with +ve real part.

Clubbing all the Eigen values with positive real part together, we can write

$$E = E_1 \oplus E_2$$

where $DF(0)|_{E_1}$ will have Eigen values with +ve real part while $DF(0)|_{E_2}$ will have Eigen values with −ve or zero real part.

We denote $DF(0)|_{E_1}$ by A and $DF(0)|_{E_2}$ by B.

Thus, $DF(0) = A \oplus B$

Consider $A \in L(E_1)$ since all the Eigen values of A have +ve real part, we can find $0 < a < \text{Re}\,\lambda$, where λ is any Eigen value of A.

Hence we can find a basis of E_1 such that with respect to the inner product and norm defined with this basis, we have

$a \,|\, x\,|^2 \le\, < Ax, x >, \forall x \in E_1$

Similarly, considering the operator B on E_2. Since all the Eigen values of B have –ve or zero real part then for any b>o, $\text{Re}\,\lambda < b$, for Eigen value λ of B.

Hence there exist a basis of E_2 such that with respect to the inner product and norm defined with this basis, we have

$< BY, Y > \, < b \,|\, Y\,|^2, \forall y \in E_2$

We choose $b>0$ in such a way that $o<b<a$.

The union of these basis of E_1 and E_2 will form a basis of E.

Further any pt. $z \in E$ will be uniquely expressed as

$z = x + y$ with $x \in E_1, Y \in E_2$

or

We may write, $z=(x, y)$

On the space E we define an inner product in such a way that it is a sum of inner product on E_1 and E_2.

Further we define a norm on E obtained from the norm of the component of E_1 and E_2.

Thus, if $z_1 = x_1 \oplus y_1 = (x_1, y_1)$

NON-LINEAR SINK

$$z_2 = x_2 \oplus y_2 = (x_2, y_2)$$

Then $<z_1, z_2> = <x_1, x_2> + <y_1, y_2>$

And $|z_1| = \sqrt{|x_1|^2 + |y_1|^2}$

As F is C' and $F(o) = 0$

We have

$F(x, y) = F(o, o) + DF(o)(x, y)$ + terms of higher

$\qquad = o + Ax + BY$ + terms of higher order

$\qquad = Ax + BY + R(x, y) + S(x, y)$

where $R(x, y)$ is the component of remainder in E_1 and $S(x, y)$ is the component of remainder in E_2.

Thus,

$$F(x, y) = \left(Ax + R(x, y), BY + S(x, y) \right)$$

$$\qquad = \left(F_1(x, y), F_2(x, y) \right), \text{ where}$$

$F_1(x, y) \equiv Ax + R(x, y), F_2(x, y) \equiv BY + S(x, y)$

i.e. $F(g) = \left(F_1(g), F_2(g) \right)$

and we shall denote $\left(R(x, y), S(x, y) \right)$ by $Q(z)$.

since $DF(o)$ is the derivative of F at o. we know that

$$\lim_{|(x,y)| \to o} \frac{|F(x, y) - DF(o)(x, y)|}{|(x, y)|} = 0.$$

i.e. $\qquad \lim_{|(x,y)| \to o} \dfrac{|Q(x, y)|}{|(x, y)|} = 0$

i.e. $\qquad \lim_{|(z)| \to o} \dfrac{|Q(z)|}{|z|} = 0$

i.e. for a given $\in> o, \exists\, \delta > o$ such that

311

$|z| \leq \delta \Rightarrow |Q(z)| < \in$

For $(x, y) \in (E_1 \times E_2)$, we define the cone C.

$C = \{z \in E \ i.e. \ (x, y) \in E_1 \times E_2 \ s.t. |x| \geq |y|\}$

We now show that we can find $\in > o$ and (hence $\delta > o$) such that if U denotes

$\{z \text{ such that } |z| \leq \delta\}$ then $\forall z \in C \cap U$, we have

$< x, F_1(x, y) > - < y, F_2(x, y) > > 0$, for $x \neq 0$

We know that

$F_1(x, y) = Ax + R(x, y), F_2(x, y) = BY + S(x, y)$

$\therefore < x, F_1(x, y) > - < y, F_2(x, y) >$

$\quad = < x, Ax > + < x, R(x, y) > - < y, BY > - < y, S(x, y) >$

Consider

$|< x, R(x, y) > - < y, S(x, y) >| \leq |< x, R(x, y) > + |< y, S(x, y) >|$

$\leq |x| \cdot |R(x, y)| + |y| \cdot |S(x, y)|$

L$\rightarrow$ by Cauchy Schwarz

$\leq |z| \cdot |R(x, y)| + |z| \cdot |S(x, y)|$

As $Q = (R, S), |Q| \geq |R|$ and $|Q| \geq |S|$

$\therefore |< x, R(x, y) > - < y, S(x, y) >| \leq 2. |z| . |Q(z)|$

$\therefore -2. |z| . |Q(z)| \leq < x, R(x, y) > - < y, S(x, y) > \leq 2 |z| . |Q(z)|$

$< x, R(x, y) > - < y, S(x, y) > \geq -2 |z| . |Q(z)|$

Now,

$< x, F_1(x, y) > - < y, F_2(x, y) > \cdot \geq < x, Ax > - < y, BY - 2 |z| | Q(z)|$

As $< Ax, x > \geq a |x|^2, < y, BY > \leq b |y|^2$

If $z \in C \cap U$, we have $|z| \leq \delta$ and hence for such a z,

NON-LINEAR SINK

$|Q(z)| < \in \cdot |z|$

$\therefore$ for $z \in C \cap U$, we have

$<x, F_1(x,y)> - <y, F_2(x,y)> \geq a|x|^2 - b|y|^2 - 2 \in |z|^2$

As $z \in C \cap U$, we have $|x| \geq |y|$

$\therefore |x|^2 = \dfrac{|x|^2 + |x|^2}{2} \geq \dfrac{|x|^2 + |y|^2}{2} = \dfrac{|z|^2}{2}$

Similarly, $|y|^2 = \dfrac{|y|^2 + |y|^2}{2} \leq \dfrac{|x|^2 + |y|^2}{2} = \dfrac{|z|^2}{2}$

$\therefore a|x|^2 \geq a\dfrac{|z|^2}{2}, -b|y|^2 \geq -b\dfrac{|z|^2}{2}$

Thus for $z \in C \cap U$.

$<x, F_1(x,y)> - <y, F_2(x,y)> \geq a\dfrac{|z|^2}{2} - b\dfrac{|z|^2}{2} - 2 \in |z|^2$

$$= \left(\frac{a}{2} - \frac{b}{2} - 2 \in \right) |z|^2$$

Since $o < b < a$ and b can be chosen arbitrarily and $\in > o$ is also arbitrary, we choose then in such a way that $\dfrac{a}{2} - \dfrac{b}{2} - 2 \in > 0$.

Let α be such that

$o < \alpha < \dfrac{a}{2} - \dfrac{b}{2} - 2 \in$.

Thus, for all $z \in C \cap U$, (where $U = \{z \| z| < \delta\}$)

$<x, F_1(x,y)> - <y, F_2(x,y)> > \alpha |z|^2$

Thus with $z \in C \cap U$, with $z = (x, y)$ and $x \neq o$. then

$<x, F_1(x,y)> - <y, F_2(x,y)> > o$ \hfill (6.1)

Further as $|z|^2=|x|^2+|y|^2\geq\left||x|^2-|y|^2\right|$

Thus for $z\in C\cap U$, we have

$$<x,F_1(x,y)>-<y,F_2(x,y)>>\alpha|z|^2\geq\alpha\left||x|^2-|y|^2\right| \tag{6.2}$$

We now define a new function $g:E_1\times E_2\to R$ as

$$g(x,y)=\frac{1}{2}\left(|x|^2-|y|^2\right),g\in C'.$$

It is clear that $g^{-1}([o,\infty)=C$ and $g'(o)=$ boundary of C.

Consider $Dg\ (x,\ y)$.

$Dg(x,\ y)$ will be given by the matrix

$$\left[<x,x'>-<y,y'>\right]$$

Therefore

$$Dg(x,y)\left(F(x,y)\right)=<x,F_1(x,y)>-<y,F_2(x,y)>$$

Thus if $(x,y)\in C\cap U$, we have

$$Dg(x,y)\left(F(x,y)\right)>0 \tag{6.3}$$

Hence if $z(t)$ is the solution curve of the diff. equation then along this solution curve in $C\cap U$

We have

$$\frac{d}{dt}g(z(t))=Dg(z(t))\cdot z'(t)=Dg(z(t))\cdot F(z)>0.$$

But in $C\cap U$,

$$g(z(t))=\frac{1}{2}\left(|x(t)|^2-|y(t)|^2\right)$$

Thus if $g(t)\in C\cap U$, then it cannot go outside $C\cap U$ while still being in U

i.e. for the solution curve z(t) to go outside $C\cap U$, it has to go outside U.

Thus if $z(t)$ is a solution curve of the diff. equation $z'=F(z)$ which

NON-LINEAR SINK

initially is in $C \cap U$ and if it is defined only over a finite interval of time say $[o, t_o]$, then as $C \cap U$ is a compact set, there must be exist some $t \in [o, t_o]$ such that $z(t) \notin C \cap U$.

That is the solution curve $z(t)$ which was initially in $C \cap U$, has to go outside $C \cap U$ at time t.

By the above argument, at t, $z(t) \notin U$.

We now consider the case where z(t) i.e. defined for $0 \le t < \infty$.

If initially, we have $z(t) \notin C \cap U$, then by inequality (2),

$$< \propto (t), F_1(x(t), y(t)) > - < y(t), F_2(x(t), y(t)) >> \alpha \mid x(t) \mid^2 - \mid y(t) \mid^2$$

But

$$< x(t), f_1(x(t), y'(t) > -, y(t), f_2(x(t), y(t)) >$$

$$= \frac{d}{df} g(x(t), y(t)) = \frac{d}{dt} \left(\frac{\mid x(t) \mid^2 - \mid y(t) \mid^2}{2} \right)$$

Thus we have

$$\frac{1}{2} \frac{d}{dt} \left(\mid x(t) \mid^2 - \mid y(t) \mid^2 \right) > \alpha \left\| x(t) \mid^2 - \mid y(t) \mid^2 \right|$$

If we consider the region in $C \cap U$, where $\mid x(t) \mid > \mid y(t) \mid$

We have, $\frac{d}{dt} \left(\mid x(t) \mid^2 - \mid y(t) \mid^2 \right) > 2\alpha \left(\mid x(t) \mid^2 - \mid y(t) \mid^2 \right)$

$$\Rightarrow \frac{\frac{d}{dt} \left(\mid x(t) \mid^2 - \mid y(t) \mid^2 \right)}{\mid x(t) \mid^2 - \mid y(t) \mid^2} > 2\alpha$$

On integration, we get

$$\mid x(t) \mid^2 - \mid y(t) \mid^2 > e^{2\alpha} \left(\mid x(o) \mid^2 - \mid y(o) \mid^2 \right)$$

As $\mid x(o) \mid^2 - \mid y(o) \mid^2 > 0$ and $\alpha > 0$ as $t \to \infty$

315

$|x(t)|^2 - |y(t)|^2$ becomes unbounded.

i.e. $|z(t)|$ becomes unbounded.

i.e. $z(t) \notin U$ after some time

we have thus shown that if we choose a solution curve in $C \cap U$ sufficiently close to origin initially then it goes outside the nbd U of an origin.

This contradicts the assumption that O is a state equilibrium.

Thus our assumption must be wrong.

Hence no Eigen value of $DF(\bar{x})$ has +ve real part.

DEFINITION

For the diff. equation $x' = f(x)$, an equilibrium pt. $\bar{x}$ is said to be hyperbolic if the derivative $DF(\bar{x})$ has no Eigen values with real part zero.

Theorem 3: A hyperbolic equilibrium point is either unstable or asymptotically stable.

Proof: If $\bar{x}$ is a hyperbolic equilibrium pt. of $x' = f(x)$ then there are two possibilities.

In one possibility, no. Eigen value of $DF(\bar{x})$ has positive real part.

In other possibility, there is at least one Eigen value of $DF(\bar{x})$ that has a +ve real part.

In the first possibility all Eigen values of $DF(\bar{x})$ will have negative real part.

Thus $\bar{x}$ is a sink i.e. asymptotically stable equilibrium.

In the other possibility $DF(\bar{x})$ has at least one Eigen value with +ve real

and as such $\bar{x}$ can not be a stable equilibrium and so not asymptotically equilibrium.

Hence $\bar{x}$ is an unstable equilibrium.

NON-LINEAR SINK

Definition: Let $\bar{x}$ be a hyperbolic equilibrium pt. of the diff. equation $x' = f(x)$. Then there are three situations:

1) If all the Eigen values of the operator $DF(\bar{x})$ have $-$ve real part then $\bar{x}$ is known as a sink.

2) If all the Eigen values of $DF(\bar{x})$ have $+$ve real part and $\bar{x}$ is said to be a source.

3) If some Eigen values have $+$ve real part and some have $-$ve real part then $\bar{x}$ is said to be a saddle point.

If $\bar{x}$ is an asymptotically stable point of equilibrium then by definition there is a nbd say N of $\bar{x}$ such that once the solution curve meets N, if not only remains in N for future time but approaches $\bar{x}$ as $t \to \infty$.

It is thus clear that any solution curve $x(t)$ of $x' = f(x)$, if meets N then $x(t) \to \bar{x}$.

The union of all solution curves which tends to $\bar{x}$ as $t \to \infty$ is called the basin of $\bar{x}$ and denoted by $B(\bar{x})$.

It is clear that any solution curve which meets N is in $B(\bar{x})$.

Conversely, any solution curve in $B(\bar{x})$ must meet N. It follows that the set $B(\bar{x})$ is open.

Further if $\bar{x}$ and $\bar{y}$ are two distinct asymptotically stable equilibrium pt. then $B(\bar{x})$ and $B(\bar{y})$ are disjoint.

Ex 1. For which of the following Linear operators A on R^n is $O \in R^n$, a stable equilibrium of $x' = Ax$.

i) $A = 0$

(ii) $\begin{bmatrix} 0 & -1 & & \\ 1 & 0 & & \\ & & 0 & 1 \\ & & -1 & 0 \end{bmatrix}$ (iii) $\begin{bmatrix} 0 & -1 & 0 & 0 \\ 1 & 0 & 0 & 0 \\ 1 & 0 & 0 & 1 \\ 0 & 0 & -1 & 0 \end{bmatrix}$ (iv) $\begin{bmatrix} 0 & 1 \\ 0 & 0 \end{bmatrix}$ (v) $\begin{bmatrix} 1 & 2 \\ 2 & -2 \end{bmatrix}$

Solution : (i) As A = 0, every $\bar{x} \in R^n$ is an equilibrium point.

Since the diff. equation is $x' = 0$, any solution curve $x(t)$ that passes through

a given pt. say $\bar{x}$ of R^n is given by

$x(t) \equiv \bar{x}, \forall t \geq 0$

Thus the solution curves are singleton sets.

If U is any nbd of zero then any solution curve $x(t)$ which initially is in U.

i.e. $x(t) = \bar{x}$, for some $\bar{x} \in U$

x(t) remains at $\bar{x}$ i.e. in U, $\forall t \geq 0$.

Thus 0 is a stable equilibrium.

As $|x(t)| - 0| = |x(0) - 0|$ remains fixed $\forall t$.

$\therefore 0$ is not asymptotically stable.

(ii) Since A is diag $\{A_1, A_2\}$

Where $A_1 = \begin{bmatrix} 0 & -1 \\ 1 & 0 \end{bmatrix}$ and $A_2 = \begin{bmatrix} 0 & 1 \\ -1 & 0 \end{bmatrix}$

The system $x' = Ax$ can be decoupled

i.e. we can write

$x_1' = -x_2 \qquad x_3' = -x_4$
$x_2' = -x_1 \qquad x_4' = -x_3$

Thus we have

$x_1 = a_1 \cos t + b_1 \sin t$
$x_2 = a_1 \sin t - b_1 \cos t$

$x_3 = c_1 \cos t + d_1 \sin t$
$x_4 = c_1 \sin t - d_1 \cos t$

NON-LINEAR SINK

Thus $(0, 0)$ a stable equilibrium point but not asymptotically equilibrium.

iii) The Eigen values of the operator A is given by

$$\begin{vmatrix} -\lambda & -1 & 0 & 0 \\ 1 & -\lambda & 0 & 0 \\ 1 & 0 & -\lambda & 1 \\ 0 & 0 & -1 & -\lambda \end{vmatrix} = 0$$

$$\therefore -\lambda[-\lambda(\lambda^2+1)] + 1[1(\lambda^2+1)] = 0$$

$$\therefore \lambda^2(\lambda^2+1) + 1(\lambda^2+1) = 0$$

$$\therefore (\lambda^2+1)(\lambda^2+1) = 0$$

$$\therefore \lambda = \pm i, \pm i$$

$\therefore$ The Eigen values are $\pm i$ with multiplicity 2.

The components of the solution will be of the form

$$c_1 \cos t + c_2 \sin t + t(c_3 \cos t + c_4 \sin t)$$

Thus $(0, 0)$ is unstable equilibrium.

iv) The Eigen values are given by

$$\begin{vmatrix} -\lambda & 1 \\ 0 & -\lambda \end{vmatrix} = 0$$

$$\Rightarrow \lambda^2 = 0$$

$$\Rightarrow \lambda = 0, 0$$

The Eigen value is 0 with multiplicity 2.

The components of the solution will be of the form.

$$c_1 + t\, c_2$$

Thus, $(0, 0)$ is an unstable equilibrium.

v) The Eigen values are given by

$$\begin{vmatrix} 1-\lambda & 2 \\ 2 & -2-\lambda \end{vmatrix} = 0$$

$\therefore (1-\lambda)(-2-\lambda)-4=0$

$\Rightarrow -2-\lambda+2\lambda+\lambda^2 -4=0$

$\Rightarrow \lambda^2 +\lambda-6=0$

$(\lambda+3)(\lambda+2)=0$

$\therefore \lambda =-3,2$

The components of the solution will be of the form

$c_1 e^{-3t} +c_2 e^{2t}$

Thus, (0, 0) is an unstable equilibrium.

Ex 2. Discuss the behaviour of equilibrium pts. of the non-linear system

$x' = f(x)$, where

i) $\begin{bmatrix} x_1 - x_1 x_2 \\ x_2 - x_1^2 \end{bmatrix}$ (ii) $\begin{bmatrix} -4x_2 + 2x_1 x_2 -8 \\ 4x_2^2 - x_1^2 \end{bmatrix}$

iii) $\begin{bmatrix} 2x_1 - 2x_1 x_2 \\ 2x_2 - x_1^2 + x_2^2 \end{bmatrix}$ (iv) $\begin{bmatrix} -x_1 \\ -x_2 + x_1^2 \\ x_3 + x_1^2 \end{bmatrix}$ v) $\begin{bmatrix} x_2 - x_1 \\ kx_1 + x_2 - x_1 x_3 \\ x_1 x_2 - x_3 \end{bmatrix}$

Solution: (i) $\begin{bmatrix} x_1 - x_1 x_2 \\ x_2 - x_1^2 \end{bmatrix}$

$f_1(x_1,x_2) = x_1 - x_1 x_2$

$f_2(x_1,x_2) = x_2 - x_1^2$

$\therefore$ Points of equilibrium are given by

$x_1 - x_1 x_2 =0 \Rightarrow x_1(1-x_2) = 0$

$x_2 - x_1^2 =0 \Rightarrow x_1 - x_1' = 0$

The equilibrium points are given by (0, 0), (1, 1) and (-1, 1).

Further with $f_1(x_1,x_2) = x_1 - x_1 x_2$

$f_2(x_1,x_2) = x_2 - x_1^2$

$$\therefore Df(x_1, x_2) = \begin{bmatrix} \dfrac{\partial f_1}{\partial x_1} & \dfrac{\partial f_1}{\partial x_2} \\ \dfrac{\partial f_2}{\partial x_1} & \dfrac{\partial f_2}{\partial x_2} \end{bmatrix} = \begin{bmatrix} 1 - x_2 & -x_1 \\ -2x_1 & 1 \end{bmatrix}$$

Then $Df(0,0) = \begin{bmatrix} 1 & 0 \\ 0 & 1 \end{bmatrix}$

$\therefore$ Eigen values are given by $\begin{bmatrix} 1-\lambda & 0 \\ 0 & 1-\lambda \end{bmatrix} = 0$

$\therefore (1-\lambda)^2 = 0 \Rightarrow \lambda = 1, 1$

$\therefore$ Eigen values are +1 source and it is unstable.

Now, $Df(1,1) = \begin{bmatrix} 0 & -1 \\ -2 & 1 \end{bmatrix}$

$\therefore$ Eigen values are given by $\begin{bmatrix} -\lambda & -1 \\ -2 & 1-\lambda \end{bmatrix} = 0$

$\therefore -\lambda(1-\lambda) - 2 = 0$

$\therefore \lambda^2 - \lambda - 2 = 0$

$\therefore (\lambda - 2)(\lambda - 1) = 0$

$\therefore \lambda = -1$ and 2.

Eigen values are -1 and 2

$\therefore (1, 1)$ is saddle point and it is unstable.

$Df(-1,1) = \begin{bmatrix} 0 & 1 \\ 2 & 1 \end{bmatrix}$

$\therefore$ Eigen values are given by $\begin{bmatrix} -\lambda & 1 \\ 2 & 1-\lambda \end{bmatrix} = 0$

$\therefore -\lambda(1-\lambda) - 2 = 0$

$\therefore \lambda^2 - \lambda - 2 = 0$

$\therefore (\lambda - 2)(\lambda - 1) = 0$

321

$\therefore \lambda = -1, 2.$

$\therefore$ Eigen values are-1 and 2.

$\therefore (1, 1)$ is saddle point & it is unstable.

iv) $\begin{bmatrix} -4x_2 + 2x_1x_2 - 8 \\ 4x_2^2 - x_1^2 \end{bmatrix}$

Here $f_1(x_1, x_2) = -4x_2 + 2x_1 x_2 - 8$

$f_2(x_1, x_2) = 4x_2^2 - x_1^2$

Points of Equilibrium are given by

$\Rightarrow -4x_2 + 2x_1x_2 - 8 = 0$

$\Rightarrow -(2x_2) - x_1)(2x_2 + x_1) = 0 = 0$

$\Rightarrow 2x_2 - x_1 = 0; \ 2x_2 + x_1 = 0 = 0$

$\Rightarrow x_1 = 2x_2 ; \ x_1 = -2x_2$

Put $\frac{1}{2}x_1 = 2x_2$ i.e. $x_2 = \frac{1}{2}x_1$ in $f_1(x_1, x_2)$, we get

$-4\left(\frac{1}{2}x_1\right) + 2x_1\left(\frac{1}{2}x_1\right) - 8 = 0$

$\Rightarrow -2x_1 + x_1^2 - 8 = 0$

$\Rightarrow x_1^2 - 2x_1 - 8 = 0$

$\therefore x_1 = \dfrac{-(-2) \pm \sqrt{4 - 4 \times | \times \in (-8)}}{2 \times 1} = \dfrac{2 \pm \sqrt{36}}{2}$

$\therefore x_1 = 4, -2$

(i) $\Rightarrow x_2 = \frac{1}{2}x_1 \Rightarrow x_2 = 2$

$x_2 = \frac{1}{2} \times (-2) \Rightarrow x_2 = -1$

When we put $x_2 = \frac{1}{2}x_1$ in $f_1(x_1, x_2)$, we get

322

$$-4\left(-\frac{1}{2}\right)x_1 + 2x_1\left(-\frac{1}{2}x_1\right) - 8 = 0$$

$$2x_1 + x_1^2 - 8 = 0$$

$$\Rightarrow x_1^2 - 2x_1 + 8 = 0$$

$$\therefore x_1 = \frac{2 \pm \sqrt{2 - 4 \times 1 \times 8}}{2} = \frac{2 \pm \sqrt{-7}}{2} = 2 \pm \sqrt{-7}$$

The pts of equilibrium are (4, 2) and (-2, -1)

Further, $f_1 = -4x_2 + 2x_1x_2 - 8$; $f_2 = 4x_2^2 - x_1^2$

$$\therefore Df(x_1, x_2) = \begin{bmatrix} \dfrac{\partial f_1}{\partial x_1} & \dfrac{\partial f_1}{\partial x_2} \\ \dfrac{\partial f_2}{\partial x_1} & \dfrac{\partial f_2}{\partial x_2} \end{bmatrix} = \begin{bmatrix} 2x_2 & -4 + 2x_1 \\ -2x_1 & 8x_2 \end{bmatrix}$$

Then $Df(4, 2) = \begin{bmatrix} 4 & 4 \\ -8 & 16 \end{bmatrix}$

$\therefore$ Eigen values are given by $\begin{bmatrix} 4 - \lambda & 4 \\ 8 & 16 - \lambda \end{bmatrix} = 0$

$$\Rightarrow (4 - \lambda)(16 - \lambda) + 32 = 0$$

$$\Rightarrow 64 - 4\lambda - 16\lambda + \lambda^2 + 32 = 0$$

$$\Rightarrow \lambda^2 - 20\lambda + 96 = 0$$

$$\therefore \lambda = 20 \pm \sqrt{\frac{400 - 4 \times 96}{2}}$$

$$= 20 \pm \sqrt{\frac{400 - 384}{2}} = \frac{20 \pm \sqrt{16}}{2} = \frac{20 \pm 4}{2}$$

$$\lambda = 12, 8$$

$\therefore$ The equilibrium pt. (4, 2) is a source and it is unstable.

Also, $Df(-2,-1) = \begin{bmatrix} -2 & -8 \\ 4 & -8 \end{bmatrix}$

$\therefore$ The Eigen values are given by $\begin{bmatrix} -2-\lambda & -8 \\ 4 & -8-\lambda \end{bmatrix} = 0$

$\Rightarrow (2+\lambda)(8+\lambda)+32=0$

$\Rightarrow 16+10\lambda+\lambda^2+32=0$

$\Rightarrow \lambda^2+10\lambda+48=0$

$\therefore \lambda = \dfrac{-10\pm\sqrt{100-4\times48}}{2} = \dfrac{-10\pm\sqrt{-92}}{2}$

$\lambda = -5\pm\sqrt{23}i$

$\therefore$ (-2, -1) is a sink and it is stable.

(iii) $\begin{bmatrix} 2x_1 - 2x_1 x_2 \\ 2x_2 - x_1^2 + x_2^2 \end{bmatrix}$

Here $f_1(x_1,x_2) = 2x_1 - 2x_1 x_2$

$\qquad\qquad f_2(x_1,x_2) = 2x_2 - x_1^2 + x_2^2$

The pts. of Equilibrium are given by

$\therefore 2x_1 - 2x_1 x_2 = 0 \Rightarrow 2x_1(x_1 - x_2) = 0$

$x_1 = 0, x_2 = 1$

Put $x_1 = 0$ in $f_2(x_1 - x_2)$, we get

$2x_2 + x_2^2 - x_1^2 = 0 \Rightarrow x_2(2+x_2)-0 = 0$

$x_2 = 0, 2+x_2 = 0$

$x_2 = 0, x_2 = -2$

$\therefore$ (0, 0), (0, -2) are equilibrium pts. for $x_1 = 0$

For $x_2 = 1$,

$f_2(x_1 - x_2) = 2x_2 - x_1^2 + x_2^2 = 0$

$\Rightarrow 2 - x_1^2 + 1 = 0$

$\Rightarrow x_1^2 = 3 \quad \therefore x_1 = \pm\sqrt{3}$

$\therefore \left(\sqrt{3},1\right)$ and $\left(-\sqrt{3},1\right)$ are equilibrium pts. for $x_2 = 1$

$\therefore$ The equilibrium pts. are $(0, 0)$, $(0, -2)$, $\left(\sqrt{3},1\right)$ and $\left(-\sqrt{3},1\right)$.

$$Df(x_1, x_2) = \begin{bmatrix} \dfrac{\partial f_1}{\partial x_1} & \dfrac{\partial f_1}{\partial x_2} \\ \dfrac{\partial f_2}{\partial x_1} & \dfrac{\partial f_2}{\partial x_2} \end{bmatrix} = \begin{bmatrix} 2 - 2x_2 & -2x_1 \\ -2x_1 & 2 + 2x_2 \end{bmatrix}$$

Then (i) $Df(0,0) = \begin{bmatrix} 2 & 0 \\ 0 & 2 \end{bmatrix}$

$\therefore$ The Eigen value are $\lambda = 2, 2$

$\therefore (0, 0)$ is a source and it is unstable.

ii) $Df(-0,-2) = \begin{bmatrix} 6 & 0 \\ 0 & -2 \end{bmatrix}$

The Eigen value are $\lambda = 6, -2$

$\therefore (0, -2)$ is a saddle pt. & it is unstable.

(iii) $Df(\sqrt{3}, 1) = \begin{bmatrix} 0 & -2\sqrt{3} \\ -2\sqrt{3} & 4 \end{bmatrix}$

Eigen values are given by $\begin{vmatrix} -\lambda & -2\sqrt{3} \\ -2\sqrt{3} & 4-\lambda \end{vmatrix} = 0$

$\therefore -4\lambda + \lambda^2 - 12 = 0$

$\Rightarrow \lambda^2 - 4\lambda - 12 = 0$

$\Rightarrow (\lambda - 6)(\lambda + 2) = 0 \quad \Rightarrow \lambda = 6, -2$

$\therefore (\sqrt{3}, 1)$ is a saddle pt. and it is unstable.

(iv) $Df(\sqrt{3}, 1) = \begin{bmatrix} 0 & 2\sqrt{3} \\ 2\sqrt{3} & 4 \end{bmatrix}$

Eigen values are given by $\begin{vmatrix} -\lambda & 2\sqrt{3} \\ 2\sqrt{3} & 4-\lambda \end{vmatrix} = 0$

$\Rightarrow \lambda^2 - 4\lambda - 12 = 0$

$\Rightarrow \lambda = 6, -2$

$\therefore (-\sqrt{3}, 1)$ is also a saddle pt. and it is unstable.

(iv) $\begin{bmatrix} -x_1 \\ -x_2 + x_1^2 \\ x_3 + x_1^2 \end{bmatrix}$ $\quad \begin{matrix} f_1 = -x_1 \\ f_2 = -x_2 + x_1^2 \\ f_3 = x_3 + x_1^2 \end{matrix}$

The pts. of equilibrium are given by

$-x_1 = 0 \Rightarrow x_1 = 0 \Rightarrow x_2 = x_3 = 0$

The only pt. of equilibrium is $(0, 0, 0)$.

Further,

$f_1(x_1, x_2, x_3) = -x_1$

$f_2(x_1, x_2, x_3) = -x_2 + x_1^2$

$f_3(x_1, x_2, x_3) = x_3 + x_1^2$

$$Df(x_1, x_2, x_3) = \begin{bmatrix} \dfrac{\partial f_1}{\partial x_1} & \dfrac{\partial f_1}{\partial x_2} & \dfrac{\partial f_1}{\partial x_3} \\ \dfrac{\partial f_2}{\partial x_1} & \dfrac{\partial f_2}{\partial x_2} & \dfrac{\partial f_2}{\partial x_3} \\ \dfrac{\partial f_3}{\partial x_1} & \dfrac{\partial f_3}{\partial x_2} & \dfrac{\partial f_3}{\partial x_3} \end{bmatrix}$$

$$= \begin{bmatrix} -1 & 0 & 0 \\ 2x_1 & -1 & 0 \\ 2x_1 & 0 & 1 \end{bmatrix}$$

Then,

$$Df(0, 0, 0) = \begin{bmatrix} -1 & 0 & 0 \\ 0 & -1 & 0 \\ 0 & 0 & -1 \end{bmatrix}$$

NON-LINEAR SINK

Eigen values are given by

$$= \begin{vmatrix} -1-\lambda & 0 & 0 \\ 0 & -1-\lambda & 0 \\ 0 & 0 & -1-\lambda \end{vmatrix} = 0$$

$$\Rightarrow (1+\lambda)(1+\lambda)(1+\lambda) = 0$$

$$\Rightarrow \lambda = -1, -1, -1$$

$\therefore (0, 0, 0)$ is a sink and it is asymptotically stable.

$$v) \begin{bmatrix} x_2 - x_1 \\ kx_1 - x_2 - x_3 \\ x_1 x_2 - x_3 \end{bmatrix}$$

Consider $(0, 0, 0)$ the pt. of equilibrium.

$$Df(x_1, x_2, x_3) = \begin{bmatrix} -1 & 1 & 0 \\ k-x_3 & -1 & -x_1 \\ x_2 & x_1 & -1 \end{bmatrix}$$

$$Df(0, 0, 0) = \begin{bmatrix} -1 & 1 & 0 \\ k & -1 & 0 \\ 0 & 0 & -1 \end{bmatrix}$$

Eigen values are given by $= \begin{vmatrix} -1-\lambda & 1 & 0 \\ k & -1-\lambda & 0 \\ 0 & 0 & -1-\lambda \end{vmatrix} = 0$

$$\therefore -(1+\lambda)(1+\lambda)^2 + k(1+\lambda) = 0$$

$$\Rightarrow (1+\lambda)[(1+\lambda)^2 + k] = 0$$

$$\Rightarrow \lambda = -1, -1, \ (1+\lambda)^2 = k$$

$$\Rightarrow 1 + \lambda = \pm\sqrt{-k}$$

$$\Rightarrow \lambda = -1 \pm \sqrt{k}i$$

If k is +ve then Eigen values are -1 and -1.

So, for k +ve, it is a sink and it is asymptotically stable.

When k = 0, $\lambda = -1, -1$

In this case, origin is a sink and it is asymptotically stable.

If k > 0, i.e. k is –ve then

$\lambda = -1, (1 + \lambda) = \pm\sqrt{-k}$

$\Rightarrow \lambda = -1 \pm \sqrt{-k}$

Eigen values are $-1, -1 + \sqrt{-k}, -1 - \sqrt{-k}$

In this case, origin is a saddle pt. and it is unstable.

vi) $\quad \dfrac{dx}{dt} = -2x(x-1)\,(2x-1);\, \dfrac{dy}{dt} = -2y$

Solution: The pts. of equilibrium are (0, 0), (1, 0) and $\left(\dfrac{1}{2}, 0\right)$

$$Df(x, y) = \begin{bmatrix} -2\left(6x^2 - 6x + 1\right) & 0 \\ 0 & -2 \end{bmatrix}$$

Then

$$Df(0, 0) = \begin{bmatrix} -2 & 0 \\ 0 & -2 \end{bmatrix}$$

Eigen values are $\lambda = -2, -2$

$\therefore (0, 0)$ is a sink and asymptotically stable.

Also, $\;Df(1, 0) = \begin{bmatrix} -2 & 0 \\ 0 & -2 \end{bmatrix}, \Rightarrow \lambda = -2, -2$

$\therefore (1, 0)$ is also a sink and asymptotically stable.

$$Df\left(\dfrac{1}{2}, 0\right) = \begin{bmatrix} -2\left(\overset{3}{6}.\dfrac{1}{42} - \overset{-3}{6}.\dfrac{1}{2} + 1\right) & 0 \\ 0 & -2 \end{bmatrix}$$

$$= \begin{bmatrix} (-3 + 6 - 2) & 0 \\ 0 & -2 \end{bmatrix} = \begin{bmatrix} 1 & 0 \\ 0 & -2 \end{bmatrix}$$

Eigen values are 1 and -2.

$\therefore (\frac{1}{2}, 0)$ is a saddle pt. and it is unstable.

LIAPUNOV FUNCTION AND STABILITY THEOREM

Let W be an open subset of R^n (on E a normed linear space) and $F : W \to R^n$ (or E) is continuously differentiable function.

We consider the diff. equation $X' = F(X)$.

Let $\bar{X}$ be a pt. of equilibrium of this diff. equation.

Let U be a nbd of $\bar{X}$ in W and V a function defined (continuously differentiable) on $U (V : U \to R)$

We define $V = DV(x) \cdot F(x)$, where the R.H.S. is the operator DV operating on F(x).

In particular, if x(t) is a solution curve of the diff. equation $X' = F(X)$ then

$\dot{V}(x(t)) = DV(x(t)) \cdot F(x(t))$

$= DV(x(t)) \cdot x(t)$ { $\therefore$ x(t) is a solution curve.

$= \dfrac{d}{dt} V(x(t))$

So, if $\dot{V} \le 0$ then $V(x(t))$ goes on decreasing (or remains constant) along the solution curve.

Theorem 4. [LIAPUNOV STABILITY THEOREM]

Let $\bar{x} \in W$ be an equilibrium for $X' = F(X)$.

Let $V : U \to R$ be a continuous function defined on a nbd $U \subseteq W$ of $\bar{x}$, differentiable on $U - \bar{x}$ such that

(a) $V(\bar{x}) = 0$ and $V(x) > 0$ if $x \ne \bar{x}$

(b) $\dot{V} \le 0$ in $U - \{\bar{x}\}$ then $\bar{x}$ is a stable equilibrium.

Furthermore, if also

(c) $\dot{V} < 0$ in $U - \{\bar{x}\}$ then $\bar{x}$ is asymptotically stable.

329

DYNAMICAL SYSTEM – A SHORT COURSE

A function V satisfying the conditions

(a) and (b) is called Liapunov Function. If condition (c) also holds then it is called strict Liapunov Function.

Proof:

Since U is a nbd of $\bar{x}$. We can find $\delta > 0$ s.t. the ball $B_\delta(\bar{x})$ along with its boundary lies entirely within U.

We consider the function V on the boundary of $B_\delta(\bar{x})$.

Obviously, v is, on this boundary and will attain its minimum say α on the boundary.

i.e. $\alpha = \min \{V(x) \, s.t \, | x - \bar{x}| = \delta\}$

we have $\alpha > 0$.

Let $U_1 = \{x \in B_\delta(\bar{x}) \, s.t \, V(x) < \alpha\}$.

Obviously U_1 is a non-empty open subset of $B_\delta(\bar{x})$.

If $x(t)$ is any solution curve in U_1, whose starting pt. is in U_1 then we show that

$x(t) \in U, \forall t \geq 0$.

Since $\dot{V} \leq 0$, it follows that for any solution curve $x(t)$ in U_1, V as a function of t is non-increasing.

Since in U_1, V < α, it follows that the solution curve $x(t)$ can not cross the boundary of $B_\delta(\bar{x})$.*

Actually it will remain in U_1. This means $\bar{x}$ is stable equilibrium.

We now assume that $\dot{V} < 0$ in $U - \{\bar{x}\}$.

We show that $\bar{x}$ is asymptotically stable equilibrium.

For this we show that if $x(t)$ is any solution curve in U_1 then $x(t) \rightarrow \bar{x}$ as

NON-LINEAR SINK

$t \to \infty$.

We assume that $x(t) \to \bar{x}$ as $t \to \infty$.

As the set $\{x(t), t \geq 0\}$ is bounded, (as the solution curve $x(t) \subseteq U_1 B_\delta(\bar{x})$)

the set $\{x(t), t \geq 0\}$ will have limit pt.

Let z_o be a limit pt. of $\{x(t), t \geq 0\}$.

Let us assume that $z_o \neq \bar{x}$.

As z_o is a limit pt. of $\{x(t), t \geq 0\}$. We can find a seq. $\{t_1, t_2, \cdots t_n, \cdots\}$ of

values of t such that $t_n \to \infty$ and $x(t_n) \to z_o$.

As V is strictly decreasing along the solution curve (as $\dot{V} < 0$ in U_1) and

$x(t_n) \to z_o$ as

$t_n \to \infty$, it follows that $V(x(t_n)) > V(z_o)$.

Since we are assuming that $z_o \neq \bar{x}$.

We consider a solution curve which starts at z_o.

If $z(t)$ denotes the pt. on this solution curve at time t, then we have

$V(z_o) > V(z(t))$.

If y_o is a pt. sufficiently close to z_o, and if $y(t)$ is the pt. on this solution

curve at time t, then (as y_o and z_o are very close), we must have

$V(z_o) > V(z(t)), \forall t \geq 0$.

Since $x(t_n) \to z_o$ as $n \to \infty$. We can choose n very large such that $x(t_n)$ is

sufficiently close to z_o.

Thus if $l(t)$ is a solution curve which starts at $x(t_n)$ then by the above

argument, we have $V(l(t)) < V(z_o), \forall t \geq 0$.

But $l(t)$ is given by $x(t_n + t)$.

Thus, we have

$$V\big(x(t_n + t)\big) < V(z_o), \forall t \geq 0$$

But we know that

$$V\big(x(t_n + t)\big) > V(z_o), \forall n$$

We thus get a contradiction.

The assumption that $z_o \neq \bar{x}$ must be wrong.

Thus $\bar{x}$ is the only limit pt. of $\{x(t), t \geq 0\}$ i.e. $x(t) \to \bar{x}$ as $t \to \infty$.

Definition:

A set P is positively invariant for a dynamical system if for each $x \in P, \phi_t(x)$ is defined & in P, $\forall t \geq 0$, where $\phi_t(x)$ denotes the flow of the system.

Definition:

An entire orbit of the system is a set of the form $\{\phi_t(x)/t \in R\}$, where $\phi_t(x)$ is defined for all $t \in R$.

Theorem 5: Let $\bar{x} \in W$ be an equilibrium of the dynamical system $X' = F(X)$. Let $V : U \to R$ be a Liapunov function for $\bar{x}$, U is a nbd of $\bar{x}$. Let $P \subseteq U$ be a nbd of $\bar{x}$, which is closed in W. Suppose that P is +vely invariant and that there is no entire orbit in $P - \{\bar{x}\}$ on which v is constant. Then $\bar{x}$ is asymptotically stable and $P \subseteq B(\bar{x})$.

Proof: We show that any solution curve $x(t)$ of the given dynamical system $X' = F(X)$ which starts at some pt. in P approaches $\bar{x}$ as $t \to \infty$.

Let us assume that $x(t)$ is a solution curve with starting pt. in P.

As P is +vely invariant $x(t)$ will lie in P, $\forall t \geq 0$.

We assume that $x(t) \to \bar{x}$ as $t \to \infty$.

NON-LINEAR SINK

Let a be a limit pt. of the set $\{x(t): t \geq 0\}$.

Thus we can find a seq. $\{t_1, t_2, \cdots\} t_n \to \infty$ such that $x(t_n) \to a$ as $n \to \infty$.

Let $\alpha = v(a)$, obviously α is the glb of $v(x(t))$, $t \geq 0$.

$\alpha = v(a) = \text{glb}\{v(x(t)), t \geq 0\}$

This follows because v is continuous, $x(t_n) \to a$ and $t_n \to \infty$ as $n \to \infty$.

Thus if a' is any other limit pt. of $\{x(t), t \geq 0\}$ i.e. if there is another sequence $t_1', t_2' \cdots$ tending to infinity such that

$a' = \lim_{n \to \infty} x(t_n')$ then $\alpha = v(a) = v(a')$

Let L be the collection of all sub-sequential limits of $\{x(t), t \geq 0\}$

i.e. L consist of all those points which are limit pts. of the set $\{x(t), t \geq 0\}$.

Obiviously, $L \subseteq P$, it follows because P is closed and element of L are Limit pts. of $\{x(t), t \geq 0\}$ a subset of P.

Let $a \in L$ and P is positively invariant, it follows that $\phi_t(a)$ lies in P for all $t \geq 0$.

We note that $a = \lim_{n \to \infty} x(t_n)$

Consider $\phi_t(x(t_n))$, we note that $\phi_t(x(t_n))$ is defined and in P for $[-t_n, 0]$.

This is because if $\phi_{-tn}(x(t_n)) = y$ then $x(t_n) = \phi_{tn}(y)$

i.e. $x(t_n)$ is a pt. on the solution curve whose starting pt. is y and is reached after time t_n.

But $x(t_n)$ is also a pt. on a pt. on a solution curve starting at say x_o (of P) such that $x(t_n)$ is reached after time t_n.

By uniqueness $y = x_o$ i.e. $\phi_{-tn}(x(t_n))$ is in P and given by x_o.

Thus $\phi_t(x(t_n)) \subseteq P$, for $[-t_n, 0]$.

This holds for all $x(t_n)$. Since $x(t_n) \to a$ as $n \to \infty$ and as $\ln \to +\infty$ with $n \to \infty$, it follows that $\phi_t(a)$ is defined in P for $t \in (-\infty, 0)$.

Thus $\phi_t(a)$ is defined and in P for $t \in (-\infty, \infty)$

i.e. the entire orbit $\phi_t(a)$ lies in P.

Consider $\phi_s(a)$, for some $S \in R$.

$$\phi_s(a) = \phi_s \left(\lim_{n \to \infty} x(t_n) \right)$$

$$= \phi_s \left(\lim_{n \to \infty} \phi_{tn}(x_o) \right)$$

$$\phi_s(a) = \lim_{n \to \infty} \phi_s \left(\phi_{tn}(x_o) \right)$$

$$= \lim_{n \to \infty} \left(\phi_{s+tn}(x_o) \right)$$

$$= \lim_{n \to \infty} \left(x(t_n + s) \right) \in L$$

$$V \left(\phi_s(a) \right) = \alpha = v(a)$$

Since $S \in R$, it follows that v is constant along the entire orbit $\phi_t(a)$ in P, which is a contradiction.

$x(t) \to \bar{x}$ as $t \to \infty$, whenever $x(t)$ is a solution curve starting in P.

$\therefore \bar{x}$ is asymptotically stable and $P \subseteq B(\bar{x})$.

Definition : The set L defined above is called the set of w-limit pts. or the w-limit set of the projector (solution curve) $x(t)$.

i.e. $L = \{ a \in W$ s.t. $a = \lim_{n \to \infty} x(t_n)$

i.e. as $t_n \to +\infty, x(t_n) \to a \}$.

Similarly, We define the set of α-limit pts. on the α-limit set of a trajectory $y(t)$ to be the set of all those pts. $b \in W$ such that $b = \lim_{n \to \infty} y(t_n)$

334

NON-LINEAR SINK

Definition: A set A in the domain W of a dynamical system is said to be invariant if for every $x \in R$, $\phi_t(x)$ is defined in A, for $t \in R$.

Theorem 6: The α-Limit set and w-limit set of any trajectory are closed invariant sets.

Proof: By definition it follows that the α-limit set and w-limit set of any trajectory are invariant sets.

Since α–limit set and w-limit set contains all α–limit pts and w-limit pts respectively.

Then the collection of all such α–limit pts and w-limit pts is a derived set and hence it is closed, since derived set is always closed.

Ex 3: consider the system $x' = 2y(z-1)$

$y' = -x(z-1)$

$z' = xy$

Solution: There are infinitely many pts. of equilibrium.

There are given by, $z = 1, x = 0; z = 1, y = 0; x = 0, y = 0; 0 = x = y = z$.

We consider the behaviour at the equilibrium point $(0, 0, 0)$.

So we find $DF(x, y, z)$:

The matrix of $DF(x,y,z) = \begin{bmatrix} 0 & 2(z-1) & 2y \\ -(z-1) & 0 & -x \\ y & x & 0 \end{bmatrix}$

Then

$$DF(0,0,0) = \begin{bmatrix} 0 & -2 & 0 \\ 1 & 0 & 0 \\ 0 & 0 & 0 \end{bmatrix}$$

Eigen values are given by $\begin{vmatrix} -\lambda & -2 & 0 \\ 1 & -\lambda & 0 \\ 0 & 0 & -\lambda \end{vmatrix} = 0$

$$\therefore -\lambda(\lambda^2 + 0) + 2(-\lambda + 0) = 0$$

$$\therefore -\lambda^3 - 2\lambda = 0$$

$$\therefore \lambda(\lambda^2 + 2) = 0$$

$$\therefore \lambda = 0, \lambda = \pm\sqrt{2}\,i$$

Eigen values are $\pm\sqrt{2}\,i, 0$.

Because the equilibrium pt. is $(0, 0, 0)$ and we want $v(0, 0, .0) = 0$ and $v(x, y, z) > 0$ deleted nbd of $(0, 0, 0)$.

We consider v to be of the type

$v(x,y,z) = ax^2 + by^2 + cz^2$ with $a, b, c \geq 0$ (all of a, b, c not vanishing simultaneously)

$$v = 2axx' + 2byy' + 2czz'$$

Since along the solution curves we have

$$x' = 2y(z-1), \ y' = -x(z-1), \ z' = xy$$

It follows that along the solution curves

$$v = 2ax(2y(z-1)) + 2by(-x(z-1)) + 2cz(xy)$$
$$= 4axy(z-1) - 2bxy(z-1) + 2cxyz$$

$\dot{v} = 0$ if b=2a and c=0

Thus, we can define v as

$v(x,y,z) = ax^2 + 2ay^2 + 0$, for $a>0$.

In particular for $a = 1$, we get

$$v(x,y,z) = x^2 + 2y^2$$

Thus $(0, 0. 0)$ is a stable equilibrium.

Furthermore along the solution curves v remains constant

($\therefore \dot{v} = 0$ along solution curves).

NON-LINEAR SINK

Ex 4. Find a strict Liapunov function for the equilibrium $(0, 0)$ of

$$x' = -2x - y^2$$

$$y' = -y - x^2$$

Solution: $Df(x, y)$ is given by the matrix $\begin{bmatrix} -2 & -2y \\ -2x & -1 \end{bmatrix}$

$$DF(0,0) = \begin{bmatrix} -2 & 0 \\ 0 & -1 \end{bmatrix}$$

Thus $(0, 0)$ is a sink (asymptotically stable equi)

To get a strict Liapunov function, we consider

$$v(x, y) = ax^2 + by^2, a, b \geq 0$$

$$\dot{v}(x, y) = 2axx' + 2byy'$$

Along the solution curves,

$$\dot{v}(x, y) = 2ax(-2x - y^2) + 2by(-y - x^2)$$

$$= -4ax^2 - 2axy^2 - 2by^2 - 2bx^2 y$$

We consider $a = b = 1$ to get $v(x, y) = x^2 + y^2$

And along the solution curves.

$$\dot{v} = -4x^2 - 2xy^2 - 2y^2 - 2yx^2$$

$$= -2x^2(2 + y) - 2y^2(x + 1)$$

It is clear that $\dot{v} < 0$ if

$2 + y > 0$ and $1 + x > 0$

i.e. $y > -2$ and $x > -1$

so, if we take U to be on open ball with centre at origin $(0, 0)$ and radius 1,

it si clear that $v(x, y) = x^2 + y^2$ is a strict Liapunov function

LIAPUNOV FUNCTION AND A BASIN FOR A PENDULUM

Consider a pendulum with a bob of mass 'm' attached to an end of massless string of length 'l'.

We assume that in addition to the weight of the bob, there is a frictional force acting on the bob. and is proportional to the velocity, with proportionality constant k.

We consider the vertically downward position of the bob as origin and measure the displacement of the bob from this vertically downward position by an angle θ which is measured in anti-clockwise direction.

The motion of the pendulum is given by a system of two simultaneous first order diff. equations namely.

$$\dot\theta = w \text{ and } \dot\omega = \frac{-g\sin\theta}{l} - \frac{k\omega}{m}$$

It is a non-Linear autonomous system.

We know that (0, 0) is a sink (asymptotically stable pt. of equilibrium).

We shall obtain a Liapunov function for the sink (0, 0).

We expert that the energy of the bob can be used as a Liapunov function.

We thus consider the total energy of the bob at any position (θ, w).

The kinetic energy (K. E) is equal to : $\frac{1}{2} m\, l^2 w^2$

(Angular velocity $\dot\theta = w$ which induces a Linear velocity $l\dot\theta$).

In order to get the expression for P.E of the bob, we measure its height from the vertically downward position which at a given value of θ is $l - l\cos\theta$.

Thus the potential energy (P.E) is : $mg(l - l\cos\theta)$.

So, at a position (θ, w) the total energy of the bob is given by

$T.E = K.E + P.E$

$$E(\theta, w) = \frac{1}{2} m\, l^2 w^2 + mgL(1 - \cos\theta)$$

NON-LINEAR SINK

$$= ml\left(\frac{1}{2} lw^2 + g(1 - \cos\theta)\right)$$

Obiviously, E(0, 0) = 0 and E(θ, w) > 0 if $(\theta, w) \neq (0,0)$

We now consider $\dot{E}(\theta, w)$

$$\dot{E}(\theta, w) = ml\left(\frac{1}{2} \times 2l\ w\dot{w} + g\sin\theta\ \dot{\theta}\right)$$

But we have

$$\dot{\theta} = w \quad \text{and} \quad \dot{w} = \frac{-g\sin\theta}{l} - \frac{kw}{m}$$

$$\dot{E}(\theta, w) = ml\left[lw\left(\frac{-g\sin\theta}{l} - \frac{kw}{m}\right) + g\sin\theta\ w\right]$$

$$= ml\left[-g\sin\theta\ w - \frac{klw^2}{m} + g\sin\theta\ w\right]$$

$$\dot{E}(\theta, w) = -kl^2 w^2$$

Certainly, $\dot{E}(\theta, w) \leq 0$.

Thus $E(\theta, w)$ is indeed a strict Liapunov function as

$$w \neq 0 \Rightarrow \dot{E} < 0$$

We now try to find subset lying in the basin of (0, 0).

For any c, $0 < c < 2mgl$.

We consider a set $P_c = \{(\theta, w) \text{ such that } E(\theta, w) \leq c \text{ and } |\theta| = \pi\}$

Obiviously, $(0,0) \in P_c$

We now show that P_c is positively invariant, closed set and $P_c - (0.0)$ does not contain an entire orbit along which $\dot{E} = 0$.

In order to show that P_c is positively invariant.

339

DYNAMICAL SYSTEM – A SHORT COURSE

We let $\alpha > 0$ and consider $0 \le t \le \alpha$ and $\{\theta(t), w(t)\}$ with $(\theta(0), w(0)) \in P_c$

We have $E(\theta(0), w(0)) \le c$ and $|\theta(0)| < \pi$

But along the solution curves $\dot{E} \le 0$

$E(\theta(0), w(t)) \le c$

We need to show that $|\theta(t)| < \pi$

If it were not true, it would mean we could find t_o,

$0 \le t_o < \alpha$ s.t $|\theta(t_o)| = \pi$

As $|\theta(0)| < \pi$ and according to an assumption $|\theta(0)|$ exceeds π.

But then for $(\theta(t_o), w(t_o))$,

$$E(\theta(t_o), w(t_o)) = ml\left(\frac{1}{2}lw(t_o)^2 + 2g\right)$$

$$\ge 2mgl > c$$

This is a contradiction as along solution curves starting with a pt. $(\theta(0), w(0))$ in P_c then for all t; $0 \le t \le \alpha, (\theta(t), w(t)) \in P_c$

But α is arbitrary. Thus P_c is positively invariant.

We next show that P_c is closed.

Let (θ_1, w_1) be a limit of P_c.

By definition, (θ_1, w_1) is a limit of seq $\{\theta^{(n)}, w^{(n)}\}$ of pts. of P_c.

As $(\theta^{(n)}, w^{(n)}) \in P_c, E(\theta^{(n)}, w^{(n)}) \le c$

In order to show that $(\theta_1, w_1) \in P_c$, we now need to how that $|\theta_1| < \pi$.

As $|\theta^{(n)}| < \pi$ and $\theta_1 = \lim \theta^{(n)}$

We have $|\theta_1| < \pi$

We show that $|\theta_1| \ne \pi$

340

If $|\theta_1| = \pi$ then $E(\theta_1, w_1) > c$ but $E(\theta_1, w_1) \le c$

$\therefore |\theta_1| < \pi$

$\therefore E(\theta_1, w_1) \in P_c$

We have show that $P_c - \{(0, 0\}$ does not contain an entire orbit (trajectory) along which E is constant.

Let us assume that there exist on entire orbit in P_c along which $\dot{E} = 0$.

But $\dot{E} = -k\, l^2 w^2$ along on orbit, thus along such an orbit, we must have w=0.

We have $\dot{\theta} = w$, $\dot{w} = -\dfrac{g \sin \theta}{l} - \dfrac{kw}{m}$

So along such an orbit/trajectory, we must have $\theta =$ constant and $\sin \theta = 0$ (does not mean $\theta = 0$, it may be $\pi, -\pi$ etc)

But in $P_c, |\theta_1| < \pi$

This along such an orbit we must have $\theta = 0$ as well as w=0.

Thus the only entire orbit in P_c along which E is constant and is the singleton set $\{(0, 0)\}$, i.e. $P_c - \{(0,0)\}$ does not contain any entire orbit along. which is constant

$\therefore P_c \subseteq B(0, 0)$

Thus for $0 < c < 2mgl$, $P_c \subseteq B(0, 0)$.

Let P= $\cup P_c$ such that $0 < c < 2mgl, P_c \subseteq B(0, 0)$

Where $P = \{\theta, w)$ such that

$E(\theta, w) < 2mgl \,\&\, |\pi| < \pi\}$

However, p is not the entire basin.

For example. The pt (π, u) where $u \ne 0$ is a pt. in B(0, 0) which is not in P.

GRADIENT SYSTEM

A gradient system on an open set $w \subseteq R^n$ is a dynamical system of the form

$$x' = -\text{grad } v(x)$$

Where, $V : U \rightarrow R$ is a c^2 – function and

$$\text{Grad } V = \left(\frac{\partial v}{\partial x_1}, \frac{\partial v}{\partial x_2}, \cdots \frac{\partial v}{\partial x_n} \right)$$

is the gradient vector field grad $V : U \rightarrow R^n$.

As $V : U \rightarrow R$ we consider Dv, a linear map from R^n to R and in the standard coordinates, its matrix is a 1xn matrix given as:

$$DV(x) = \left(\frac{\partial v}{\partial x_1}, \frac{\partial v}{\partial x_2}, \cdots \frac{\partial v}{\partial x_n} \right)$$

As such for any $y \in R^n$,

$$DV.Y = \left(\frac{\partial v}{\partial x_1}, \frac{\partial v}{\partial x_2}, \cdots \frac{\partial v}{\partial x_n} \right) \begin{pmatrix} y_1 \\ y_2 \\ \vdots \\ y_n \end{pmatrix}$$

$$= \frac{\partial v}{\partial x_1}.y_1 + \frac{\partial v}{\partial x_2}.y_2 + \cdots + \frac{\partial v}{\partial x_n} y_n$$

$$= \sum_{i=1}^{n} \frac{\partial v}{\partial x_i}.y_i$$

$DV.Y = <\text{grad } v, y>$

For the gradient system $x' = -\text{grad } v$, we define

$$\dot{V} = \frac{d}{dt}\big(v(x(t))\big)$$

Theorem 7: $\dot{V}(x) \leq 0, \forall x \in U$ and $\dot{V}(x) = 0$ if and only if x is an equilibrium of the system $x' = -\text{grad } v(x)$.

NON-LINEAR SINK

Proof: By the definition,

$$\dot{V} = \frac{d}{dt} v\big(x(t)\big)\big|_{t=0}$$

$$= DV.x' \text{ (where, } x' = \frac{d}{dt} x(t) \equiv \dot{x})$$

But along the trajectories (solution curves) of the $x' = $-grad v.

Along the trajectories,

$$\dot{V}(x) = <\text{grad v, -grad v} >.$$

$$\left(\because \forall y \in R^n, DV.Y = < grad\ v, y > \right)$$

Along the trajectories, we have

$$\dot{V}(x) = - < \text{grad v, grad v} >.$$

$$= -| grad\ v |^2$$

$$\dot{V}(x) \leq 0$$

Also, $\dot{V}(x) = 0$ iff grad v(x) $= 0$

$\Rightarrow \dot{V}(x) = 0$ iff x is a pt. of equilibrium of the system.

Theorem 8: Let $\bar{x}$ be an isolated minimum of v then $\bar{x}$ is an asymptotically stable equilibrium of the gradient system $x' = $-grad v(x).

Proof: Since $\bar{x}$ is an isolated minimum of v.

It is clear that grad $v(\bar{x}) = 0$ i.e. $\bar{x}$ is an equilibrium point.

In a nbd of $\bar{x}$ which does not contain any other minimum of v, we define a function of g as

$$g(x) = v(x) - (\bar{x})$$

Obiviously, g is differentiable, $g(\bar{x}) = 0$ and in the above nbd of $\bar{x}$ at $x \neq \bar{x}, g(x) > 0$

$(\because v(x) > v(\bar{x}), \forall x$. In the nbd with $x \neq \bar{x})$

343

Further $\dot{g}(x) = \dot{v}(x) - |grad\, v|^2$ and as $\bar{x}$ is the isolated minimum.

We must have $\dot{g}(x) < 0$ in the nbd with $x \neq \bar{x}$.

g is thus a strict Liapunov function for the equilibrium pt. $\bar{x}$.

As such $\bar{x}$ is an asymptotically stable equilibrium point.

LEVEL SURFACES

For the function $V : U \to R$ and any $c \in R$ we consider the subset $V^{-1}(c)$ of U. it is known as the Level surfaces of v taking the constant value c at every pt. of it.

If u is a pt. on the level surface $V^{-1}(c)$, s.t $(grad\, v)_u \neq 0$, then we say that u is a Regular point of the level surface.

If $(gradV)_u \neq 0$, we have at least one of the components

$$= \frac{\partial v}{\partial x_1}, \frac{\partial v}{\partial x_2}, \cdots, \frac{\partial v}{\partial x_n} \text{ at u non-zero.}$$

If say $\left.\dfrac{\partial v}{\partial x_j}\right|_u \neq 0$, then by implicit function thm,

We can express the Jth component x_j in terms of the remaining components

$$x_1, x_2, \cdots, x_{j-1}, x_{j+1}, \cdots x_n$$

As say $x_j = g\left(x_1, x_2, \cdots, x_{j-1}, x_{j+1}, \cdots x_n\right)$

to get

$$V\left(x_1(u), \cdots x_{j-1}(u), g\left(x_1, x_2, \cdots, x_{j-1}, x_{j+1}, \cdots x_n\right)^{(u)}, x_{j+1}(u), \cdots, x_n(u)\right) = c$$

This is an n-1 dimensional surface in R^n.

The tangent to this surface at any regular pt. u on it is given by the collection of all the vectors y s.t. *DV(u). Y= 0.*

NON-LINEAR SINK

But $DV(u). Y = 0 \Leftrightarrow <grad\ v(u), y> = 0$

i.e. $grad\ V(u) \perp Y$, where y is a tangent vector to the level surface.

i.e. grad v(u) is orthogonal to the level surface at the regular pt. u on it.

Thus, at regular points, the vector field-grad v(x) is let to the level surface of v.

Theorem 9: Let $x' = -grad\ v(x)$ be a gradient system. At regular pts., the trajectories cross level surfaces orthogonally Non-regular pts. are equilibrium of the system. Isolated minima are asymptotically stable.

Proof: Since the solution curves are solution of the system

$x' = -grad\ v(x)$

It follows that the tangent to the trajectories are parallel to the vector field $-grad\ v(x)$.

Since $-grad\ v(x)$ is orthogonal to the Level surfaces at regular pts. it follows that the trajectories of the gradient system cut the level surfaces orthogonally at regular pts.

Ex 5. $V : R^2 \to R^2$ be the function $v(x,y) = x^2(x-1)^2 + y^2$

Find the pts. of equilibria and discuss their nature for the corresponding gradient system.

Solution: The gradient system will be given by

$$x' = \frac{\partial v}{\partial x}, y' = \frac{\partial v}{\partial y}$$

$$\frac{\partial v}{\partial x} = 2x(x-1)^2 + 2x^2(x-1)$$

$$= 2x(x-1)[x-1+x]$$

$$= 2x(x-1)(2x-1)$$

$$\frac{\partial v}{\partial y} = 2y$$

Thus, the gradient system is:

$$x' = -2x(x-1)(2x-1)$$

$$y' = -2y$$

It is a non-Linear autonomous system. The pts. of equilibria are $(0, 0)$,

$(1, 0)$ and $\left(\dfrac{1}{2}, 0\right)$

To study the nature of the pts. of equilibria, we consider the operator Dv and is given by

$$\begin{bmatrix} \dfrac{\partial f_1}{\partial x} & \dfrac{\partial f_1}{\partial y} \\ \dfrac{\partial f_2}{\partial x} & \dfrac{\partial f_2}{\partial y} \end{bmatrix} = \begin{bmatrix} -2(6x^2 - 6x + 1) & 0 \\ 0 & -2 \end{bmatrix}$$

At $(0, 0)$, the matrix $A = \begin{bmatrix} -2 & 0 \\ 0 & -2 \end{bmatrix}$

Both the Eigen values are –ve (-2).

$\therefore (0, 0)$ is a sink (asymptotically statically stable equilibrium).

At $(1, 0)$

$A = \begin{bmatrix} -2 & 0 \\ 0 & -2 \end{bmatrix}$, in this case, $(1, 0)$ is a sink.

At $\left(\dfrac{1}{2}, 0\right)$

$A = \begin{bmatrix} 1 & 0 \\ 0 & -2 \end{bmatrix}$, Eigen values are +1 and -2.

$\therefore \left(\dfrac{1}{2}, 0\right)$ is an unstable equilibrium pt. (saddle0.

Ex 6. $v(x, y) = x^2 + 2y^2$

Solution : The gradient system will be given by

$$x' = -\frac{\partial v}{\partial x}, y' = -\frac{\partial v}{\partial y}$$

$$\frac{\partial v}{\partial x} = 2x, \frac{\partial v}{\partial y} = 4y$$

Thus, the gradient system is:

$$x' = -2x, \quad y' = -4y$$

Here (0, 0) is the equilibrium point.

$$\begin{bmatrix} \dfrac{\partial f_1}{\partial x} & \dfrac{\partial f_1}{\partial y} \\ \dfrac{\partial f_2}{\partial x} & \dfrac{\partial f_2}{\partial y} \end{bmatrix} = \begin{bmatrix} -2 & 0 \\ 0 & -4 \end{bmatrix}$$

At (0, 0), the matrix $A = \begin{bmatrix} -2 & 0 \\ 0 & -4 \end{bmatrix}$

$\therefore$ Eigen values are -2 and -4.

$\therefore$ (0, 0) is a sink (asymptotically stable equi.).

Ex 7. $v(x, y) = y \sin x$

Solution: The gradient system will be given by

$$x' = -\frac{\partial v}{\partial x}, y' = -\frac{\partial v}{\partial y}$$

$$x' = -y \cos x, \quad y' = -\sin x$$

The pts of equilibrium are (0, 0), $(n\pi, 0)$,

Consider the pt. of equilibrium $(n\pi, 0)$. The matrix A at (x, y) is:

$$A = \begin{bmatrix} y \sin x & -\cos x \\ -\cos x & 0 \end{bmatrix}$$

At $(n\pi, 0)$; $\begin{bmatrix} 0 & (-1)^{n+1} \\ (-1)^{n+1} & 0 \end{bmatrix}$

Eigen values are :

$$\begin{vmatrix} -\lambda & (-1)^{n+1} \\ (-1)^{n+1} & -\lambda \end{vmatrix} = 0$$

$$\Rightarrow \lambda^2 - (-1)^{n+1}(-1)^{n+1} = 0$$

$$\Rightarrow \lambda^2 - 1 = 0$$

$$\therefore \lambda = \pm 1$$

$\therefore (n\pi, 0)$ is unstable pt. of equilibrium (saddle)

Theorem 10: Let z be an α-limit pt. or an w-limit pt. of a gradient flow. Then z is an equilibrium Pt.

Proof: By the definition, there exist a trajectory $x(t)$ of the gradient system

$$x' = -grad\ v(x)$$

And seq. $\{t_n\}, t_n \to \infty$ s.t. $x(t_n) \to z$ and z is known as w-limit pt. of the trajectory x(t).

We know that along a trajectory $\dot{v} \le 0$

We must have

$$v(z) = glb\{v(x(t)),\ \forall t\ \text{for which x(t) is defined}\}.$$

We know that if z_1 is any pt. on the trajectory x(t), then

For some $s \in R$, we have $z_1 = \phi_s(z)$

Where $\phi_t(.)$ denotes the flow of the given gradient system.

But as $z = \lim_{n \to \infty} x(t_n)$, we have

$$z_1 = \lim_{n \to \infty} x(t_n + s)$$

As $t_n \to +\infty$ with $n \to \infty, t_n + s \to \infty$ as $n \to \infty$

By the above argument, $v(z_1) = glb\{v(x(t))\}$

$$\therefore v(z_1) = v(z)$$

NON-LINEAR SINK

But z_1 is arbitrary pt. on the trajectory.

V is constant along the trajectory

$\therefore \dot{v} = 0$

In particular, $\dot{v}(z) = 0$

$\therefore z$ is a pt. of equilibrium.

Thus every w-limit pt. of the gradient system is an equilibrium point.

Let z be an α-limit pt. of a trajectory of the gradient system

$x' = -grad\ v(x)$.

By the definition there exist a trajectory say x(t) of the given gradient

system and a seq. $\{t_n\}$ with $t_n \to +\infty$ as $n \to \infty$ s.t. $z = \underset{n \to \infty}{\lim}\ x(t_n)$.

If we change t to –t, then α-limit pt. can be consider as an w-limit point.

Thus an α-limit pt. z of the gradient system,

$x' = -grad\ v(x)$

Can be considered as an w-limit pt. of the gradient system

$x' = +grad\ v(x)$

$x' = -grad\ (v(x))$

By the previous argument, -v remains constant along the trajectory of z.

$\therefore -\dot{v}(z) = 0$

$\Rightarrow \dot{v}(z) = 0$

z is also a pt. of equilibrium.

CHAPTER 7

DUAL SPACES

Let E be a finite dimensional real vector space

i.e. $E \cong R^n$, for some $n \neq 1$

We consider Linear maps from E to R and the collection of all such linear maps is denoted by E^*.

Thus, $E^* = L(E, R)$.

E^* itself is a real vector space.

Theorem 1: E^* is isomorphic to E and thus has the same dimension.

Proof: We construct a linear map from E to E^* which is one-one and onto.

As E is n-dimensional, let $\{e_1, e_2, \cdots, e_n\}$ be a basis of E.

With respect to this basis, we define an inner product on E as

$$< e_i, e_j > = \delta_{ij}$$

and extend it on the entire space E.

we now define a map $u : E \to E^*$ as $x \to u_x$, where u_x is s.t.

for every $y \in E$,

$$< u_x(y) = < x, y >$$

We show that the map $x \to u_x$ is well defined, linear, one-one and onto.

Let $x_1, x_2, y_1, y_2 \in E, \alpha \in R$

Consider

$$u_x(y_1 + y_2) \ = < x, y_1 + y_2 >$$

$$= < x, y_1 > + < x, y_2 >$$

$$= u_x(y_1) + u_x(y_2)$$

$$u_x(\alpha y) = < x, \alpha y > = \alpha < x, y > = \alpha \, u_x(y)$$

$\therefore u_x \in E^*$

Thus $x \to u_x$ is indeed a map from E to E^*.

We now show that $x \to u_x$ is a linear map

Consider $u_{x_1+x_2} = u_{x_1} + u_{x_2}$

For any

$$y \in E, u_{x1+x2}(y) = <x_1 + x_2, y>$$

$$= <x_1, y> + <x_2, y>$$

$$= u_{x1}(y) + u_{x2}(y)$$

$$= (u_{x1} + u_{x2})(y)$$

$\therefore u_{x1+x2} = u_{x1} + u_{x2}$

Further,

$$u_{\alpha x}(y) = <\alpha x, y> = \alpha <x, y> = \alpha u_x(y)$$

For $y \in E$.

$$u_{\alpha x} = \alpha u_x$$

The map $x \to u_x$ is a linear map.

To show that $x \to u_x$ is 1-1, we note that x is in the kernel of the map iff

u_x is a zero functional on E.

iff $u_x(y) = 0$, for $y \in E$

iff $u_x(x) = 0$

iff $<x, x> = 0 \Leftrightarrow \| x \|^2 = 0$

$\Leftrightarrow x = 0$

In order to show that the map $x \to u_x$ is onto.

DUAL SPACES

Let $v \in E^*$ we show that $x \in E$ s.t $u_x = v$.

As $\{e_1, e_2, \cdots, e_n\}$ is a basis of E, let $v(e_i) = l_i$, for some $l_i \in R$.

We let $x = \sum(e_i) = \,< x, e_i > \, = \,< \sum_{j=1}^{n} l_j e_j, e_i > \, = l_i$ $\qquad (\because < e_i, e_j > = \delta_{ij})$

But $v(e_i) = l_i$

Thus, for $1 \le i \le n, U_x(e_i) = v(e_i)$

Let $y \in E$ then we have

$$y = \sum_{j=1}^{n} y_j \cdot e_j \text{, for real } y_1, y_2, \cdots y_n$$

$$u_x(y) = \,< x, y >$$

$$= \left\langle \sum_{j=1}^{n} l_i \cdot e_j, \sum_{k=1}^{n} y_k e_k \right\rangle$$

$$= \sum_{i=1}^{n} l_i \cdot y_i$$

Also,

$$v(y) = v\left(\sum_{k=1}^{n} y_k e_k \right)$$

$$= \sum_{k=1}^{n} y_k \cdot v(e_k) = \sum_{k=1}^{n} y_k \cdot l_k$$

Thus the map $x \to u_x$ from E to E^* is a linear bijection i.e. an isomorphism.

$E \cong E^*$

Dim E^* = dim E = n.

NOTE

We can indeed get a basis of E^* from the basis of E.

We define e_1^* as an element of E^* such that

$$e_1^*(e_1) = 1, e_1^*(e_2) = e_1^*(e_3) = \cdots = e_1^*(e_n) = 0.$$

In general, $e_1^*(e_j) = \delta_{ij}$.

We thus get n distinct elements $e_1^*, e_2^*, \cdots, e_n^*$ of E^* .

We show that they are Linearly Independent.

Let $\alpha_1, \alpha_2, \cdots, \alpha_n \in R$ such that

$$\alpha_1 e_1^* + \alpha_2 e_2^* + \cdots + \alpha_n e_n^* = 0$$

i.e. $\forall y \in E, \left(\sum_{i=1}^{n} \alpha_i e_i^* \right) y = 0$

in particular with $y = \in_1$, we get

$$\left(\sum_{i=1}^{n} \alpha_i \cdot e_i^* \right) \cdot e_1 = 0$$

$$\sum_{i=1}^{n} \alpha_i \left(e_1^*(e_1) \right) = 0$$

$\alpha_1 = 0 \quad \left(\because e_1^*(e_1) = \delta_{i1} \right).$

Similarly, we can show that

$\alpha_2 = \alpha_3 = \cdots = \alpha_n = 0$

Thus, $\left\{ e_1^*, e_2^*, \cdots, e_n^* \right\}$ is a basis of E^*.

It is known as basis dual to the basis $\left\{ e_1, e_2, \cdots, e_n \right\}$.

Let E be an n-dimensional real vector space with an inner product and

let E^* be its dual space.

We define a map $\phi : E \to E^*$ as

$x \to \phi(x), \forall x \in E$

DUAL SPACES

Where $\phi(x)$ is defined as

$\phi(x)y = <x, y>, \forall y \in E$

With this definition for every $x \in E$, $\phi(x)$ is indeed a functional on E

i.e. $\phi(x): E \to R$.

Further we can show that the map $\phi: x \to \phi(x)$ is an isomorphism.

Let w be an open subset of E and $v: W \to R$ a continuously differentiable real valued function for each $x \in W$, we have DV(x) a function on E.

Thus DV can be considered as a map from w to E^*. This type of map is known as **1-form**.

Since for every $x \in W$, grad $v(x) \in E$ and $Dv(x) \in E^*$ and $\phi: E \to E^*$ is an **isomorphism**.

As we know that in the standard basis system,

$\forall y \in E, \forall x \in W, (Dv(x))(y) = <grad\ v(x), y>$

We can thus define

grad $v(x) = \phi^{-1}(Dv(x))$

This definition is consistent with the result that

$Dv(x)\ (y) = <grad\ v(x), y>$

Theorem 2: Let $v: W \to R$ be a C^2-function on an open W of E (as n-dimensional real vector space with an inner product).

i). $\bar{x}$ is an equilibrium pt. of the diff. Equation

$x' = -\text{grad}\ v(x)$ iff $Dv(\bar{x}) = 0$

ii). If $x(t)$ is a solution of $x' = -\text{grad}\ v(x)$ then

$\dfrac{d}{dt}v(x(t)) = -|\ grad\ v(x)|^2$

iii). If $x(t)$ is not constant, then $v(x(t))$ is a decreasing function of t.

355

Proof: i). $\bar{x}$ is an equilibrium pt. of the gradient system.

$x'grad\ v(x) = -$ iff grad $v(\bar{x}) = 0$

But grad $v(\bar{x}) = \phi^{-1}(Dv(\bar{x}))$

As ϕ^{-1} is an isomorphism,

$\phi^{-1}(Dv(\bar{x})) = 0$ iff $Dv(\bar{x}) = 0$

ii). If x(t) is the trajectory of the gradient system

$x' = -$ grad $v(x)$

Then along the trajectory x(t), we have

$\dfrac{d}{dt}v(x(t)) = Dv(x)\cdot\dfrac{d}{dt}x(t)$.

$= Dv(x)\cdot(x'(t))$

$= Dv(-\ grad\ v(x))\quad (\therefore x' = -\ grad\ v(x(t))$

$\dfrac{d}{dt}v(x(t)) = Dv(x(t))\cdot(-\ grad\ v(x(t)))$

$= <grad\ v(x(t)), -grad\ v(x(t))>$

$= -\mid grad\ v(x(t))\mid^2$

iii). Obviously, if $x(t)$ is not constant, then $\dot{v} < 0$.

Definition: Let T be an operator on real vector space (with an inner product). We define the Adjoint operator T^* on E as

$<T_x Y> = <x, T^* Y>, \forall x, Y \in E$

T is said to be **self Adjoint** iff $T = T^*$.

Let $\{e_1, e_2, \cdots, e_n\}$ be an orthonormal basis of the real vector space E with an inner product $<\cdot, \cdot>$.

Let T be a self Adjoint operator on E.

Let $A = \left(a_{ij}\right)^n_{i,j=1}$ be the matrix of A w.r.t. the basis $\{e_1, e_2, \cdots, e_n\}$.

We show that the matrix A is symmetric.

By the definition of the matrix of T w.r.t. the basis $\{e_1, e_2, \cdots, e_n\}$

We know that

$$T_{ei} = \sum_{k=1}^{n} a_{ki} \cdot e_k, \text{ for } i = 1, 2, \cdots n$$

$$< T_{ei}, e_j > = a_{ji}$$

But $< T_{ei}, e_j > = < e_i, T^* e_j > = < e_i, T_{ej} >$

$\therefore$ T is self Adjoint

But $T_{ei} = \sum_{k=1}^{n} a_{kj} \cdot e_k$

$\therefore < e_i, T_{ej} > = a_{ji}$

Thus $a_{ij} = a_{ji}, 1 \le i, j \le n$.

i.e. the matrix of T w.r.t. the orthonormal basis is symmetric.

Conversely,

If T is an operator on E and has a symmetric matrix w.r.t. an orthonormal basis then we must have

$$a_{ji} = < T_{ei}, e_j > = a_{ij} = < e_i, T_{ej} >$$

But $< T_{ei}, e_j > = < e_i, T^* e_j >$

This holds for $1 \le i, j \le n$

Theorem 3: Let E be a real vector space with an inner product and let T be a self Adjoint operator on E. Then the Eigen values of T are real.

Proof: Since Eigen value of T are to be obtain from the characteristic polynomial of T and since this poly may have in general complex roots, we have to consider complexification of E_c of the space E and T_c of the operator T to consider complex Eigen values.

Let E_c be the complexification of E and T_c be the complexification of T then

$$T_c(a+ib) = T(a) + iT(b), \forall a, b \in E$$

We extend the inner product on E to E_c in the following way:

$$< a+ib, x+iy > = < a+ib, x > + < a+ib, iy >$$
$$= < a, > + i < b, x > - i < a, y > + < b, y >$$

For $a, b, x, y \in E$

It is clear that for any $a \in E_c, < a, a > = 0 \Leftrightarrow a = 0$

Also for any complex number λ and $a, b \in E_c$

$$< \lambda a, b > = \lambda < a, b >$$
$$< a, \lambda b > = \bar{\lambda} < a, b >$$

We next show that for any $a, b \in E$

$$< T_c a, b > = < a, (T^*)_c b >$$

For this let $a = a_1 + ia_2$ and $b = b_1 + ib_2$

$\therefore$ L.H.S. $= < T_C(a_1 + ib_2), b_1 + ib_2 >$

$$= < Ta_1 + ib_2, b_1 + ib_2 >$$
$$= < Ta_1, b_1 > + i < Ta_2, b_1 > - i < Ta_1, b_2 > + Ta_2, b_2 >$$

R.H.S. $= < a_1 + ia_2, T_c^*(b_1 + ib_2) >$

$$= < a_1 + ia_2, T^*b_1 + iT^*b_2 >$$
$$= < a_1, T^*b_1 > + i < a_2, T^*b_1 > - i < T^*b_2 > + < a_2, T^*b_2 >$$

DUAL SPACES

$$= <Ta_1,b_1> +i<Ta_2,b_1> -i<Ta_1,b_2> +<a_2,b_2>$$

Thus $\forall a,b \in E_c, <T_ca,b> = <a,(T^*)_cb>$

If (T_c^*) is Adjoint of T_c, then by the definition

$$<T_ca,b> = <a,(T_c)^*b>$$

Thus $\forall a,b \in E_c, <a,(T^*)_cb> = <a,(T_c)^*b>$

i.e. $(T^*)_c = (T_c)^*$

Since T is self –Adjoint, we have $T=T^*$

$\therefore \forall a,b \in E_c, <T_ca,b> = <a,T_cb>$ (i.e. T_c is also self Adjoint)

Let λ be an Eigen value of T_c,

By the definition, there exists $a \in E_c$ such that

$$T_c a = \lambda a$$

$$\therefore <T_c a,a> = <\lambda a,a> = \lambda <a,a>$$

But $<T_c a,a> = <a,T_c a> = <a,\lambda a> = \bar{\lambda}<a,a>$

$$\lambda<a,a> = \bar{\lambda}<a,a>$$

As a sis a non-zero in E_c.

We have $\lambda = \bar{\lambda}$

i.e. λ is real.

Thus all Eigen values of T_c are real

i.e. all Eigen values of T are real.

Theorem 4: A symmetric real n x n matrix has real Eigen value. (obviously true)

Theorem 5: At an equilibrium of a gradient system, the Eigen values are real.

Proof: Consider a gradient system

$$x' = -grad\ v(x), \text{ where } x = (x_1, x_2, \cdots x_n)$$

Thus the system is given by

$$x_1' = -\frac{\partial v}{\partial x_1}$$

$$x_2' = -\frac{\partial v}{\partial x_2}$$

...

$$x_n' = -\frac{\partial v}{\partial x_n}$$

Let $\bar{x} = (\bar{x}_1, \bar{x}_2, \cdots \bar{x}_n)$ be a pt. of equilibrium.

To study the behaviour of the system at this pt. $\bar{x}$ of equilibrium, we have o consider the linear part i.e. we have to consider D(-grad v(x)).

We know that this operator is given by the matrix

$$\begin{bmatrix} \dfrac{-\partial^2 v}{\partial x_1.\partial x_1} & \dfrac{-\partial^2 v}{\partial x_2.\partial x_1} & \cdots & \dfrac{-\partial^2 v}{\partial x_n.\partial x_1} \\ \dfrac{-\partial^2 v}{\partial x_1.\partial x_2} & \dfrac{-\partial^2 v}{\partial x_2.\partial x_2} & \cdots & \dfrac{-\partial^2 v}{\partial x_n.\partial x_2} \\ \vdots & \vdots & & \\ \dfrac{-\partial^2 v}{\partial x_1.\partial x_n} & \dfrac{-\partial^2 v}{\partial x_2.\partial x_n} & \cdots & \dfrac{-\partial^2 v}{\partial x_n.\partial x_n} \end{bmatrix}$$

Since $\dfrac{\partial^2 v}{\partial x_i.\partial x_j} = \dfrac{\partial^2 v}{\partial x_j.\partial x_i}$, the above matrix is the Eigen values will be real.

Theorem 6: Let E be a real vector space with an inner product. Then any self Adjoint operator on E can be diagonalised.

Proof: Let T be a self Adjoint operator on the real inner product space E. As T is self Adjoint, the Eigen values of T are all real.

DUAL SPACES

Consider an Eigen value say α_1 of T. By the definition, we can find a non-zero element say e_1 of E such that $T(e_1) = \alpha_1 e_1$

Let $E_1 = \{x \in E \text{ such that} < x, e_1 >= 0\}$

$\therefore E_1$ is a subspace of E.

We show that E_1 is invariant under T.

Let $x \in E_1$. By definition $< x, e_1 >= 0$

$< Tx, e_1 > = < x, T^* e_1 > = < x, Te_1 > (\therefore T$ is self Adjoint)

$$= < x, \alpha_1 e_1 >$$

$$= \alpha_1 < x, e_1 >= 0$$

$\therefore T_x \in E_1$

Thus E_1 is invariant under T i.e. $T|_{E_1}$ is an operator on E_1

Let $T_1 \equiv T|_{E_1}$

Using the inner product on E, E_1 is also an inner product space.

We now consider the next Eigen value of T say α_2.

We can find a non-zero element say $e_2 \in E_1$

Such that $Te_2 = \alpha_2 e_2$ i.e. $T_1 e_2 = \alpha_2 e_2$

We then consider $E_2 = \{x \in E_1 \text{ such that } < x, e_2 >= 0\}$

We can show that E_2 is invariant under T_1 so we denote by T_2 the restriction of T_1 to E_2.

$$T_2 = T_1|_{E_2} = T|_{E_2}$$

We can proceed in this way to get a collection of elements $\{e_1, e_2, \cdots e_n\}$ (the above process must stop at some finite stage because E is finite dimensional space).

Since $\{e_1, e_2, \cdots e_n\}$ is an orthogonal collection the elements $e_1, e_2, \cdots e_n$ are Linear independent. Hence they will form a basis.

Since we have, $T(e_i) = \alpha_i e_i, 1 \le i \le n$

The matrix of T with respect to this basis is a diagonal matrix.

Since we can replace $\dfrac{e_i}{|e_i|}$, the sequence $\{e_1, e_2, \cdots e_n\}$

can be considered as an orthonormal w.r.t. which T is diagonal.

Theorem 7: At an equilibrium of a gradient flow. The Linear part of the vector field is a diagonalisable by an orthonormal basis.

Ex 1. Find an orthonormal diagonalising basis for each of the following operators:

(a) $\begin{bmatrix} 2 & 1 \\ 1 & 1 \end{bmatrix}$ (b) $\begin{bmatrix} 0 & -2 \\ -2 & 0 \end{bmatrix}$ (c) $\begin{bmatrix} 0 & 0 & 1 \\ 0 & 1 & 2 \\ 1 & 2 & -1 \end{bmatrix}$

Solution: Consider the operator with the matrix $\begin{bmatrix} 0 & -2 \\ -2 & 0 \end{bmatrix}$

The Eigen values are given by

$\begin{bmatrix} -\lambda & -2 \\ -2 & -\lambda \end{bmatrix} = 0 \Rightarrow \lambda^2 - 4 = 0$

$\Rightarrow \lambda^2 = 4$

$\therefore \lambda = \pm 2$

Consider the Eigen value +2

We find the Eigen space belonging to Eigen value +2.

If will be given by (x, y) such that

$\begin{bmatrix} 0-2 & -2 \\ -2 & 0-2 \end{bmatrix} \begin{bmatrix} x \\ y \end{bmatrix} = 0$

$\Rightarrow -2x - 2y = 0 \Rightarrow x + y = 0$

DUAL SPACES

$\therefore x = -y$

Thus the Eigen space belonging to $\lambda = +2$ is :

$\{(x,y) \in R^2 \text{ such that } x + y = 0\}$

We can thus choose $e_1 = \left(\dfrac{1}{\sqrt{2}}, \dfrac{-1}{\sqrt{2}} \right)$ we have made $|e_1| = 1$

It is clear that the space E_1 which is orthogonally complement to E_1 is :

$\{(x,y) \; s.t \; x = y\}$

We next consider the 2nd Eigen value namely -2.

The Eigen space belonging to the Eigen value -2 is given by

$\begin{bmatrix} 2 & -2 \\ -2 & 2 \end{bmatrix} \begin{bmatrix} x \\ y \end{bmatrix} = \begin{bmatrix} 0 \\ 0 \end{bmatrix}$ i.e. $x = y$

Thus the eigen space belonging to the eigen value -2 is same as the space E_1

We can thus choose $e_2 = \left(\dfrac{1}{\sqrt{2}}, \dfrac{1}{\sqrt{2}} \right)$

Thus a $\{e_1, e_2\}$ is the orthonormal basis w.r.t. which the matrix will be

$\begin{bmatrix} 2 & 0 \\ 0 & -2 \end{bmatrix}$

Ex 2. Let A be a self Adjoint operator. If x and y are eigen vectors belonging to different eigen values then $<x, y> = 0$

Solution: Let x and y be such that x belongs to eigen value say α and y belongs to eigen value say β of A with $\alpha \neq \beta$ then we have

$Ax = \alpha x, \; Ay = By$

So consider $<Ax, y> = <\alpha x, y> = \alpha <x, y>$

But $<Ax, y> = <x, A^* y> = <x, Ay> = <x, \beta y> = \beta <x, y>$

$\therefore \alpha <x, y> = \beta <x, y>$

$\therefore (\alpha - \beta) < x, y > = 0$

But $\alpha \neq \beta$

We must have, $< x, y > = 0$

*RLC EQUATIONS

We consider a circuit consisting of some resistance, inductance and capacitance with changing values the objective is to find out the electric current and voltage in such arm of the circuit.

Applying Kirchhoff's law one get a relation for these electric quantities. This relation is a system of diff. Equation namely

$$\frac{dx}{dt} = y - f(x)$$

$$\frac{dx}{dt} = -x$$

It is known as Lienard's equation.

If $f(x) = x^3 - x$ then this equation is known as van der pol's equation.

LIENARD'S EQUATION

It represents an RLC circuit equation and typically it is given by a system of 1st order diff. equation

$$\frac{dx}{dt} = y - f(x)$$

$$\frac{dx}{dt} = -x$$

We first consider a special case, where

$$f(x) = +kx, \forall k > 0$$

This corresponds to an ordinary resistance in the circuit.

The system becomes,

$$\frac{dx}{dt} = y - kx$$

$$\frac{dx}{dt} = -x$$

The system has an equilibrium point (0, 0) writing $z = (x, y)$ it becomes

$z' = Az$, where $A = \begin{bmatrix} -k & 1 \\ -1 & 0 \end{bmatrix}$

The eigen values are given by $\begin{vmatrix} -k - \lambda & +1 \\ -1 & -\lambda \end{vmatrix} = 0$

i.e. $\lambda(k + \lambda) + 1 = 0$

$\Rightarrow \lambda^2 + \lambda k + 1 = 0$

$\therefore \lambda = \dfrac{-k \pm \sqrt{k^2 - 4}}{2}$

Since $\left| \sqrt{k^2 - 4} \right| \leq |k|$, it follows that even if both the roots are real they are

–ve, other-wise for complex roots the real part is –ve.

Thus the equilibrium pt. (0, 0) is a sink i.e. an asymptotically stable equilibrium point.

In case, $0 < k < 2$, it is a spiral sink.

We now consider the general case.

i.e. $x' = y - f(x)$ and $y' = -x$

it has a unique pt. Of equilibrium $(0, f(0))$.

In order to consider the behaviour of the system at this pt. of equilibrium, we consider the matrix of the linear part of the R.H.S at the equilibrium point.

It is $\begin{bmatrix} -f'(0) & 1 \\ -1 & 0 \end{bmatrix}$ {At any pt. (x, y) the matrix is $\begin{bmatrix} -f'(x) & 1 \\ -1 & 0 \end{bmatrix}$}

365

DYNAMICAL SYSTEM – A SHORT COURSE

The eigen values are given by

$$\begin{vmatrix} -f'(0)-\lambda & 1 \\ -1 & 0-\lambda \end{vmatrix} = 0$$

$$\Rightarrow \lambda(f'(0)+\lambda)+1=0$$

$$\Rightarrow \lambda^2 + f'(0)\cdot\lambda +1=0$$

$$\Rightarrow \lambda = \frac{-f'(0)\pm\sqrt{(f'(0))^2-4}}{2}.$$

The unique equilibrium pt. $(0, f(0))$ is a sink if $f'(0)>0$ (is +ve and source if $f'(0)<0$.

We now consider van der pol's Equation namely we let $f(x)=x^3-x$.

In this case, the unique equilibrium pt. Is (0, 0).

As $f'(0)=-1$, it is a source.

GEOMETRIC ANALYSIS OF VAN DER POL'S EQUATION

We first consider the function $f(x)=x^3-x$

It has three zeros 0, +1, -1 and $f'(x)=3x^2-1$

Thus immediately to the right of (0, 0), the graph will go below the x-axis, will attain minimum at $x=\frac{1}{\sqrt{3}}$, will cross x-axis at x = +1 and then extend to infinity.

Immediately to the left of (0, 0), the graph will go above the x-axis, attains maximum $x=\frac{-1}{\sqrt{3}}$ and then will fall down to meet x-axis at

x = -1 and then extend to -∞.

We draw four boundaries in the R^2 –plane as:

1) $v+=\{(x,y)s.t.\ x=0, y>0\}$
2) $g+=\{(x,y)s.t.\ x>0, y=x^3-x\}$

366

DUAL SPACES

3) $v- = \{(x, y) s.t. \; x = 0, y < 0\}$

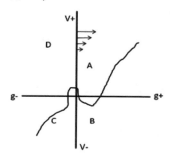

4) $g- = \{(x, y) s.t. \; x < 0, y = x^3 - x\}$

Along with origin (0, 0), these four curves divide

The entire plane into four disjoint parts A, B, C, D.

Where,

A is bounded by v+ and g+, (0, 0)

B is bounded by (0. 0), g+ and v-

C is bounded by (0. 0), v- and g-

D is bounded by (0,0) g- and v+

We not that $y' = 0$ only at $v + Uv - U(0,0)$

Similarly, $x' = 0$ only at $g + Ug - U(0,0)$

We now consider the geometric behaviour of the tangent vector $(x'(t), y'(t))$ at various pt. of R^2.

At pts of V+ i.e. $(x(t), y(t)) \in V +$

i.e. x(t) =0 and y(t) >0

we have $y'(t) = 0$ and $x'(t) = y(t) > 0$

thus at points of V+, the direction of tangent is along the horizontal.

367

At pts. of $g+,\left(x(t)>0, y(t)=x(t)^3 - x(t)\right)$.

We have, $x'(t)=0, y'(t)<0$

Thus the direction of the tangent $\left(x'(t), y'(t)\right)$ to the trajectory $\left(x(t), y(t)\right)$ on g+ is vertically downwards.

At pts. on V-, $y'=0, x'<0$. Thus the direction of the tangent is horizontal but in the negative direction.

At pts. on $g-, x'=0, y'>0$, vertically upward.

At pts in the region A, (where $x>0, y>x^3 - x$)

We have $x'>0, y'<0$

Thus in the region A, the direction is towards the four quadrant.

At pts in the region $B(x>0, y<x^3 - x)$.

We have $x'<0, y'<0$

i.e. the direction of the tangent is towards the 3rd quadrant.

At pts in the region $C, (x<0, y<x^3 - x)$

We have $x'<0, y'<0$

i.e. the direction is pointing towards the 2nd quadrant.

At pts in the region $D, (x<0, y>x^3 - x)$

$x'>0, y'>0$

i.e. the direction is towards the 1st quadrant.

Theorem 8: Any trajectory starting on $v+$ entries A. Any trajectory starting in A meets $g+$ furthermore, if meets $g+$ before it meets $V-, g-$ or $V+$.

Proof: Consider a trajectory $\left(x(t), y(t)\right), t\geq 0$ such that $\left(x(0), y(0)\right)\in V+$

But $\left(x(0), y(0)\right)\in V+$ we have

$x'(0) > 0$ and $y'(0) = 0$

Thus for initially small +ve values of t, $x(t)$ will increase but then with this non-zero x(t), $y'(t)$ will become −ve.

As $x'(0)$ was +ve, it can not immediately become equal to zero.

Thus for some small values of $t>0$, we shall have $x'(t) > 0, y'(t) < 0$

As this behaviour is characteristic of the region A, it follows that the trajectory will enter the region.

Let us assume that the trajectory is in the region.

A goes out of the region A.

We show that it will meet g+ before it can meet

V+, g-, V-

The region A is characterised by $x' > 0, y' < 0$

The boundaries of the region A are V+, g+ and (0, 0).

As (0, 0) is a source, the trajectory can not go to (0, 0).

Thus for the trajectory in A, to come out of A, it has to meet V+ or g+.

But in A, $x' > 0$ i.e. x goes on increasing and as such −x goes on decreasing i.e. y' goes on decreasing.

Thus y goes on decreasing.

Thus the R.H.S. of $y' = y - (x^3 - x)$ goes on decreasing.

Thus we can have $x' \to 0, y' < 0$. The meeting of the trajectory with V+ is ruled out, because for this we shall require $y' = 0$

i.e. y' should increase from its −ve value to zero which is not possible.

Thus the trajectory in A if it goes out of A, it will go out through g+

We now show that the trajectory with initial pt. in A will go out of A after some time and thus met g+.

DYNAMICAL SYSTEM – A SHORT COURSE

Let us assume that this is not true, we thus consider a trajectory $(x(t), y(t))$ with $(x(0), y(0)) \in A$.

We draw a line parallel to the x-axis through the pt. $(x(0), y(0))$.

Let P denote the compact set bounded by V+, g+ and the above line (including the boundaries).

We first show that the trajectory $(x(t), y(t))$ is defined $\forall t \geq 0$

Since the curve is initially at the pt. $(x(0), y(0))$ in P.

It is clear from the geometry of P and the fact that in $x' > 0, y' < 0$, that if $(x(t), y(t))$ does not meet g+, for some t, then $(x(t), y(t))$ will remain in P.

If the solution curve $(x(t), y(t))$ is defined only over a finite interval, we have a contradiction. (as if a solution curve z(t) is defined over a finite maximal interval (α, β) then for any compact set Q, $\exists t \in (\alpha, \beta)$ s.t.

$z(t) \notin Q$)

It therefore follow that $(x(t), y(t))$ is defined for all $t \geq 0$.

Let $a = x(0) > 0$ we can write

$$y(t) - y(0) = \int_0^t y'(s)\, ds$$

But as $a = x(0)$ and $x' > 0$.

We have, $x(t) \geq a, \forall t$

But $y' = -x$

$$\therefore y'(t) \leq -a \Rightarrow y(t) - y(0) \leq \int_0^t -a\, ds = -at$$

$$\therefore y(t) = y(0) - at$$

But we are assuming that the solution is defined for all $t \geq 0$.

Thus, we can let $t \to +\infty$. As $t \to \infty$

This is a contradiction as we are assuming that

$(x(t), y(t))$ remains in P, for all time.

Thus, any solution curve with starting pt in A must meet g+.

Theorem 9: Every trajectory is defined (at least for all $t \geq 0$. Except for (0,

0) each trajectory respectively crosses the curves v+, g+, v-, g- in clock

wise order passing among the region A, B, C, D in clockwise order.

VAN DER POL'S EQUATION

Consider the van-der pol's equation, we know that if the entire R^2 plane is

divided into four disjoint regions A, B, C, D bounded by curves v+, g+, v-,

g- and the equilibrium pt. (0, 0), any solution curve starting at same pt on

v+ will be defined for all $t \geq 0$ and as t goes on increasing it will move into

the region A then cross g+, enter the region B, cross v-, enter the region C,

cross g-, will enter the region D and again crossing v+ and so on.

Thus if we take a solution curve starting at pt. P on v+ there will exists

smallest time $t \geq 0$ such that at time t_1 the solution curve meets v+ again.

The map $t \to t_1(p)$ is a continuous function.

We now define map $\sigma : v+ \to v+$ as $\sigma(p) = \phi_{t1}(p)$

Where, ϕ_t is the flow of the diff. Equation & $\phi_t(p)$ will denote the point

on V+ where the solution curve with starting pt at P meets V+ for the first

time.

The function σ is well defined, continuous and one-one (as the solution

curves do not cross each other). The map σ is known as a solution map.

Theorem 10: Let p∈v+. Then p is a fixed pt. of σ (i.e. $\sigma(p) = p$) if and

only if p is on a periodic solution of van der pol's equation moreover every periodic solution curve meets v+.

Proof: Let $p \in v+$ be such that $\sigma(p) = p$

By the definition, $\phi_{t_1}(p) = p$

$\phi_{2t_1}(p) = \phi_{t_1}(\phi_{t_1}(p)) = \phi_{t_1}(p) = p$

$\phi_{3t_1}(p) = p,\ldots\ldots,\phi_{nt_1}(p) = p, n = 1,2,\cdots$

This means that the solution curve starting at p reaches p after a time

interval of t_1 i.e. the solution curve is a periodic orbit with period t_1 .

Since it is passing through the point p on V+, it is a non-trivial orbit

(a trajectory).

We now assume that the pt. p on V+ is such that $\sigma(p) \neq p$.

Since V+ is a +ve part of the y-axis, we can identify every pt. (0, y) on v+

to the real no. Y.

With the help of this identification, we can define a natural order on V+ as

follows. We shall say that

$(0, y_1) < (0, y_2)$ *iff* $y_1 < y_2$

With the help of this order, we get a complete order on V+.

As $\sigma(p) \neq p$, we have with

$\sigma(p) < p$ or $\sigma(p) > p$

Suppose $\sigma(p) > p$

As the solution curves cannot cross each other, it follows that

$\sigma(\sigma(p)) > \sigma(p)$

($\therefore$ the solution curve starting at P meets V+ at $\sigma(p)$, a point above P.

Hence a solution curve starting at $\sigma(p)$ will have to meet V+ at a point

above $\sigma(p)$ otherwise it will cross the previous part of solution curve.

Thus if $p < \sigma(p)$ then $p < \sigma(p) < \sigma^2(p) < \cdots$

DUAL SPACES

(By Repeating the argument)

Since $\sigma(p), \sigma^2(p), \sigma^3(p), \cdots$ are the pts where the solution curve starting from p meets V+, it follows that the solution cure never meets V+ at P.

Thus the solution curve through P cannot be periodic

By the similar argument if $p > \sigma(p)$ then we have

$p > \sigma(p) > \sigma^2(p) > \cdots$ and thus the curve never meets V+ at P.

Thus if $p \in v+$ is such that $\sigma(p) \neq 0$

Then the solution curve through P is not a periodic solution curve.

Since any solution curve starting at a pt. on V+ (we can take pt. g+, v-, g- but for simplification we take v+) meets v+, it is obvious for periodic solution cure.

Theorem 11: There is one non-trivial periodic solution of the van der pol's equation and every non equilibrium solution tends to this periodic solution.

Proof: We consider the union v+, v- and (0, 0), the entire y-axis.

Thus every pt. of this set is uniquely associated with a real number (its y-coordinate).

Thus if $(0, y) \to y$ then we can define a natural order on $V+UV-U$ (0, 0) as.

$(0, y_1) \leq (0, y_2)$ iff $y_1 \leq y_2$

This is a total order on $V+UV-U$ (0, 0).

We now define a map α from $V+$ to $V-$.

Since the solution curve starting at V+ must meet V- after some finite time, for every p∈V+, we can find smallest t denoted by t_2 s.t $\phi_{t_2}(p) \in V-$.

We define α as $\alpha(p) = \phi_{t_2}(p), \alpha$ is well-defined, Continuous, one-one.

Again using the fact that solution Curves do not cross each other, it follows that if $p_1 < p_2$ on V+ then

$\alpha(p_1) > \alpha(p_2)$.

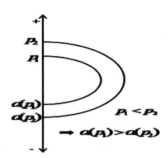

Consider the pt. (1, 0), where g+ meets x-axis. We can find a unique point P_o on V+ s.t the solution curve starting at P_o meets g+ at (1, 0) after some finite interval of time (to get the pt. P_o. We just reverse the process $t \rightarrow -t$ and starting from the pt. (1, 0).)

Let $r = |P_o|$.

We define a function δ from V+ to R as for every

$P \in V+,\ \delta(p) = |\alpha(p)^2|\ p|^2$

We now use the result, $\delta(p) > 0$ if $0 < |P| < r$

$\delta(p)$ decreases monotonically to -∞ as $|p| \mapsto \infty$ with $|p| \geq r$.

As the function δ is continuous, it follows that there will exists a unique pt. q_o is above the pt. P_o on V+.

We recollect that the van der Pol's equation is given by

$x' = y - (x^3 - x)$

$y' = -x$

$\therefore t \rightarrow (x(t), y(t))$ is a solution curve

$\Rightarrow x'(t) = y(t) - \left(x^3(t) - x(t)\right)$

$y'(t) = -x(t)$

$\therefore -x'(t) = -y(t) - \left((-x(t))^3 - (-x(t))\right).$

$-y'(t) = (-x(t))$

i.e. $\dfrac{d}{dt}(-x(t)) = +(-y(t)) - \left((-x(t))^3 - (-x(t))\right)$

$\dfrac{d}{dt}(-y(t)) = -(-x(t))$

i.e. if $t \rightarrow (-x(t), y(t))$ is a solution then $t \rightarrow (-x(t), y(t))$ is also a solution of the van der Pol's equation.

It means that if $\phi_t(a) = b$ then $\phi_t(-a) = -b$.

As $\delta(q_o) = 0$, by definition, $|\alpha(q_o)|^2 - |q_o|^2 = 0$

i.e. $|\alpha(q_o)| = |q_o|$

$\therefore \alpha(q_o) = -q_o$

i.e. $\phi_{t_2}(q_o) = q_o$

$\therefore \phi_{t_2}(-q_o) = -(q_o) = q_o$

Thus we have $\phi_{t_2}(q_o) = -q_o$ and $\phi_{t_2}(-q_o) = q_o$

$\therefore \phi_{2t_2}(q_o) = q_o$

Thus with $t_1 = 2t_2$, we have $\phi_{t_1}(q_o) = q_o$

i.e. $\sigma(q_o) = q_o$

by the (previous) proposition, it follows that the trajectory/solution curve through q_o is a periodic orbit.

If is thus a non-trivial periodic orbit. Also because of the choice of the function δ and α it follows that this periodic orbit is unique.

We now show that every other non-trivial orbit approaches this periodic orbit.

Let $p \in V+$ be a point such that $p \neq q_o$

We shall show that the trajectory through the pt. p finally converge to this periodic orbit.

Let us assume that $p > q_o$.

As $\sigma(q_o) = q_o$, we must have $\sigma(p) > q_o$

We shall show that $p > \sigma(p)$

Since $p > q_o$ and $\delta(q_o) = 0$ we must have $\delta(p) < 0$

$|\alpha(p)|^2 - |p|^2 < 0$

$|\alpha(p)|^2 < |p|^2$

$\therefore |\alpha(p)| < |p| = |-p|$

$\therefore \alpha(p) > -p$

We now define a function β from V- to V+ as

For $p \in V-$, $\beta(p)$ is the pt. on V+, where the solution curve through the pt. P meets V+ for the first time.

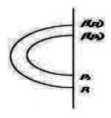

It is clear that β is well defined, Continuous, one-one.

For pts. p_1, p_2 on V-, $p_1 < p_2$, it is clear that

$\beta(p_1) > \beta(p_2)$, it follows that

$\beta(p) = -\alpha(-p), \forall p \in v_-$

Also, $\sigma = \beta\alpha$, we have $\alpha(p) > -p$

We apply β on this to get

$\beta(\alpha(p)) < \beta - p) = -\alpha(p)$

i.e. $\sigma(p) < p$

thus we have shown that if $p > q_o$ then

$\sigma(p) > q_o$ and $p > \sigma(p) > q_o$

Repeated application of σ on this will give

$p > \sigma(p) > \sigma^2(p) > q_o$

i.e. $p > \sigma(p) > \sigma^2(p) > \sigma^3(p) > \ldots\ldots > q_o$.

Thus $p, \sigma(p), \sigma^2(p), \sigma^3(p)\ldots\ldots$ is a decreasing seq. On V+ which is

bounded below by q_o. This seq. will converge to a pt. say q_1.

We show that $\sigma(q_1) = q_1$

∴ By definition, $q_1 = \overset{\lim}{\underset{n\to\infty}{}} \sigma^n(p)$

∴ $\sigma(q_1) = \sigma\left(\overset{\lim}{\underset{n\to\infty}{}} \sigma^n(p)\right) = \overset{\lim}{\underset{n\to\infty}{}}\left(\sigma\sigma^n(p)\right) = q_1$

Thus $\sigma(q_1) = q_1$ i.e. the trajectory through the pt. q_1 is a periodic orbit.

But there is only one periodic orbit through q_o on V+.

∴ $q_1 = q_o$ i.e. $\sigma^n(p) = q_o$

We can use similar argument for $p > q_o$

Thus every non-trivial trajectory converges to the periodic orbit.

Let $R^2 \to R$ be defined by

$$w(x, y) = \frac{1}{2}\left(x^2 + y^2\right).$$

Theorem 12: Let $z(t) = (x(t), y(t))$ be a solution curve of Lienard's equation $x' = y - f(x)$, $y' = -x$

Then $\dfrac{d}{dt} w(z(t)) = -x(t) \cdot f(x(t))$

Proof: By the definition, $\dfrac{d}{dt} w(z(t)) = DW \cdot z'(t)$

As w: $w: R^2 \to R,\ DW \in L(R^2, R)$. In the standard basis Dw is given by the row matrix

$$\left(\frac{\partial w}{\partial x}, \frac{\partial w}{\partial y}\right) = (x, y)$$

Thus along the trajectory of the Lienard's equation,

We have

$$\frac{d}{dt} w(z(t)) = (x(t), y(t))\begin{pmatrix} x'(t) \\ y'(t) \end{pmatrix}$$

$$= (x(t), y(t))\begin{pmatrix} y(t) - f(x(t)) \\ -x(t) \end{pmatrix}$$

$$= (x(t).y(t)) - x(t)f(x(t)) - x(t)\,y(t)$$

$$\therefore \frac{d}{dt} w(z(t)) = -x(t)\,f(x(t))$$

Thus if the product $x(t). f(x(t)) > 0$, then along the trajectories of the Lienard's equation w goes on decreasing.

Definition: A resistor is said to be passive resistor if its characteristics given by the function f is of the following type

DUAL SPACES

HOPF BIFURCATION

Particular case of van der Pol's Equation Consider a particular type of van-der Pol's equation given as

$$x' = y - (x^3 \mu x)$$

$$y' = -x$$

We let the parameter μ to vary in [-1, 1]

For this system, (0, 0) is the only equilibrium point.

The behaviour of this equilibrium pt. is given by the expression $-f'(0)$, where $f(x) = x^3 - \mu x$ i.e. $+\mu$

So, if $-1 \le \mu < 0$, in this case the equilibrium pt. (0, 0) is asymptotically stable (i.e. a sink).

For $0 < \mu \le 1$, the pt. of equilibrium (0, 0) is an unstable equilibrium (i.e. a source) .

For $\mu = 0$, the pt. of equilibrium may either be a centre or a focus (follows from Dulac's theorem).

Thus in the transition from negative values to +ve value of μ, the behaviour of the system at its pt. of equilibrium changes drastically.

Thus we say that $\mu = 0$ is a Bifurcation point (value). No geometrical interpretation for $\mu = 0$.

Thus if the resistor in the circuit of Lienard's equation given by

$f(x) = x^3 - \mu x$ is such that $-1 \le \mu < 0$

Then (0. 0) being a sink, every trajectory will approach the value (0, 0) after an elapse of time.

Since (x, y) denote current and voltage in the circuit, it mean that for such a resistor, the circuit dies down after certain time. Therefore such type of resistors are known as passive resistors.

For $0 < \mu \le 1$, (like van der Pol's Equation) one can show the existence of a non-trivial periodic orbit to which all other orbit approach.

Thus in such cases the current circuit is very much alive.

Hopf proved that for a fairly general one parameter families of equation $x' = f_{\mu}(x)$, there must be a closed orbit for $\mu > \mu_o$ if the eigen value character of on equilibrium change suddenly at μ_o from a sink to a source.

Ex 3. Discuss the bifurcation point for the system

$$x' = -y + x\left(\mu - x^2 - y^2\right)$$
$$y' = -x + y\left(\mu - x^2 - y^2\right)$$

Solution : Here (0, 0) is the only pt. of equilibrium.

The matrix for the linear part of the vector field $\left(f_1(x,y), f_2(x,y)\right)$ with

$$f_1(x,y) = -y + x\left(\mu - x^2 - y^2\right)$$
$$f_2(x,y) = x + x\left(\mu - x^2 - y^2\right)$$

is:

$$\begin{bmatrix} \dfrac{\partial f_1}{\partial x} & \dfrac{\partial f_1}{\partial y} \\ \dfrac{\partial f_2}{\partial x} & \dfrac{\partial f_2}{\partial y} \end{bmatrix} = \begin{bmatrix} \mu - 3x^2 - y^2 & -1 - 2xy \\ 1 - 2xy & \mu - x^2 - 3y^2 \end{bmatrix}$$

at pt. (0. 0), the matrix of the linear part is:

$$\begin{bmatrix} \mu^2 & -1 \\ 1 & \mu \end{bmatrix}$$

The eigen values are given by

$$\begin{vmatrix} \mu - \lambda & -1 \\ 1 & \mu - \lambda \end{vmatrix} = 0$$

$\therefore (\mu - \lambda)^2 + 1 = 0$

$(\mu - \lambda)^2 = -1 \Rightarrow \mu - \lambda = \pm i$

$\therefore \lambda = \pm i$

Thus if $\mu < 0$, the pt. of equilibrium (0, 0) is a sink and for $\mu > 0$ it is a source

Thus $\mu = 0$ is the bifurcation value.

Ex 4. $\dfrac{dx}{dt} = \mu x + y, \dfrac{dx}{dt} = x - 2y$

Solution: Here the pt. of equilibrium is (0, 0)

It is a plane autonomous system with the pt. of equilibrium (0, 0) and the matrix A is given by

$$\begin{bmatrix} \mu & 1 \\ 1 & -2 \end{bmatrix}$$

Eigen values are given by

$$\begin{vmatrix} \mu - \lambda & 1 \\ 1 & -2 - \lambda \end{vmatrix} = 0$$

$\Rightarrow -(\mu - \lambda)(2 + \lambda) - 1 = 0$

$\Rightarrow (\mu - \lambda)(2 + \lambda) + 1 = 0$

$\Rightarrow 2\mu + \mu\lambda - 2\lambda - \lambda^2 + 1 = 0$

$\therefore -\lambda^2 + (\mu - 2)\lambda + (2\mu + 1) = 0$

i.e. $\lambda^2 - (\mu - 2)\lambda - (2\mu + 1) = 0$

$\therefore \lambda = \dfrac{(\mu - 2) \pm \sqrt{[-(\mu - 2)]^2 - 4 \times 1 \times (-(2\mu + 1))}}{2 \times 1}$

$\therefore \lambda = \dfrac{(\mu - 2) \pm \sqrt{(\mu - 2)^2 + 4(2\mu + 1)}}{2}$

So, for $2\mu+1>0$ both roots are real and of opposite sign.

Thus (0, 0) is unstable equilibrium.

If $2\mu+1>0$, then certainly $\mu-2<0$

Then both the roots will be –ve and thus we get a sink at (0, 0).

Thus $2\mu+1=0$ i.e. $\mu=-\dfrac{1}{2}$ is a bifurcation value.

CHAPTER 8

LIMIT SETS

Let w be an open subset of a normed Linear (finite dimensional) space E. consider a vector field defined on w namely say F and the corresponding dynamical system given by the system of diff. equation $x' = f(x)$

Let ϕ_t denote the flow of this diff. equation (on the vector field f)

For $x \in W$, we say that y is an w-limit point of x if there exist a seq. $\{t_n\}, t_n \to \infty$ such that

$$Y = \lim_{n \to \infty} \phi_{tn}(x)$$

Collection of all w-limit pts. of x is denoted by $L_w(x)$ and is known as w-limit set of the pt. x.

In the same way, y is known as an α-limit point of the pt. x if there exist a seq. $\{t_n\} \to -\infty$ s.t.

$$Y = \lim_{n \to \infty} \phi_{tn}(x)$$

The collection of all α-limit pts. of x is denoted by $L_\alpha(x)$ and is known as α-limit set of x.

By a limit set we mean either $L_\alpha(x)$ or $L_w(x)$, for some x.

Ex 1. Let $\bar{x}$ be an asymptotically stable equilibrium.

It follows that if y is any pt. in its basin then any trajectory through y will be tending to $\bar{x}$

i.e. $\phi_t(y) \to \bar{x}$ as $t \to +\infty$.

Thus $\bar{x}$ is the only w-limit pt. of every point in the basin of $\bar{x}$.

Ex 2. Let $\bar{x}$ be an equilibrium pt.

Then by the definition of an equilibrium point

$$L_w(\bar{x}) = L_\alpha(\bar{x}) = \{\bar{x}\}$$

(The trajectory at $\bar{x}$ remains at $\bar{x}$ for all time $t \in R$)

Ex 3. Let x be a closed trajectory of a dynamical system.

Let x_o be a pt. on x. since the trajectory is closed it follows that the pt. x_o will be attained infinitely many time for positive as well as negative t.

$x_o \in L_w(x_o)$ as well as $x_o \in L_\alpha(x_o)$

If y_o is only other point on x, let $\phi_s(y_o) = x_o$

If $\phi_{t_1}(x_o) = x_o$, for some $t_1 > 0$ then

$\phi_{s+t_1}(y_o) = \phi_{s+2t_1}(y_o) = \dots \dots = x_o$

$\therefore x_o \in L_w(y_o)$. Similarly $x_o \in L_\alpha(y_o)$

Since y_o is any point of x, it follows that every pt. of x belongs to L_w and L_α of every point of x.

Ex 4. Consider the van der Pol's equation

In this case, (0, 0) is an unstable equilibrium pt. (source) and there exists a non-trivial periodic orbit.

It then follows that this periodic orbit is w-limit of every other non-trivial orbit.

Since (0, 0) is the source for any point x lying in the region bounded buy the periodic orbit, $(0, 0) \in L_\alpha(x)$

However, if x lies outside the periodic orbit, $\in L_\alpha(x)$ will be empty.

Ex 5. Consider a phase portrait of a dynamical system with three equilibrium pts. A, B, C.

The pts. A and B are sources while B is just an equilibrium pt. there is a closed trajectory given by the figure of eight. This figure of eight is in the

LIMIT SETS

w-limit sets of all pts. lying outside it.

It follows that A is the α-limit pt. of every pt. in he left half of figure of eight and C is α–limit pt. of every pt in the right half of figure of eight.

Theorem 1: a) If x and z are an the same trajectory, then $L_w(x) = L_w(z)$. Similarly for α-limits.

b) If D is a closed positively invariant set and $z \in D$, then $L_w(z) \subseteq D$. Similarly, for negatively invariant sets and α-limit sets of every pt. in it.

Proof: (a) Since x and z are on the same trajectory, we have

For some $s \in R, \phi_s(x) = z$

Let $y \in L_w(x)$ so by the definition, $\exists$ a seq. $\{t_n\}$ with $t_n \to \infty$ such that

$$Y = \lim_{n \to \infty} \phi_{tn}(x)$$

But $\phi_s(x) = z \Rightarrow x = \phi_{-s}(z)$

$$\therefore Y = \lim_{n \to \infty} \phi_{tn}(\phi_{-s}(z))$$

$$= \lim_{n \to \infty} \phi_{tn-s}(z)$$

Let $t'_n = t_{n-s}$. As $n \to \infty$, $t'_n \to +\infty$

Thus $Y = \lim_{n \to \infty} \phi_{t'n}(z)$, where $t'_n \to +\infty$ as $n \to +\infty$

$\therefore y \in L_w(z)$

$\therefore L_w(x) \subseteq L_w(z)$

Similarly, we can show that $L_w(z) \subseteq L_w(x)$

$\therefore L_w(x) = L_w(z)$

On the similar lines, $L_\alpha(x) = L_\alpha(z)$

(b) Let D be a closed and positively invariant set.

385

DYNAMICAL SYSTEM – A SHORT COURSE

Let $z \in D$. Let $y \in L_w(z)$.

By the definition, we can find a seq. $\{t_n\}$ with $t_n \to \infty$ as $n \to \infty$ s.t.

$$Y = \lim_{n \to \infty} \phi_{tn}(z).$$

Since $z \in D$ and D is +vely invariant,

$$\phi_{tn}(z) \in D, \forall n$$

Y is thus a limit of a seq. of pts. in D.

But D is closed

$$\therefore y \in D$$

$$\therefore L_w(z) \subseteq D$$

Similarly, if D is dosed and –vely invariant and

$z \in D$ then $L_a(z) \subseteq D$

(c) Since the set is invariant, it is both positively and negatively invariant.

Hence, For every z in it, $L_a(z) \subseteq D$, $L_w(z) \subseteq D$

LOCAL SECTION AND FLOW BOX

Let E be an n-dimensional normed linear space and w an open subset of it.

By a hyper plane we mean a subspace of E with dimension n-1.

Let f be a vector field on W. let S be an open subset of a hyper plane (we can also assume that $a \in S$).

We say that f is transverse to S when for every $x \in S$, $f(x) \notin S$ (i.e. in particular $f(x) \neq 0$).

A local section of a (assume that $o \in W$) of F is an open set S containing in a hyper plane $H \subseteq E$ which is transverse to f.

We can use the local section to obtain a flow box of any pt. which is not an equilibrium pt. of the given dynamical system.

LIMIT SETS

We consider the flow field and a pt. in this flow field we assume that the flow is not stationary. We can take a small nbd of this point and assume that the stream lines are parallel to each other in this nbd.

We can take a section of the stream tube through the pt. a s.t. the stream lines are not parallel to this section i.e. the stream lines are either coming out or going in the section.

If we take any point other than a in this small stream tube, it will lie on a unique stream line which will meet the above at some point.

Thus the given pt. can be described as the one which lies on a stream line (passing through a unique pt. in the section) and which is attained along this stream line after an elapse of some time.

This is a very simple description of all pts in a small stream tube surrounding the pt. a this can also be done in the field trajectories of a given dynamical system where stream lines will be replaced by trajectories and stream tube is replaced by what is known as a flow box.

A diffeomorphism is a map from an open subset of a normed linear space to an open subset of another normed linear space such that the map and its inverse are both differentiable.

If the map and its inverse are k-times differentiable for $k \geq 1$, the map is said to be $C^k -$ diffeomorphism.

Two diffeomorphism $f, g : w \rightarrow w$ are said to be topologically (or $C^\infty -$) conjugate if there is a homomorphism $h : w \rightarrow w$ s.t.

$h.f = g.h$

Two flows ϕ_t and $\psi_t : W \rightarrow W$ are said to be topologically conjugate if there exists a homeomorphism $h : W \rightarrow W$ such that

$h \cdot \phi_t = \psi_t \cdot h, \forall t \in R$

i.e. $\phi_t(x) \xrightarrow{h} \psi_t(h(x)), \forall t \in R$ and $x \in W$

i.e. each trajectory of ϕ_t is mapped onto one and only one trajectory of ψ_t and vice-versa.

If h in the above definition is a C^k – diffeomorphism with $k \geq 1$ then we say that ϕ_t and ψ_t are C^k – conjugate.

C^k – Conjugacy of ϕ_t and ψ_t corresponds to there being a k-times differentiable change of coordinates h, which transforms the differential equation,

$\dot{x} = x(x)$ of ϕ_t into $\dot{y} = y(y)$ of ψ_t.

Since we have $h(\phi_t(x)) = \psi_t(h(x))$ and h is diffeomorphism .

We differential this equation w.r.t. t and evaluate at $t=0$ to get

$$\left\{ Dh(\phi_t(x)) \cdot \frac{d}{dt} \phi_t(x) \right\}_{t=0} = \frac{d}{dt} \psi_t(h(x)) \bigg|_{t=0}$$

i.e. $Dh(x) \cdot x(x) = Y(h(x))$ $\{\because \phi_o(x) = x\}$

let us denote the change of coordinates i.e. $h(x)$ by Y i.e. $y(x) = h(x)$ then

$y(x(t)) = h(x(t))$

$\dot{y} = \dot{h}(x(t)) = Dh(x) \cdot \dot{x}$

$\quad = Dh(x) \cdot X(x) = Y(h(t)) = Y(y)$

i.e. $\dot{y} = Y(y)$

thus if h is the diffeomorphism through which the flow fields ϕ_t and ψ_t of the dynamical system

$\dot{x} = X(x)$ and $\dot{y} = Y(y)$

Then the coordinate transformation given by h transform the dynamical system $\dot{x} = X(x)$ into $\dot{y} = Y(y)$

LIMIT SETS

Let x_o be a regular point (which is not an equilibrium pt) of a dynamical system $\dot{x} = X(x)$.

Let S be a local section the point x_o. We can choose S in such a way that $X(x_o)$ is $\perp$er to S

i.e. the trajectory of the flow field through x_o is $\perp$er to the plane S.

There is a nbd V of x_o such that any pt. $x \in V$ can be written as

$x = \phi_u(y)$, for some $y \in S$

In other words, we can use the trajectory of the flow to define new coordinates on V.

This new coordinates are best related to local coordinates at x_o.

By shifting of origin, we can consider x_o to be the origin of the coordinate system.

Suppose we can choose a basis in R^n which has $X(0)$ as its first vector. Then the first coordinate of every point Y in S is zero and S defines a nbd $\widetilde{S}$ of O in R^{n-1}.

Each pt. of $\widetilde{S}$ can be specified by $\xi \in R^{n-1}$ and every pt. x of v can be written as

$x = \phi_u(0, \xi) = \phi(u, (0, \xi)) = h(u, \xi)$

By definition, $h : R^n \rightarrow R^n$ in c^1. Further $h|_{\widetilde{S}}$ is an identity and

$Du(h(0)) = X(0) \quad (\because x = 0)$

Thus Det. $Dh(0) \neq 0$ and so h^{-1} exists.

In the new coordinates, the trajectories of the flow are simply the lines of constant ξ

i.e. $\psi_t(u,\xi) = (u+t,\xi)$

To show that ϕ_t and ψ_t are conjugate.

We note that $h(\psi_t(u,\xi)) = h(t+u,\xi)$

But $h(t+u,\xi) = \phi(t+u,(u,\xi))$

$$= \phi_t(\phi_u(0,\xi))$$

$$= \phi_t\phi(u,(0,\xi))$$

i.e. $h(\psi_t(u,\xi)) = \phi_t(h(u,\xi))$

MONOTONE SEQUENCE IN PLANAR DYNAMICAL SYSTEM

Let $x_o, x_1, \cdots$ be a finite or infinite seq. of distinct pts. on the solution curve

$$C = \{\phi_t(x_o) : 0 \le t \le \alpha\}$$

We say that the sequence $\{x_o, x_1, \cdots\}$ is monotone along the trajectory if

$\phi_{tn}(x_o)x_n$ with $0 \le t_1 \le t_2 \le t_3 \le \cdots \le \alpha$

Let $y_o, y_1, \cdots$ be a finite or infinite sequences of pts. On a line segment I of R^2.

We say that the seq. is monotone along I if the vector $y_n - y_o$ is a scalar multiple of $y_1 - y_o$ as

$y_n - y_o = \lambda_n(y_1 - y_o)$ with $\lambda_n > 1, \lambda_1 < \lambda_2 < \lambda_3 \cdots$.

The seq. $\{\lambda_o, \lambda_1, \lambda_2 \cdots\}$ is monotone along a segment I means the pts $\lambda_1, \lambda_2 \cdots$ are on the same side of y_o on I and y_n lies in between y_{n-1} and y_{n-1}.

SOME PROPERTIES OF W-LIMIT SETS (α-LIMIT SETS)

Consider a trajectory Γ of a dynamical system let x and y be any two pts.

on Γ.

We know that $L_w(x) = L_w(y)$

Similarly, $L_\alpha(x) = L_\alpha(y)$. i.e. the w-limit set/$\alpha$-limit set remains same for any pt. on a given trajectory.

Thus we can talk of $w(\Gamma)$ and $\alpha(\Gamma)$

i.e. w limit set of pts of Γ and α limit set of pts of Γ.

i.e. $w(\Gamma) = L_w(x)$, where x is any pt. on Γ.

i.e. $\alpha(\Gamma) = L_\alpha(x)$,

Theorem 2: Consider a dynamical system in a normed linear space E and let Γ, $w(\Gamma)$ and $\alpha(\Gamma)$ denote the trajectory, w-limit set and α-limit set of the trajectory of the dynamical system. Then

a) $w(\Gamma)$ is closed in E ($\alpha(\Gamma)$ is a closed subset of E).

b) If Γ lies in a compact subset of E, then $w(\Gamma)$ (and $\alpha(\Gamma)$ is non-empty, connected and compact subset of E.

Proof: a) We show that $w(\Gamma)$ is a closed se in E.

Let $\{p_1, p_2 \cdots\}$ be a seq. of pts. In $w(\Gamma)$ which converges to say a pt. $p \in E$. We show that $p \in w(\Gamma)$

As $p_1 \in w(\Gamma), p_1 \in L_w(x), x \in \Gamma$

By the definition, we can find a seq. $\{t_1^{(1)}, t_2^{(1)}, t_3^{(1)}, \cdots\}$

Such that $t_n^{(1)} \to +\infty$ as $n \to \infty$ and

$\phi(t_n^{(1)}, x) \to p_1$ as $n \to \infty$

Where, ϕ denotes the flow of the Dynamical system.

As $\phi_2 \in L_w(x)$, we can find a seq. $\{t_1^{(2)}, t_2^{(2)}, \cdots\}$

Such that $t_n^{(2)} \to +\infty$ as $n \to \infty$ and

$\phi\left(t_n^{(2)}, x\right) \to p_2$ as $n \to \infty$

While choosing a seq. $\left\{t_1^{(2)}, t_2^{(2)}, \cdots\right\}$ we choose the elements of the seq. in such a way that

$t_1^{(1)} < t_2^{(2)}, t_2^{(1)} < t_2^{(2)}$ and so on.

(this is always possible as both the seq. are tending to $+\infty$).

We repeat this process to get a seq. $\left\{t_k^{(n)}\right\}$ s.t. $t_k^{(n)} \to +\infty$ as $k \to \infty$ and

$\phi\left(t_k^{(n)}, x\right) \to p_n$ as $k \to \infty$

and $t_k^{(n-1)} < t_k^{(n)}, n = 1, 2, \cdots, k = 1, 2, \cdots$

since $\phi\left(t_k^{(n)}, x\right) \to p_n$ as $k \to \infty$

then for $\in = \dfrac{1}{n}$, there exist a number $k^{(n)}$ s.t.

$\forall k \geq k^{(n)}, |\phi\left(t_k^{(n)}, x\right) - p_n| < \dfrac{1}{n}$

We can again choose the numbers $k^{(1)}, k^{(2)}, \cdots$

Such that $k^{(1)} < k^{(2)} < k^{(3)} < \cdots$

We now define $t_n = t_{k^{(n)}}^{(n)}$, it is clear that as $n \to \infty, t_n \to \infty$

We consider

$|\phi(t_n, x) - p| \leq |\phi(t_n, x) - p_n| + |p_n - p|$ by triangle inequality.

$\leq \dfrac{1}{2} + |p_n - p|$

But R.H.S $\to 0$ AS $n \to \infty$ $(\because P_N \to P)$

Thus if p is a limit pt. of the seq. $\{p_1, p_2, \cdots\}$

It is possible for us to find a seq. $\{t_1, t_2, \cdots\}$ s.t. $t_n \to \infty$ as $n \to \infty$ and

LIMIT SETS

$|\phi(t_n, x) - p| \to 0$ as $n \to \infty$

Thus p is an w-limit pt. of x.

$\therefore p \in L_w(x) = w(\Gamma)$

$\therefore w(\Gamma)$ is a closed subset of E.

b) We assume that the trajectory Γ lies within a compact subset of E.

We show that $w(\Gamma) \neq \phi$

Since the entire trajectory Γ lies within a compact set, it must be defined for t > 0 i.e. for $t \in [0, \infty)$

Thus if $x \in \Gamma$ then $\phi_t(x) \in \Gamma, \forall t \geq 0$.

Consider the seq. $\{\phi_o(x), \phi_1(x), \cdots\}$. There pts. Are on Γ and Γ lies in a compact set.

As such the seq. $\{\phi_o(x), \phi_1(x), \cdots\}$ will have a limit pt i.e. seq. (if required we may replace it by appropriate sub seq.) $\phi_n(x)$ converges to a point.

Obiviously this limit pt. will be in $L_w(x)$

i.e. $L_w(x) \neq \phi$ i.e. $w(\Gamma) \neq \phi$.

Since $w(\Gamma)$ will be contained in the compact set and it is closed, it follows that $w(\Gamma)$ is a compact set (when Γ lies entirely within, the compact set).

We now show that $w(\Gamma)$ is a connected set.

Let us assume that $w(\Gamma)$ is not connected.

Let A, B be a separation of $w(\Gamma)$ i.e. A and B are closed subset of $w(\Gamma)$.

i.e. $A \cap B = \phi, A \cup B = w(\Gamma)$

since A and B are closed disjoint sets.

$\therefore d(A, B) > 0$

Let $\delta = d(A,B)$. Let $x \in \Gamma$.

Any pt. of A is an w-limit pt. of x and as such we can find $t_1^{(1)}$ such that

$$d\left(\phi(t_1^{(1)}, x), A\right) < \frac{\delta}{2}$$

Similarly, any pt. of B is an w-limit pt. of x and we can find $t_2^{(1)}$ such that

$$d\left(\phi(t_2^{(1)}, x), B\right) < \frac{\delta}{2}$$

Since $d(A,B) = \delta$, it follows that $d\left(\phi(t_2^{(1)}, x), A\right) > \frac{\delta}{2}$.

If we fix the point x and set A, we use the fact that the distance is a continuous function of $t_1^{(1)}$ and $t_2^{(1)}$ and hence we can find an intermediate value say t_1 such that $d\left(\phi(t_1, x), A\right) > \frac{\delta}{2}$.

We repeat the process by getting values $t_1^{(2)}, t_2^{(2)}$ s.t

$$d\left(\phi(t_1^{(2)}, x), A\right) < \frac{\delta}{2}, d\left(\phi(t_2^{(2)}, x), A\right) > \frac{\delta}{2}$$

and therefore for an intermediate value say t_2

$$d\left(\phi(t_2, x), A\right) > \frac{\delta}{2}$$

In this way we get a seq. of numbers $t_1, t_2, t_3, \cdots$ such that

$$\forall n, \ d\left(\phi(t_n, x), A\right) > \frac{\delta}{2} \ \text{and} \ t_n \to \infty \ \text{as} \ n \to \infty.$$

Since the trajectory Γ lies in a compact set, the seq. $\{\phi(t_n, x)\}$ will have a limit pt. say p.

Since $p = \lim_{n \to \infty} \phi(t_n, x)$ and $t_n \to +\infty$ as $n \to \infty$, it follows that $p \in L_w(x)$ i.e. $p \in w(\Gamma)$.

LIMIT SETS

As $d\big(\phi(t_n,x),A\big)=\dfrac{\delta}{2}$ and $p=\displaystyle\lim_{n\to\infty}\phi(t_n,x)$

We have $d(P,A)=\dfrac{\delta}{2}$, we have $p\notin A$.

We use the triangle inequality namely,

$d(A,B)\le d(P,A)+d(P,B)$

$\therefore d(P,B)\ge d(A,B)-d(P,A)$

$$\ge\delta-\frac{\delta}{2}=\frac{\delta}{2}.$$

Thus $d(P,B)\ge\dfrac{\delta}{2}$,

$\therefore p\notin B$

Thus $p\notin A$, $p\notin B$, $p\in w(\Gamma)$ and $w(\Gamma)=A\cup B$

This is a contradiction.

$\therefore w(\Gamma)$ is connected.

Theorem 3: Let $y\in w(\Gamma)$ then the trajectory through y lies in $w(\Gamma)$.

Similar result for α-limit set.

Proof: Since $y\in w(\Gamma)$, we can write

$y\in L_w(x)$, some $x\in\Gamma$

So by the definition of w limit pt., we can find a seq. say $t_1,t_2,\cdots$ s.t.

$t_n\to+\infty$ as $n\to\infty$ and

$\phi(t_n,x)\to y$ or $\phi_{t_n}(x)\to y$ as $n\to\infty$.

Let z be a point on a trajectory through y.

$z=\phi_s(y)$, for some $s\in R$

Consider the seq. $\{t_{1+s},t_{2+s},\cdots\}$ and consider

DYNAMICAL SYSTEM – A SHORT COURSE

$$\phi_{tn+s}(x) = \phi_{s+tn}(x) = \phi_s\big(\phi_{tn}(x)\big)\phi_s \cdot \phi_{tn}(x)$$

$$\therefore \lim_{n \to \infty} \phi_{tn+s}(x) = \phi_s\Big(\lim_{n \to \infty} \phi_{tn}(x)\Big) = \phi_s(y) = z$$

$\therefore \phi$ is continuous.

Further as $s \in R$ and $t_1, t_2, \cdots \to +\infty$ as $n \to \infty$.

The seq. $t_{1+s}, t_{2+s}, \cdots \to +\infty$ as $n \to \infty$.

$\therefore z \in L_w(x) = w(\Gamma)$

Theorem 4: The α-limit set and w-limit set of a trajectory is invariant in the sense if $\phi(,)$ denote the follow of the dynamical system then

$$\phi_t\big(w(\Gamma)\big) \subseteq w(\Gamma)$$

$$\phi_t\big(\alpha(\Gamma)\big) \subseteq \alpha(\Gamma)$$

PLANE DYNAMICAL SYSTEM

We consider $E = R^2$ and the dynamical system.

$x' = f(x)$, Here $x \in R^2$

Similarly, $f(x) : R^2 \to R^2$

Since $E = R^2$, it follows that local section must be line segment.

We consider a trajectory of a plane dynamical system and also a line segment.

Suppose the line segment cuts the trajectory in a seq. of pts.

If the seq. of pts. Is monotone along the trajectory does it necessarily mean that the seq. is monotone along the line segment.

In case (a), the seq. $\{y_o, y_1, y_2\}$ is monotone along both the trajectory and the line segment.

In case (b), the seq. $\{y_o, y_1, y_2\}$ is monotone along both the trajectory and the line segment.

In case (c), the seq. $\{y_o, y_1, y_2\}$ is monotone along the trajectory but not along the line segment.

We note that in case (a), the line segment is not a local section as the direction of the trajectory is reversing.

In case (b), the line segment can be a local section.

In case (c), the line segment cannot be a local section.

Thus along a line segment one may have a seq. of pts. which are monotone along the trajectory but may or may not be monotone along the segment (case (a) and (c)).

However if the line segment is also a local section if a seq. is monotone along a trajectory then it must be monotone along the local section also.

Theorem 5: Let S be a local section of a C^1 – planar dynamical system and $y_o, y_1, y_2 \cdots$ a seq. of distinct pts. Of S that are on some solution curve C.

If the seq. is monotone along C, it is also monotone along S.

Proof: We shall first consider three pts. y_o, y_1 and y_2 which are monotone along a trajectory and then show that they are also monotone along the local section S.

We consider the pts. y_o and y_1 on the trajectory and then show that y_2 is located on the local section S in such a way that y_o, y_1, y_2 is monotone along S.

Let Σ denotes the closed curve consisting of the trajectory starting at y_o and ending at y_1 (including both y_o and y_1) and the part T of the local section S that lies between y_o and y_1 excluding y_o and y_1.

Let (1) denote the bounded closed region bounded by Σ.

DYNAMICAL SYSTEM – A SHORT COURSE

Since the pt. y_1 is on S, a local section, the trajectory at y_1 can not be along S i.e. the trajectory must cross S. at y_1. (Because at pts. of a local section the vector field f is transverse to S).

Thus at y_1 their t are two possibilities i.e. at y_1 the trajectory either enter D or exits D.

Let us assume that at y_1, the trajectory exits D.

If we take any pt. on T, then using the fact that T is a part of local section, the trajectory through any pt. of T will either be entering D or will be exiting D.

We denote by T- the set of all those pts. On T at which the trajectories exit D and we denote by T+ the set of all those pts. on T at which the trajectories enter D.

Thus $T = T\text{-}U\ T\text{+}$, a disjoint union. It further follows that both T+ and T- are open subsets of T.

Further at y_1, the direction of the trajectory is from right to left and hence in the immediate vicinity (nbd) of y_1 same must be the direction of the trajectory. Then $T_- \neq \phi$

Thus we have $T = T\text{-}\ U\ T\text{+}$, a disjoint union of open sets.

But I is connected and $T_- \neq \phi$. Hence $T\text{+} = \phi$.

Thus, at every pt. of T the trajectories will be exiting D.

Hence if we take any pt. in $R^2 - D$. It follow that its trajectory cannot enter D and thus will remain in $R^2 - D$.

Since at y_1, the trajectory is exiting D and as $y_2 = \phi_{t_1}(y_1)$, for some $t > 0$.

It follows that $y_2 \in R^2 - D$. But it is also on S. since the portion T of S lies in D, it follows that $y_2 \in S - T$.

LIMIT SETS

$S-T$ is a disjoint union of two half open intervals say I_1, (with end pt.

y_1) and I_2 with end pt. y_o.

But I_2 lies within D. Hence y_2 lies on S beyond y_1.

Hence y_o, y_1, y_2 are monotone along S.

We next consider the pts. y_o, y_1, y_2 and repeat the argument to get

y_o, y_1, y_2, y_3 monotone along S.

Repeating this we show that $\{y_o, y_1, \cdots\}$ is monotone along S.

Theorem 6: Let $y \in L_w(x) \cup L_a(x)$. Then the trajectory of y crosses any local section at not more than one point.

Proof: Let us assume that there exist a local section say S such that a trajectory of y meets S in two pts. Say y_1 and y_2.

We further assume that $\{y_o, y_1, y_2 \cdots\}$ is monotone along the trajectory,

Without loss of generality, we assume that $y \in L_w(x)$. It follows that both y_1 and y_2 are in $L_w(x)$.

We consider tow flow boxes around y_1 and y_2 obtained by taking two small intervals say J_1 and J_2 on S around the pts. y_1 and y_2 respectively.

We can assume that J_1 and J_2 are disjoint.

As $y \in L_w(x)$ and y_1, y_2 are along the trajectory of y, both $y_1, y_2 \in L_w(x)$

It therefore follows that the trajectory through x will be passing through the flow boxes y_1 and y_2 respectively at infinitely many times and therefore the trajectory will be meeting the parts J_1 and J_2 of the local section at infinitely many pts. Say $a_1, b_1, a_2, b_2, \cdots$ are pts. on J_1 while

$a_1, b_2, \cdots$ are pts. on J_2.

Sine $\{a_1, b_1, a_2, b_2, \cdots\}$ is monotone along the trajectory, then by previous theorem, the seq. $\{a_1, b_1, a_2, b_2, \cdots\}$ must be monotone along S.

However, this is not possible as

$\{a_1, b_2, \cdots\} \subseteq J_1, \{b_1, b_2, \cdots\} \subseteq J_2$ and J_1, J_2 are on the same local section S and $J_1 \cap J_2 = \phi$.

Thus the assumption that a trajectory through y meets.

A local section S at two pts. Leads to a contradiction.

So, any trajectory through Y will meet any local section in of the most one point.

Definition: By a closed orbit of a dynamical system we mean the image of a non-trivial periodic solution. Thus a trajectory r is a closed orbit of r is not an equilibrium and $\phi_p(x) = x$, for $x \in r, p \neq 0$

It follows that $\phi_{np}(y) = y, y \in r, n = 0, \pm 1, \pm 2, \cdots$

Theorem 7. [POINCARE-BENDIXSON THEOREM]

A non-empty compact Limit set of a $C'-$planar dynamical system, which contains no equilibrium pt. if a closed orbit.

Proof: Without loss of generality, we consider a non-empty compact limit set $L_w(x)$, for some x.

We further assume that $L_w(x)$ does not contain any equilibrium point.

Let $y \in L_w(x)$. Every point on the trajectory of y_1 is again a pt. in $L_w(x)$.

Since $L_w(x)$ is a compact set, $L_w(y)$ is non-empty.

Since pts. of $L_w(y)$ are Limit pts of the trajectory through y and entire trajectory lies in $L_w(x)$ a compact set, it follows that $L_w(y)$ is non-empty

LIMIT SETS

subset of $L_w(x)$.

Let $z \in L_w(y)$. Let S be any local section through the pt. z.

As $y \in L_w(x)$, it follows by the previous thm that the trajectory through y will meet S in not more than one point.

But as z is a w-limit pt. of y, there will be infinitely many pts. on the trajectory through y which will be sufficiently closed to z and hence will meet S.

Thus we can certainly find two numbers r and s (distinct) such that $\phi_r(y)$ and $\phi_s(y)$ will lie on S.

If therefore follows that the trajectory through y will meet S in exactly one pt. so

$$\phi_r(y) = \phi_s(y)$$

As $r \neq s$, we have

Either $r < s$ or $r > s$.

If $r < s$ then $\phi_{-r}(\phi_r(y)) = \phi_{-r}(\phi_s(y))$

i.e. $y = \phi_{s-r}(y)$

i.e. $y = \phi_p(y)$, where p= s-r

If $r > s$ then $y = \phi_{r-s}(y)$

Thus in any case, we can find p > 0 s.t $y = \phi_p(y)$

Thus the trajectory through y is a closed orbit say r.

But entire trajectory through y is contained in $L_w(x)$.

Thus, $r \subseteq L_w(x)$

We now show that $r = L_w(x)$

To this end we show that $\lim_{t\to\infty} d(\phi_t(x),r)=0$

We note that pt. of $L_w(x)$ are limit pts. of the type $\lim_{t\to\infty}\phi_t(x)$.

So, $d(\phi_t(x),r)\to 0$ as $t\to\infty$ will mean the w-limit pt. of x are the limit pt. of r.

But r is closed. Thus we shall get $L_w(x)\subseteq r$.

Let z be a pt. on r. let S be a local section through z taken small enough s.t. $S\cap r=\{z\}$.

But as $z\in r$ and r is a trajectory through y, $z\in L_w(x)$.

If $r_\in$ denotes a flow box at z with the local section S. then a trajectory from x will enter this flow box infinitely many times and as such will be meeting the local section S.

Thus we can find a seq. $\{t_1,t_2,\cdots\}, t_n\to\infty$ such that

$\phi_{tn}(x)\in S$, for sufficiently large n. $\phi_{tn}(x)\to z$ as $n\to\infty$

and $\phi_{tn}(x)\notin S$ for $t_n<t<t_{n+1}$, for sufficiently large n.

since the seq. $\{t_n\}$ tends to infinity as $n\to\infty$

we shall show that $t_{n+1}-t_n$ remains bounded.

We note that $z\in R$, a closed orbit, so we can find a number say $p>0$ s.t.

$\phi_p(z)=z$

We shall denote $\phi_{tn}(x)$ by x. we know that $x_n\to z$ as $n\to\infty$.

Thus for every large n, we can consider $\phi_p(x_n)$.

Since x_n is sufficiently close to z and $\phi_p(z)=z$, it follow that $\phi_p(x_n)$ is also sufficiently close to z and as such we can assume that $\phi_p(x_n)\in v_e$.

LIMIT SETS

Since V_e is a flow box determined by the section S, it follows that the pt $\phi_p(x_n)$ will be on some trajectory through S.

Thus we can find t, $t \in [-\in, \in]$ s.t. $\phi_{p+t}(x_n) \in S$.

We now claim that $t_{n+1}t_n \leq p + \in$, because otherwise

$t_{n+1}t_n > p + \in$

$t_{n+1} > t_n + p + \in$

But we have for some $t \in [-\in, \in]$,

$\phi_{p+t}(x_n) \in S$

$\therefore \phi_{p+t+t_n}(x) \in S \ \{\because x_n = \phi_{t_n}(x)\}$

But we know that $\phi_t(x) \notin S$, for $t_n < t < t_{n+1}$.

We thus get a contradiction and hence we have

$t_{n-1}t \leq p + \in$

Let $\beta > 0$ be chosen arbitrarily.

From the regularity result of the solution of dynamical system, it follows that $\exists \delta > 0$ S.t if $|x_{n-u}| < \delta$ and $t | \leq p + \in$ then $|\phi_t(x_n 0 - \phi_t(u)| < \beta$.

Since $x_n \to z$, for sufficiently large value of n.

We can assume $d(x_n, z) < \delta$ and thus for $t | \leq p + \in$.

$d(\phi_t(x_n), \phi_t(z)) < \beta$

Thus for very large value of n, we choose t s.t. $t_n \leq t \leq t_{n+1}$.

Consider $d(\phi_t(x), \phi_{t-t_n}(z)) = d(\phi_{t-t_n}(x_n), \phi_{t-t_n}(z)) \ \{\because \phi_{t_n}(x) = x_n\}$

Since $z \in R$ and r is a closed orbit,

$\phi_{t-t_n}(z) \in r$.

Further as $t_n \le t \le t_{n+1}, t - t_n \le t_{n+1} - t_n \le p + \in$

We thus have, $d\big(\phi_{t-t_n}(x_n), \phi_{t-t_n}(z)\big) < \beta$

Thus for very large value of t,

$$d\big(\phi_t(x_n), \phi_{t-t_n}(z)\big) < \beta$$

But $\phi_{t-t_n}(z)$ is a point on r.

Thus for very large values of n and t s.t $t_n \le t \le t_{n+1}$

We have $d\big(\phi_t(x), r\big) \le d\big(\phi_t(x), \phi_{t-t_n}(z)\big) < \beta$.

$$\therefore \lim_{t \to \infty} d\big(\phi_t(x), r\big) = 0$$

APPLICATION OF POINCARE BENDIXSON THEOREM

Definition : A Limit cycle is closed orbit r such that

$r \subseteq L_w(x)$ or $r \subseteq L_\alpha(x)$, for some $x \notin r$.

In case a limit cycle r is such that $r \subseteq L_w(x)$. Then it is known as w-limit cycle.

Similarly, if a limit cycle r is such that $r \subseteq L_\alpha(x)$ then it is known as α-limit cycle.

From the Poincare-Bendixson theorem, it follows that if r is an w-limit cycle then for some $x \notin r$

$$\lim_{t \to \infty} d\big(\phi_t(x), r\big) = 0$$

i.e. the trajectory $\phi_t(x)$ moves closer and closer to the limit cycle r as $t \to \infty$

Similarly, if r is an α-limit cycle, it follows that for some $x \notin r$,

$$\lim_{t \to \infty} d\big(\phi_t(x), r\big) = 0.$$

Suppose r is an w-limit cycle and $d\big(\phi_t(x), r\big)$ tends to zero as $t \to \infty$.

We consider a pt. say z on v and a local section through z and assume that

LIMIT SETS

the local section meets r only at z.

It follows that we can find values of t say to $< t_1$ such that $\phi_{t_o}(x)$ and

$\phi_{t_1}(x)$ and the portion of the local section between $\phi_{t_o}(x)$ and $\phi_{t_1}(x)$.

We can assume that for $t_o < t < t_1, \phi_t(x)$ does not lie on the local section this region is invariant since the trajectory in this region cannot cross the local section between $\phi_{t_o}(x)$ and $\phi_{t_1}(x)$. Also it cannot cross the other part of the boundary of the region as it consist of trajectories.

Thus any trajectory in this region will tend towards r.

Theorem 8: Let r be an w-limit cycle. If $r = L_w(x), x \notin r$ then x has a nbd v such that $r = L_w(y), \forall y \notin v$

In other words, the set, $A = \{y \mid r = L_w(y)\}$ is open.

Proof: Let x be such that $r = L_w(x)$.

By considering a pt. z on r and a small local section at z, we can find values say t_o and t_1 such that the region enclosed by r on one side and part of the trajectory between $\phi_{t_o}(x), \phi_{t_1}(x)$ and $\phi_{t_1}(x)$ and $\phi_{t_1}(x)$ is + vely invariant.

It is such that any trajectory in this region will tends to r as $t \to \infty$.

It is clear that for large t, specially $\phi_{t_1}(x)$ will be in this region.

By the regularity of solution of dynamical system, it follows that if y is sufficiently close to x (i.e if y is in a small nbd of x) then for sufficiently large values of t, $\phi_t(y)$ will also be in this region.

Then $\lim\limits_{t \to \infty} \phi_t(y)$ will approach r i.e $r = L_w(y)$

405

Theorem 9: A non-empty compact set k that is +vely or negatively invariant contains either a limit cycle or an equilibrium point.

Proof: Let us assume that the compact set k is positively invariant and $x \in K$.

Thus $L_w(x)$ is non-empty compact set.

Then by Poincare-Bendixson theorem, $L_w(x)$ is either an equilibrium pt. or closed orbit.

Thus $L_w(x)$ is either an equilibrium point or an w-limit cycle.

Similarly, we can prove for α(-ve invariant k)

Theorem 10: Let r be a closed orbit and suppose that the domain W of the dynamical system includes the whole open region U enclosed by r. then U contains either an equilibrium or a limit cycle.

Proof: We consider the compact set u Ur

For any $x \in U, L_w(x)$ as well as $L_\alpha(x)$ are non-empty.

We further note that the set U is invariant as its boundary is a closed trajectory r.

Thus Let $x \in U$. We assume that u does not contain an equilibrium pt. or a closed orbit.

We must then have $L_w(x) = L_\alpha(x) = r$.

We take some $z \in r$, since $z \in L_w(x)$ as well as $L_\alpha(x)$.

We take a local section at z and thus we get a seq. $t_1, t_2, \cdots \to +\infty$ such that $\phi_{ti}(x) \in S, i \geq 1$

We note that $\phi_{t_1}(x)$ will be monotone along S.

Since $z \in L_\alpha(x)$, we can get a seq. $t_i', t_i' \to -\infty$ as $i \to \infty$ such that

$\phi_{t_i}(x) \in S$ and $\phi_{t_i'}(x)$ also need to be monotone along S.

Since we have $t_k' < t_{k-1}' < \cdots t_1' < t_1 < t_2 < \cdots$

We cannot have the seq.

$$\phi_{t_k}(x), \phi_{t_{k-1}}(x), \cdots, \phi_{t'_1}(x), \phi_{t_2}(x), \cdots$$

Monotone along S. We thus a get a contradiction.

Thus U must contain either an equilibrium pt. or a closed orbit.

Theorem 11: Let r be a closed orbit enclosing an open set U contained in the domain W or the dynamical system. Then U contains an equilibrium point.

Proof: To prove the theorem, let us assume that U does not contain any equilibrium point so U must contain closed orbits.

Since any closed orbit in U will enclose some non-zero area.

Let A denote the glb of the areas enclosed by closed orbits in U.

Let $A_1, A_2, \cdots$ be a seq. of areas enclosed by closed orbits in U such that

$$A = \lim_{n \to \infty} A_n$$

Let $\alpha_1, \alpha_2, \cdots$ be closed orbits in U such that area enclosed by α_i is A_i

Let x_i be any point on α_i. We thus get a seq. $\{x_1, x_2, \cdots\}$ of pts. in U.

As $U \bigcup r$ is a compact set, we thus a get seq. $\{x_1, x_2, \cdots\}$ of the pts. in a compact set.

Let x be a limit point.

We show that x must lie on a closed orbit.

If x does not lie on a closed orbit, then it follows that $\phi_t(x)$ will approach the boundary r.

But as $x_n \to x$ for sufficiently large values of n, x_n is sufficiently close to x

DYNAMICAL SYSTEM – A SHORT COURSE

and therefore it then follows that $\phi_t(x_n)$ will also tend to r.

However this is not possible because x_n lies on a closed orbit α_n in U.

Thus x must lie on a closed orbit.

Since the pts. x_n are approaching to x, using a local section, we can show that the closed orbit α_n will approach the closed orbit of x.

It therefore follows that the area of the closed orbit of x will be limit of A_n as $n \to \infty$

Thus the area of the closed orbit of x is in A.

By the definition A is the glb of areas enclosed by closed orbits in U.

If we consider the region enclosed by the closed orbit of x, then this region will neither contain an equilibrium pt nor a closed orbit.

(Because if it contains a closed orbit, it will also be for understanding purpose closed orbit in U and thus the area enclosed by if will be at least the area enclosed by the orbit of x which is not possible.)

This contradicts the previous thm.

$\therefore$ U must contain an equilibrium point.

Theorem 12: Let H be a first integral of a planar c^1 – dynamical system (i.e H is real valued function which is constant on trajectories). If H is not constant on any open set, there are no limit cycles.

Proof: Let us assume that there are limit cycles.

Let r be a limit cycle. It follows that H is constant on r.

Since by the definition

For some $x \notin v$, $\lim\limits_{t \to \infty} d(\phi_t(x), r) = 0$

By the continuity of H, H must be constant (same value as on r) on the trajectory $\phi_t(x)$.

408

LIMIT SETS

But we know that the set $A = \{y \mid r = L_w(y)\}$ is an open set.

Thus, if $\lim_{t \to \infty} d(\phi_t(y), r) = 0$ then H is same on $\phi_t(x)$ as well as on $\phi_t(y)$ and this holds on a nbd of x.

Thus H remains constant on an open nbd of x.

Thus, if there exists a limit cycle then H remains constant in an open set.

ECOLOGY

Ecology is a scientific study of interrelationship with the environment.

SINGLE SPECY MODEL

Let $y(t)$ denote the population (number) of a specy at time t. let $y + \Delta y$ denote its population at time $t + \Delta t$.

Thus the rate of change of the specy at time t is $\dfrac{\Delta y}{\Delta t} \cdot \dfrac{1}{y(t)}$

Taking limit as $\Delta t \to 0$, we get instantaneous average rate of change of population of a specy given by

$\dfrac{y'(t)}{y(t)}$ i.e. $\dfrac{d}{dt}(\log y(t))$

It is also known as average rate of growth of the specy.

In the simplest model we consider the average rate of growth of the specy to be given by a positive constant say α.

Thus we have $\dfrac{d}{dt}(\log y(t)) = \alpha$

$\Rightarrow y(t) = y(0) \cdot \exp(\alpha t)$

However, this model is unrealistic because as $t \to \infty$, $y(t) \to \infty$ and in practice no specy can grow infinitely.

We modify the model in such a way that the dependence of growth of a specy on the availability of food is incorporated.

DYNAMICAL SYSTEM – A SHORT COURSE

We thus consider the model

$$\alpha = a(\sigma - \sigma_o)$$

Where α is the average rate of growth.

Here σ_o denotes the necessary quantity of food for the growth of the specy and σ denotes the actual availability of food.

Thus if $\sigma < \sigma_o$, the population will decrease and $\sigma > \sigma_o$, the population will increase.

Thus we have $\dfrac{d}{dt}(\log y(t)) = \alpha(\sigma - \sigma_o)$ with $a > 0$

i.e. $y(t) = y(0) \cdot \exp\left(\alpha(\sigma - \sigma_o)t\right)$

However, this model is also not realistic one as $\sigma > \sigma_o$ will mean that the population will go on increasing indefinitely with time. Even though there is sufficient food available, population of a specy do not go on increasing indefinitely, there are other factors in addition to availability of food which determine the growth of a specy.

It is found that the growth of a specy is positive only up to a limit beyond which it starts declining.

This observation is incorporated in a model where we consider the rate of growth α to be a function of population itself.

We consider $\alpha = a(\eta - y)$

Thus we get a model, $\dfrac{d}{dt}(\log y(t)) = \alpha(\eta - y)$

i.e. $\dfrac{y'(t)}{y(t)} = \alpha(\eta - y(t))$

i.e. $\dfrac{d}{dt}y(t) = \alpha(\eta - y(t))y(t)$

for the equation $\dfrac{d}{dt} y(t) = a\eta y - ay^2$.

Here are two pts. Of equilibria namely $y = 0$ & $y = \eta$.

It is a non-linear dynamical system.

To consider the behaviour of these equilibria, we consider the linear part of the function $a\eta y - ay^2$.

It is given by $a\eta - 2ay$.

At $y = 0$, it is $a\eta > 0$.

So, the eigen value is $a\eta(+ve)$ and thus $y = 0$ is an unstable (i.e. source) equilibrium.

At $y = \eta$, the Linear part is $-a\eta$ which is $-ve$ no.

Thus $y = \eta$ is an asymptotically stable equilibrium i.e. if the population of the specy is $< \eta$, it will go on increasing and if it is $> \eta$, then it will decrease. This is again a simplistic model.

Usually we consider the model given by

$y' = M(Y) \cdot y$ for some value $\eta, M(\eta) = 0$, for $y < \eta, M(Y) > 0$ & $y > \eta, M(Y) > 0$

In the previous model, $M(Y) = a(\eta - y)$.

PREDATOR AND PREY MODEL

We consider the species. One specy is serving as a food for other specy. Thus we get a predator-prey model.

Let $x(t)$ and $y(t)$ denote the population of prey and predator specy at a given instance t.

Since x(t) serves as food supply for $y(t)$, it is clear that for the positive growth of $y(t)$, $x(t)$ must be above a base value.

Thus we have $y' = (C_x - D)y(t)$

This equation is obtainable from the general expression

$$y' = a(\sigma - \sigma_o)y$$

Where, σ is the quantity of food available and

σ_o is the minimum quantity of food necessary for the survival of the specy.

Since $x(t)$ denotes the food supply, we have

$y'(t) = (C_x - D)y$, for the predator

Since the food given by x(t) is itself a specy, it itself has its rate of growth.

Since the growth in the predator specy may lead to the decrease in the prey specy, the growth model for the prey specy is given by

$$x'(t) = (A - BY)x$$

Thus the predator prey model is given by a system of two simultaneous differential equations namely

$$x'(t) = (A - BY(t))x(t)$$

$$y'(t) = (C_x(t) - D) \cdot y(t)$$

This system is known as equations of Voltera and Lotka with A, B, C, D > 0.

There are two pts. of equilibrium namely (0, 0) and $\left(\dfrac{D}{C}, \dfrac{A}{B} \right)$.

The volterra-Lotka equations represents a non-linear dynamical system.

The Linear part is given by the Jacobian matrix with

$$f_1(x, y) = A_x - Bxy$$

$$f_2(x, y) = Cxy - Dy$$

The matrix is : $\begin{bmatrix} A - By & -Bx \\ Cy & Cx - D \end{bmatrix}$

LIMIT SETS

At $(0, 0)$, the matrix is $= \begin{bmatrix} A & 0 \\ 0 & D \end{bmatrix}$

Obiviously, the equilibrium pt. $(0, 0)$ is a saddle and hence unstable equilibrium.

For the equilibrium point $\left(\dfrac{D}{C}, \dfrac{A}{B} \right)$.

The matrix is : $\begin{bmatrix} 0 & -BD/C \\ Ac/B & 0 \end{bmatrix}$

The eigen values are given by

$\begin{vmatrix} -\lambda & -BD/C \\ Ac/B & -\lambda \end{vmatrix} = 0$

i.e. $\lambda^2 + \dfrac{BD}{C} \cdot \dfrac{AC}{B} = 0$.

$\lambda^2 = -AD$

$\therefore \ \lambda = \pm i - \sqrt{AD}$,both imaginary.

Q. Prove that every trajectory of Voltera-Lotka equations is a closed orbit, except the equilibrium and the coordinate axes.

Solution: The predator-Prey model given by the Voltera–Lotka equation is given by the dynamical system

$x'(t) = (A - BY(t))x$

$y'(t) = (C_x(t) - D)y$

with $A, B, C, D > 0$.

It is non-linear dynamical system with two pts. Of equilibrium $(0, 0)$, $\left(\dfrac{D}{C}, \dfrac{A}{B} \right)$, where $(0, 0)$ is a saddle.

Since x(t) and y(t) are the populations of living species, we consider only

413

the region $x(t) \geq 0$ and $y(t) \geq 0$

We draw lines $x = \dfrac{D}{C}$ and $y = \dfrac{A}{B}$

Region $x(t) \geq 0$, $y(t) \geq 0$ into four regions.

We first show that the +ve parts of the x-axis as well as the y-axis are trajectories of the volterra-Lotka equations.

If we have initial pt. say $(x_o, 0)$ on the +ve x-axis $(x_o > 0)$ then for this paint, we have

$x' = Ax$ i.e. $x(t) = x_o e^{At}$

$y' = 0,\ y(t) = 0$

i.e. the solution curve moves along the x-axis.

Similarly, if the initial value is $(0, y_o)$, for some $y_o > 0$

Then the equations become $x' = 0$ and $y' = -DY$

i.e. $x(t) = 0,\ y(t) = y_o e^{-Dt}$.

I) We shall show that the trajectories of the dynamical system (other than the equilibrium pt and coordinate axis i.e. x & y axis) rotate (not necessarily along a circle) in anticlockwise direction around the point

$$z = \left(\frac{D}{C}, \frac{A}{B} \right).$$

II) We next show that the pt. $z = \left(\dfrac{D}{C}, \dfrac{A}{B} \right)$ is a stable equilibrium.

III) We finally show that the trajectories mentioned above are closed orbits.

Proof: of I : We shall denote the four quadrants into which the region $x(t) \geq 0$, $y(t) \geq 0$ is divided by

LIMIT SETS

The line $x = \dfrac{D}{C}$ and $y = \dfrac{A}{B}$ as quadrants I, II, III & IV,

Where In quadrant I, $x > \dfrac{D}{C}$, $y > \dfrac{A}{B}$

In quadrant II, we have $x < \dfrac{D}{C}$, $y > \dfrac{A}{B}$ and so on.

Since we have,

$$x' = (A - BY)x$$
$$y' = (Cx - D)y$$

In the 1st quadrant, we have $x' < 0, y' < 0$

We now consider a point say *(u, v)* in the 1st quadrant and consider the trajectory through it.

Thus we have

$$u > \dfrac{D}{C}, \ v > \dfrac{A}{B}$$

$$\therefore A - Bv = -r < 0$$

$Cu - D = s > 0$, for some +ve r and s.

Since we are consider a trajectory with initial pt. *(u, v)* in quadrant I, it will remain in that quadrant for some time.

Thus, Let t be such that $(0, t)$ is the maximal interval over which the trajectory remains in quadrant I.

We shall show that $\tau \neq \infty$

For all $t \in [0, \tau)$, we have

$$x' < 0, y' > 0$$

Also we have $x' = (A - BY)x$

$$y' = (Cx - D)y$$

i.e. $\dfrac{x'}{x} = A - BY$

$\dfrac{y'}{y} = Cx - D$

Since in the 1st quadrant $x' < 0, y' > 0$, y goes on increasing.

Therefore the expression A-BY will decrease from its maximum value A-Bv =-r.

Similarly, as $x' < 0$, x is decreasing, therefore Cr-D will go on decreasing from its max. value Cu-D=s.

Thus for all $t \in [0, \tau)$, we have

$\dfrac{x'}{x} = A - BY \le -r$

$\dfrac{y'}{y} = Cx - D \le s$

i.e. $\dfrac{d}{dt}\left(\log x(t)\right) \le -r$

$\dfrac{d}{dt}\left(\log y(t)\right) \le s$

On integrating.

$x(t) \le x_o \, e^{-rt}$

$y(t) \le y_o \, e^{st}$

i.e. $\forall t \in [0, \tau), x(t) \le u \, e^{-rt}$ & $y(t) \le v \, e^{st}$

since the trajectory is in the 1st quadrant, we have

$\dfrac{D}{C} < x(t) \le u \, e^{-rt}$

$\dfrac{A}{B} < y(t) \le v\,e^{st}, \forall t \in [0,\tau)$

Since $\dfrac{D}{C} < x(t) \le u\,e^{-rt}$ and as $u\,e^{-rt} \to 0$ as $t \to \infty$.

Thus the trajectory will remain in the 1st quadrant only for a finite time t.

Thus we shall have $\dfrac{D}{C} < x(t) \le u$ and $\dfrac{A}{B} < y(t) \le v\,e^{st}, \forall t \in [0,\tau)$

Thus at $t = \tau$, we shall have

$x(t) = \dfrac{D}{C}$ and $y(t) = v\,e^{st}$.

Once the trajectory is on the line $x = \dfrac{D}{C}$ and $y(t) > \dfrac{A}{B}$, it will move into second quadrant as $x' < 0$.

The same argument can be applied to show that the trajectory will remain in 2nd quadrant only for a finite time.

Thus, it will enter quadrant III, quadrant IV and quadrant I and so on.

Proof of II: We now show that the pt. $z = \left(\dfrac{D}{C}, \dfrac{A}{B}\right)$ is a stable equilibrium.

We shall show this by obtaining an appropriate Liapunov function.

We consider a differentiable function $H=H(x, Y)$.

We further assume that $H(x, y) = F(x) + G(Y)$.

We consider the variation of H along the trajectories.

Thus consider $\dfrac{d}{dt}H(x,y) = \dfrac{d}{dt}F(x) + \dfrac{d}{dt}G(y)$

$\dfrac{d}{dt}F(x)\cdot x'\dfrac{d}{dy}G(y)\cdot y'$

But along the trajectories, we have

$x' = (A - By)x, y' = (Cx - D)y$

417

Thus along the trajectories, we have

$$\frac{d}{dt}H(x,y) = \frac{d}{dt}F(x) \cdot x(A - By) + \frac{d}{dt}G(y) \cdot y(Cx - D)$$

We further require that along the trajectories

$$\frac{d}{dt}H(x,y) = 0$$

Thus this will imply

$$\frac{d}{dt}F(x) \cdot x(A - By) + \frac{d}{dy}G(y) \cdot y(Cx - D) = 0$$

i.e.
$$\frac{x\dfrac{d}{dt}F(x)}{(Cx - D)} = \frac{y\dfrac{d}{dy}G(y)}{By - A}$$

but as the two sides are independent of each, other, the above will happen only when both the expression are constant.

We choose the constant equal to 1 we the get

$$\frac{x \cdot \dfrac{d}{dt}F(x)}{Cx - D} = 1 = \frac{y\dfrac{d}{dy}G(y)}{By - A}$$

i.e. $\dfrac{d}{dt}F(x) = C - \dfrac{D}{x}; \dfrac{d}{dy}G(y) = B - \dfrac{A}{y}$

on integrating, w get

$$F(x) = Cx - D\log x; \ G(y) = By = By - A\log y$$

Thus, $H(x,y) = Cx - D\log x + By - A\log x$

We shall show that H attains its min at pt. $z = \left(\dfrac{D}{C}, \dfrac{A}{B}\right)$

We note that

LIMIT SETS

$$\frac{\partial H}{\partial x} = 0 = \frac{\partial H}{\partial y} \text{ at z}$$

$$\frac{\partial^2 H}{\partial x \partial y} = 0.$$

Further, $\dfrac{\partial^2 H}{\partial x^2} = \dfrac{D}{x^2} > 0; \dfrac{\partial^2 H}{\partial y^2} = \dfrac{A}{y^2} > 0.$

Thus $z = \left(\dfrac{D}{C}, \dfrac{A}{B}\right)$ is the pt of absolute minimum of the function H.

We define $V(x,y) = H(x,y) - H(z)$

It is clear that $\dot{V} \le 0$ (Actually $\dot{V} = 0$)

$\therefore v$ is a Liapunov function i.e. z is a point of stable equilibrium.

We know that the differentiable function

$$H(x,y) = Cx - D\log x + By - A\log y$$

Is constant along the trajectories.

From the expression for functin H, it is clear that H is not constant in an open set.

Hence there are no Limit cycles.

Proof of III: We show that trajectories in the region x(t) >0, y(t) >0 other than the point $z = \left(\dfrac{D}{C}, \dfrac{A}{B}\right)$ are closed orbits.

Let w be any point in this region s.t. $W \ne z$.

Consider a trajectory through w. If $\phi_t(w)$ denotes the flow of the diff. equations then various pts. on the trajectory through w will be given by $\phi_t(w)$.

It possible let us assume that the trajectory through w is not a closed orbit.

DYNAMICAL SYSTEM – A SHORT COURSE

Considering the trajectory $\phi_t(w)$, it is clear that for infinitely many values

oft say $t_o, t_1, t_2, \cdots t_n \rightarrow +\infty$

We have $\phi_{ti}(w)$ meeting the portion of $x = \dfrac{D}{C}$ above $z = \left(\dfrac{D}{C}, \dfrac{A}{B}\right)$

Similarly, taking negative variation of t, it follows that for the values of t

say $t_{-1}, t_{-2}, t_{-3}, \cdots \phi_{ti}(w)$

Meets the above portion of $x = \dfrac{D}{C}$, for $i = -1, -2, -3, \cdots$.

Since we are assuming that the trajectory is hot a closed orbit

$\phi_{ti}(w), i = 0, \pm 1, \pm 2, \cdots$

Are all distinct pts on the above portion of the line $x = \dfrac{D}{C}$.

Since we have $t_{-n} < t_{-n+1} < t_{-n+2} < t_o < t_1 < t_2 < \cdots$ the pts. $\phi_{ti}(w)$ are monotone along the trajectory.

A s the line $x = D/C$ can be considered as a local section, there points must

be monotone along the above portion of the line $x = \dfrac{D}{C}$.

Since as t increases, the monotonicity implies the pts. on $x = \dfrac{D}{C}$ to be on

the same side of $\phi_{to}(w)$, it follows that either $\phi_{t-1}(w) < \phi_{to}(w) < \phi_{t1}(w) < \cdots$

or $\phi_{t1}(w) < \phi_{to}(w) < \phi_{t-1}(w) < \phi_{t-2}(w) < \cdots$

In the first case we have $\dfrac{A}{B} < \cdots \phi_{t-2}(w) < \phi_{t-1}(w) \cdots$

We have $\phi_{sj}(w)$ a strictly decreasing seq. bounded below by $\dfrac{A}{B}$, where

$S_j = t_{-j}, j = 0, 1, 2, \cdots$

We must have limit of y coordinate of $\phi_{tn}(w) \rightarrow \dfrac{A}{B}$ as $n \rightarrow \infty$.

LIMIT SETS

Similarly, considering the other case we get limit of y coordinate of

$$\phi_{tn}(w) \to \frac{A}{B} \text{ as } n \to \infty..$$

Thus we have either $\phi_m(w) \to z$ as $n \to +\infty$.

or

$\phi_{tn}(w) \to z$ as $n \to -\infty$.

However H is constant on the trajectory $\phi_t(w)$. Using the continuity of H,

it follows that $H(z)$ is the constant value of H on the trajectory $\phi_t(w)$.

This is a contradiction as H has absolute minimum at z.

Thus the trajectory through w must be a closed orbit.

Thus for the predator prey model given by the Volterra- Lotka equation whatever the number of prey and predator species neither species will die out, nor will it grow indefinitely.

On other hand, except for the state z, the populations will not remains constant.

MODIFIED PREDATOR-PREY MODEL

The predator –Prey (Model) equations of species with limited growth are given by

$$x' = (A - By - \lambda x)x$$
$$y' = (Cx - D - \mu y)y$$

Where, the constants A, B, C, D, λ and μ are all +ve

To study this modified model, we again consider the relevant region namely $x(t) > 0, y(t) > 0$.

We draw the lines $L: A - By - \lambda x = 0$

$M: Cx - D - \mu y = 0$

In the above region.

There are two cases depending on whether the lines L and M meet each other in the above region or not.

We first consider the case where L and M do not meet each other in the above region.

We now consider the signs of x' and y' in the various parts into

In this case, it is not possible for both the prey and predator population to increase simultaneously. If the prey is above its limiting population namely A, it must decrease and after a while the predator population also starts decreasing (when the trajectory crosses M). After that point the prey can never increase past $\dfrac{A}{\lambda}$.

We can consider the case where lines L and M meets each other at $z = (\bar{x}, \bar{y})$ in the above region (i.e. x > 0, y > 0).

Since z is a pt. of intersection of L and M, we have

$$A - B\bar{y} - \lambda\bar{x} = 0;\ C\bar{x} - D - \mu\bar{y} = 0$$

$$\Rightarrow \bar{x} = \frac{A - B\bar{y}}{\lambda};\ \bar{x} = \frac{D + \mu\bar{y}}{C}$$

On equating, we get

$$\Rightarrow \bar{x} = \frac{A - B\bar{y}}{\lambda} = \frac{D + \mu\bar{y}}{C}$$

$$\Rightarrow AC - BC\bar{y} - \lambda D + \mu\lambda\bar{y}$$

$$\Rightarrow AC - BC\bar{y} - \lambda D - \lambda\mu\bar{y} = 0$$

$$\Rightarrow \bar{y} = \frac{D\lambda - AC}{\lambda\mu + BC}$$

$$\therefore \bar{x} = A - B\frac{(D\lambda - AC)}{\lambda(\lambda\mu + BC)}$$

LIMIT SETS

$$\bar{x} = \frac{\mu A + BD}{\mu x + BC}$$

The pt. $z = (\bar{x}, \bar{y})$ is an equilibrium point.

The linear part of the R.H.S of the diff. equation at $(\bar{x}, \bar{y})$ is:

$$\begin{bmatrix} A - BY - 2\lambda x & -Bx \\ CY & Cx - D - 2\mu y \end{bmatrix}$$

At pt. $(\bar{x}, \bar{y}), = \begin{bmatrix} -\lambda\bar{x} & -B\bar{x} \\ C\bar{y} & \mu\bar{y} \end{bmatrix}$

Eigen values are given by

$$\begin{vmatrix} \dfrac{-\lambda(\mu A + BD)}{\mu x + BC} & \dfrac{-B(\mu A + BD)}{\mu x + BC} \\ \dfrac{C(AC - \lambda D)}{\lambda\mu + BC} & \dfrac{-\mu(AC - \lambda D)}{\lambda\mu + BC} \end{vmatrix}$$

Since the eigen values have –ve real part. The equilibrium pt.

$z = (\bar{x}, \bar{y})$ is asymptotically stable.

The point P, where the line L meets the x-axis is also a pt. of equilibrium.

It is a saddle

From the model of predator and prey with limited growth, it follows that the population eventually settles down to either a constant or a periodic solution.

There are absolute upper bounds that no population can exceed in the long run no matter what the initial populations are.

COMPETING SPECIES

In this model we consider two species whose population is given by variables x and y which are functions of time.

It is further assume that there is a single source of food for both the species.

Thus the species will have to compete with each other for survival.

As such the ratio of growth of both the species will depend will depend on the population of both the species.

The equation for model are thus given as

$$x' = M(x,y)x$$

$$y' = N(x,y)y$$

We make the following assumption for the reasonableness of this model

1) If species increases, the growth rate of the other goes down.

i.e. $\dfrac{\partial M}{\partial N} < 0, \dfrac{\partial M}{\partial N} < 0$

2) If either population is very large neither species can multiply.

Hence $\exists\, k > 0$ s.t.

$M(x,y) \leq 0$ and $N(x,y) \leq 0$ if $x \geq k$ or $y \geq k$.

3) In the absence of either species, the other has a positive growth rate up to a certain population and negative growth beyond it.

Therefore there are constant $a > 0$, $b > 0$ such that

M(x, 0) > 0, for x<a

M(x, 0) < 0, for x>a

N(0, y) >0, for y <b

N(0, y) <0, for y >b

ASYMPTOTIC STABILITY OF CLOSED ORBITS

Let $f : W \to R^n$ be a C^1 − vector field on an open set $W \subseteq R^n$. We denote by ϕ_t the flow of the differential equation $x' = f(x)$.

Let $r \subseteq W$ be a closed orbit which is nontrivial. We say that r is

LIMIT SETS

asymptotically stable if for every open set $U_1 \subseteq W$ with $r \subseteq U_1$ the is an open set $U_2, r \subseteq U_2 \subseteq U_1$ such that

$\phi_t(U_2) \subseteq U_1$, for all $t>0$ and

$$\lim_{t \to \infty} d(\phi_t(x), r) = 0$$

We know that for the Vander pol's equation , there exist a nontrivial closed orbit which is such that every nontrivial trajectory approaches it.

Thus it is an example of an asymptotically stable closed orbit.

However, closed orbit of Harmonic oscillator Is not asymptotically stable.

Definition: A point $x \in W$ is said to be of asymptotic period λ if

$$\lim_{t \to \infty} | \phi_{\lambda+t}(x) - | \phi_t(x) | = 0.$$

Theorem 13: Let r be an asymptotically stable closed orbit of period λ. Then r has a nbd $\cup \subseteq W$ such that every pt. of $\cup$ has asymptotic period λ.

Proof: For the open set W (we have $r \subseteq W$), there will exist an open set U containing r such that $r \subseteq U \subseteq W$,

$\phi_t(U) \subseteq W, t \geq 0$ and

$$\lim_{t \to \infty} d(\phi_t(x), r) = 0, \forall x \in U.$$

We shall show that every $x \in U$ has asymptotic period λ.

Let $\in > 0$ be chosen arbitrarily.

We know that the flow namely ϕ_t is a continuous function for $t \geq 0$.

In particular ϕ_λ is continuous.

Thus for any $z \in r$, we can find $\delta > 0$ such that

$$\forall y \in W, |y - z| < \delta \Rightarrow |\phi_\lambda(y) - \phi_\lambda(z)| < \frac{\in}{2}$$

425

We choose δ so that $0 < \delta < \dfrac{\epsilon}{2}$

Also, for $x \in U$, we know that $\lim\limits_{t \to \infty} d\big(\phi_t(x), r\big) = 0$.

Hence, we can find $t_o > 0$ such that

For $t > t_o , d\big(\phi_t(x), r\big) < \delta$

i.e. there exist a pt. say $z_t \in r$ such that

$|\phi_t(x) - z_t| < \delta$, for $t > t_o$

We now show that $|\phi_{\lambda+t}(x) - \phi_t(x)| < \epsilon,$, for $t > t_o$

Consider, $\phi_{\lambda+t}(x) - \phi_t(x) = \phi_{\lambda+t}(x) - \phi_\lambda(z_t) - \phi_t(x)|$

$\therefore |\phi_{\lambda+t}(x) - \phi_t(x)| \le |\phi_{\lambda+t}(x) - \phi_\lambda(z_t)| + \phi_\lambda(z_t) - \phi_t(x)|$

We note that r is a closed orbit of period λ.

As $z_t \in r, \phi_\lambda(z_t) = z_t$

Since $|\phi_t(x) - z_t| < \delta$ and $z_t \in r$, we have

$|\phi_\lambda\big(\phi_t(x)\big) - \phi_\lambda(z_t)| < \dfrac{\epsilon}{2}$

i.e. $|\phi_{\lambda+t}(x) - \phi_\lambda(z_t)| < \dfrac{\epsilon}{2}$

thus, we have

$|\phi_{\lambda+t}(x) - \phi_t(x)| \le |\phi_{\lambda+t}(x) - \phi_\lambda(z_t)| + |z_t - \phi_t(x)|$

$$< \dfrac{\epsilon}{2} + \delta < \dfrac{\epsilon}{2} + \dfrac{\epsilon}{2} = \epsilon .$$

Thus for a given $\epsilon > 0, \exists$ to such that $t > t_o$

$|\phi_{\lambda+t}(x) - \phi_t(x)| < \epsilon$

$\therefore d\big(\phi_{\lambda+t}(x), \phi_t(x)\big) \to 0$ as $t \to \infty$.

LIMIT SETS

As $x \in U$ is arbitrary, the result is proved

Definition: An operator $T \in L(E)$ is said to be a contraction if for all $x \in E$, $\lim\limits_{n \to \infty} T^n x = 0$

Theorem 14: The following statements are equivalent:

(a) T is a linear contraction.

(b) The eigen value of T have absolute values less than 1 (strictly less than 1)

(c) There is a norm on E and $\mu < 1$ such that

$|Tx| \leq \mu |x|, \forall x \in E$

(a) $\Rightarrow$ (b): We show that $\sim$ (b) $\Rightarrow \sim$ (a)

Thus let us assume that there is an eigen value λ of T such that $|\lambda| \geq 1$.

If λ is real, we can find $x \in E$, $x \neq 0$ such that $Tx = \lambda x$

$T^2 x = T(\lambda x) = \lambda T(x) = \lambda(\lambda x) = \lambda^2 x$

$T^n x = \lambda^n x$

$\therefore |T^n x| = |\lambda|^n \cdot |x|$

Certainly, as $x \neq 0$ and $|\lambda| \geq 1$.

$|\lambda|^n |x| \not\to 0$ as $n \to \infty$.

$\therefore T^n x \not\to 0$ as $n \to \infty$ for this particular $x \in E$.

i.e. T is not a linear contraction.

$$(c) \Rightarrow (a)$$

We are given that

$|Tx| \leq \mu |x|$

$\therefore |T^2 x| = |T(Tx)| \leq \mu |Tx|$

$\qquad \leq \mu.\mu |x| = \mu^2 |x|$

427

$|T^2 x| \leq \mu^2 |x|$

In general, $|T^n x| \leq \mu^n |x|$

But $0 < \mu < 1, \therefore \mu^n \to 0$ as $n \to \infty$

$\therefore |T^n x| \to 0$ as $n \to \infty$

$\therefore T^n x \to 0$ as $n \to \infty, \forall x \in E$

$\therefore$ T is a linear contraction.

(b) $\Rightarrow$ (c)

Let $\lambda_1, \lambda_2, \cdots \lambda_k$ be the distinct values of T.

We are given that $|\lambda_i| < 1$, for $1 \leq i \leq k$

Let t_{λ_i} be the generalized eigen space of T belonging to the eigen value λ_i.

We have $E = E_{\lambda_1} \oplus E_{\lambda_2} \oplus \cdots \oplus E_{\lambda_k}$

Further each of them is invariant under T.

If Ti is the restriction of T to E_{λ_i} then we have

$T = T_1 \oplus T_2 \oplus \cdots \oplus T_k$

Since $E = E_{\lambda 1} \oplus E_{\lambda 2} \oplus \cdots \oplus E_{\lambda k}$, we have

$x = x_1 + x_2 + \cdots + x_k$, for unique $x_i \in a\, E_{\lambda i}$.

If we take a max. norm then the thm will be proved if it is proved on $E_{\lambda i}$

for the operator $T_i, 1 \leq i \leq k$.

Since T_i is an operator on $E_{\lambda i}$ with only one eigen value namely λ_t.

Without loss of generality, we can assume that T has only one eigen value

say λ on the space E with $|\lambda| < 1$.

We can find a basis of E w.r.t. which the matrix of T on E is given by λ-

Jordan block.

LIMIT SETS

As a Jordan black consists of various elementary Jordan blocks put along the diagonal, we can assume that the matrix of T w.r.t. a basis $\{e_1, e_2, \cdots e_n\}$ of E is an elementary λ-Jordan block.

$$\begin{bmatrix} \lambda & & & & \\ 1 & \lambda & 0 & & \\ & & \ddots & & \\ & 0 & & \lambda & \\ & & & 1 & \lambda \end{bmatrix}$$

i.e.
$$\begin{aligned} Te_1 &= \lambda e_1 + e_2 \\ Te_2 &= \lambda e_2 + e_2 \\ &\vdots \\ Te_n &= \lambda e_n \end{aligned}$$

since $|\lambda| < 1$, we can choose $\in$ small enough so that $|\lambda| + \in < 1$

for this $\in, \bar{e}_i = \dfrac{e_i}{\in^{i-1}}, 1 \le i \le n$.

It is clear that $\bar{e}_1, \bar{e}_2, \cdots \bar{e}_n$ is a new basis and the matrix of T w.r.t. this basis can be obtained from the following equations:

$$T\bar{e}_1 = T(e_1) = \lambda e_1 + e_2 = \lambda \bar{e}_1 + \in \bar{e}_2$$

$$T\bar{e}_2 = T\left(\frac{e_2}{e}\right) = \frac{1}{\in} T(e_2) = \frac{1}{\in}[\lambda e_2 + e_3]$$

$$= \frac{1}{\in} \lambda \in \bar{e}_2 + \frac{1}{\in} \cdot e^2 \bar{e}_3$$

$$= \lambda \bar{e}_2 + \in \bar{e}_3 \text{ and so on.}$$

Thus the matrix of T w.r.t. the basis $\{e_1, e_2, \cdots e_n\}$.
is:

$$\begin{bmatrix} \lambda & & & \\ \in & \lambda & 0 & \\ & & \ddots & \\ 0 & & \lambda & \\ & & \in & \lambda \end{bmatrix}$$

Hence for any $x \in E, x = x_1\bar{e}_1 + x_2\bar{e}_2 + \cdots + x_n\bar{e}_n$

We have

$$Tx = T\left(\sum_{i=1}^{n} x_i\bar{e}_i\right)$$

$$= x_1 T(\bar{e}_1) + x_2 T(\bar{e}_2) + \cdots + x_n T(\bar{e}_n)$$

$$= x_1(\lambda\bar{e}_1 + \in\bar{e}_2) + x_2(\lambda\bar{e}_2 + \in\bar{e}_3) + \cdots + x_{n-1}(\lambda\bar{e}_{n-1} + \in\bar{e}_n)$$

$$= (x_1\lambda)\bar{e}_1 + (x_1 \in +x_2\lambda)\bar{e}_2 + (x_3 \in +x_3\lambda)\bar{e}_3 + \cdots + (x_{n-1} \in +\lambda x_n)\bar{e}_n$$

Since we are using max, norm,

$$|Tx| = \max.\{|x_1\lambda|, |x_1 \in +\lambda|, \cdots, |x_{n-1} \in +\lambda x_n|\}$$

$$\leq |\lambda||x| + \in|x| = (|\lambda| + \in)|x|$$

If $\mu = |\lambda| + \in$, then we have

$|Tx| \leq \mu|x|$ with $\mu < 1$

DISCRETE DYNAMICAL SYSTEM

If f is a continuously differentiable function on an open set $w \subseteq E$ then we know that $\phi_t(\cdot)$ is a flow of the dynamical system $x' = f(x)$.

When $\phi_t(x)$ denotes the solution of the differential equation at time t when initially it is at x.

In this case, $t \in R$ a continuous variable. However, for some situations, a continuous variable may not be definable and one may have to deal with variable which takes only discrete values.

430

LIMIT SETS

For such situations, the usual concept of dynamical system is not useful.

We therefore need to modify the definition of dynamical system for a discrete variable.

For example, let g be a continuously differentiable function from $W \rightarrow W$ such that g^{-1} is also C^1 (g is a diffeomorphism).

We may visualize w as a state of space of some kind. Then g(w) can considered as a new state of space after a 'unit of time' (unit because g(w) is obtained from w by applying once). $g^2(w)$ can be considered as the state of space after two units of time and so on. In this way, we get a concept of a discrete dynamical system.

Definition: A discrete dynamical system is a C^1-map, $g : w \rightarrow E$ where w is an open subset of E. if $w \neq E$, it is possible that g^2 is not defined at all pts. of E or even at points on w.

EQUILIBRIUM POINT

For a usual dynamical system $x' = f(x)$, we know that $\bar{x}$ is an equilibrium point iff

$$\phi_t(\bar{x}) = \bar{x}, \forall t \in R$$

Similarly, if $g : w \rightarrow E$ is a discrete dynamical system then a fixed point $\bar{x}$ of g is known as an equilibrium pt. of the discrete dynamical system.

It follows that as $\bar{x} = g(\bar{x})$, we have

$$\bar{x} = g^n(\bar{x}), \text{ for } n = 0, 1, 2, \cdots$$

Definition: We define an equilibrium point $\bar{x}$ of a discrete dynamical system to be asymptotically stable if for any open set U_1 of W containing $\bar{x}$, there exists an open set U_2 containing $\bar{x}$ such that $g(U_2) \subseteq U_1$ and $g^n(x) \rightarrow \bar{x}$ as $n \rightarrow \infty$.

DYNAMICAL SYSTEM – A SHORT COURSE

Definition: An equilibrium point $\bar{x}$ is a discrete dynamical system is said to be a sink of It provided all the eigen values of $Dg(\bar{x})$ have absolute value less than one.

Theorem 15: Let $\bar{x}$ be a fixed point of a discrete dynamical system $g : W \to E$. If the eigen values of $Dg(\bar{x})$ are less than 1 in absolute value, then $\bar{x}$ is asymptotically stable.

Proof: Without loss of generality, we can assume that $\bar{x} = 0$.

Thus we have $g(o) = 0$ and further all eigen values of $Dg(o)$ have absolute value less than 1.

Therefore we can define a norm on E and find a +ve constant $u < 1$ such that

$$|Dg(o)x| \leq \mu |x|, \forall x$$

Let $\in$ be chosen in such a way that $o < \in < 1 - u$

Using the definition of differentiability of g at o, we have

$$\lim_{|x| \to 0} \frac{|g(x) - g(o)x|}{|x|} = 0$$

Therefore for the above $\in > o$, we can find $\delta > o$

Such that $o < |x| < \delta$.

$$\Rightarrow \frac{|g(x) - Dg(o)x|}{|x|} < \in$$

i.e. $|x| < \delta \Rightarrow |g(x) - Dg(o)x| < \in |x|$

as $g(x) = g(x) - Dg(o)x + Dg(o)x$

$|g(x)| \leq |g(x) - Dg(o)x| + |Dg(o)x|$

$\qquad \leq \in |x| + \mu |x| = (\mu + \in)|x|$ with $\mu + \in < 1$

Let U_1 be any open set containing o.

LIMIT SETS

We consider U_2 to be an open set given by the intersection of δ ball $B_\delta(o)$ and U_1.

We show that $g(U_2) \subseteq U_1$ and for all $x \in U_2, g^n(x) \to o$.

If $x \in U_2$ then $|x| < \delta$ and therefore

$|g(x)| \le (\mu + \in)|x|$

$\Rightarrow (\mu + \in)|x| < \delta$ i.e. $|g(x) < \delta$

Thus for $x \in U_2$, we have $g(x) \in B_\delta(o)$

We can take small enough so that $B_\delta(o) \subseteq U_1$

Further for $x \in U_2$,

$|g^2(x)| \le (\mu + \in)^2 |x|$

$\vdots$

$|g^n(x)| \le (\mu + \in)^n |x|$

However, $\mu + \in < 1 \Rightarrow (\mu + \in)^n \to 0$ as $n \to \infty$

$\therefore g^n(x) \to 0$ as $n \to \infty \; \forall x \in U_2$

POINCARE MAP

Consider a flow ϕ_t of a C^1 – vector field $f : W \to E$.

Let $r \subseteq W$ be a closed orbit and suppose $o \in r$.

Consider a section S at O. if $\lambda > 0$ is a period of r, then we know that

$\phi_\lambda(o) = 0$

i.e. $\phi_{n\lambda}(o) \in S, n = 1, 2, \cdots$ as $\phi_{n\lambda}(o) = 0$

if x is sufficiently near zero, then there will be a time $\tau(x)$ near λ when

$\phi_{\tau(x)}(x)$ crosses S.

in this way we get a map $g : U \to S$ (where U is the above nbd of o) defined as

DYNAMICAL SYSTEM – A SHORT COURSE

$g(x) = \phi_{\tau(x)}(x)$.

Let $S_o = S \cap U$. We define a C^1 – map $g: S_o \to S$ as:

$g(x) = \phi_{\tau(x)}(x)$.

This map is known as a Poincare map.

Further g gives rise to a discrete dynamical system on S_o with O with as its equilibrium point. (Note that $\tau(o) = \lambda$ and $\phi_\lambda(o) = 0$).

We know that for the small nbd U of o in E (n-dimensional space) with $x \in U$, we can associate the image $\phi_{\tau(x)}(x)$, where $\phi_{\tau(x)}(x)$ is a point on the section S. this will give us a map from U (in n-dimensional space) to S (in n-1 dimensional space).

In order to define the Poincare map we instead take S_o to be intersection of U and S and define the map $x \to \phi_{\tau(x)}(x)$ a map from an open set in n-1 dimensional space to an open set S in an n-1 dimensional space.

We denote this map by g a map from an open set in n-1 dimensional space to an open set in n-1 dimensional space.

If $g(x) \in S_o, g^2(x)$ will also be defined and so on, g gives a discrete dynamical system.

Theorem 15: Let $g: S_o \to S$ be a Poincare map for r as above. Let $x \in S_o$ be such that $\lim_{n \to \infty} g^n(x) = 0$ .then $\lim_{n \to \infty} d(\phi_t(x), r) = 0$,

Proof: Let us denote $g^n(x)$ by x_n and x by x_o.

Similarly, let λ_n denote $\tau(x_n)$.

Since $g^n(x)$ is defined for $n = 1,2,3,\cdots$, it follows that $x_n \in S_o, n = 0,1,2,\cdots$

Since $x_n \to 0$ as $n \to \infty$, using the continuity of

LIMIT SETS

$\tau, \tau(x_n) \to t(o) = \lambda$, we get $\lambda_n \to \lambda$.

As the seq. λ_n a seq. of non-negative real nos. is convergent, it is bounded.

Let r be the upper bound for nay $s \in [o, r]$

$\phi_s(x_n) \to \phi_s(o)$ (by continuity of the flow ϕ_s and the fact that $x_n \to 0$)

i.e. $|\phi_s(x_n) - \phi_s(o)| \to 0$ as $n \to \infty$.

Actually we have

$|\phi_s(x_n) - \phi_s(o)| \to 0$ uniformly on [o, r]

For $t \geq 0$. Consider $\phi_t(x)$.

Since $\phi_o(x) = x \equiv x_o, \phi_{\lambda o}(x) = x_1, \phi_{\lambda 1}(x_1) = x_2, \cdots$

For the given $t \geq 0$, we can find n such that $\phi_t(x)$ lies between x_n and

$x_{n(t)+1}$ along the trajectory, we must have

$\phi_t(x) = \phi_s(x_{n(t)})$, where $0 \leq s < \lambda$.

and as such $s \in [o, r]$. Since n depends on t, we also write S as S(t).

Then for $t \geq 0$, we can find $s(t) \in [o, r]$ and n(t) such that

$\phi_t(x) = \phi_{s(t)}(x_{n(t)})$

As $t \to \infty, n(t) \to \infty$ and $s(t) \in [o, r]$

Consider $d(\phi_t(x), r)$.

By the definition, $d(\phi_t(x), r) \leq |\phi_t(x) - 0|$

But $\phi_t(x) = \phi_{s(t)}(x_{n(t)})$ (as o∈r)

$o \leq d(\phi_t(x), r) \leq |\phi_{s(t)}(x_{n(t)}) - \phi_{s(t)}(0)|$ ($\therefore \phi_{s(t)}(0) \in r$)

But as $t \to \infty, n(t) \to \infty$ and $x_{n(t)} \to 0$.

$\therefore |\phi_{s(t)}(x_{n(t)}) - \phi_{s(t)}(0)| \to 0$ as $t \to \infty$

435

DYNAMICAL SYSTEM – A SHORT COURSE

Theorem 16: If o is a sink for g, then r is asymptotically stable.

Proof: We not e that r is to be asymptotically stable with respect to the flow ϕ_t corresponding to the usual dynamical system $x' = f(x)$, for some vector field

Thus given a nbd U_1 of r, we need to find a nbd say U of r such that $r \subseteq U \subseteq U_1$ and $\phi_t(U) \subseteq U_1$, for each $x \in U$,

$d(\phi_t(x), r) \to 0$ as $t \to \infty$.

So Let U_1 be some nbd of r. Let λ be the period of the closed orbit r.

Let N be a very small nbd of r in U_1 such that for all $x \in N$ and $0 \le t < 2\lambda$, $\phi_t(x)$ lies in U_1, $\phi_t(x)$ lies in U_1.

As O is a sink for g, the eigen values of $Dg(o)$ will have absolute values less than 1.

We can find a norm on the hyper plane H (containing the section S) and a small nbd say V of o in H s.t.

$\forall x \in V, |g(x) \le \mu |x|$, for some $\mu < 1$.

We choose $\delta > 0$ to be sufficiently small so that the ball $B_\delta(o)$ (in H) lies entirely in $V \cap H$ so that $\forall x \in B_\delta(o)$, we have

$|g(x) \le \mu |x|$ and $\tau(x) < 2\lambda$ (this follows from the definition of N).

We now define $U = \{\phi_t(x), x \in B_\delta(o), t \ge 0\}$

Obiviously $r \subseteq U$.

We next show that $U \subseteq U_1$, Let $y \in U$.

By definition, $Y = \phi_t(x)$, for some $x \in B_\delta(o)$

Since $x \in B_\delta(o)$, $g(x)x \in B_\delta(o) (\therefore |g(x) \le \mu |x|)$

We can thus get $g^n(x), n = 0, 2, 3, \cdots$

LIMIT SETS

We denote $g^n(x)$ by x_n and $\tau(x_n)$ by λ_n.

For given $t \geq 0$, we can find n(t) such that $\phi_t(x)$ lies between $x_{n(t)}$ and

$x_{n(t)+1}$

on the trajectory.

$\therefore \exists s(t), 0 \leq s(t) < \tau\left(x_{n(t)}\right) = \lambda_{n(t)}$ such that

$\phi_t(x) = \phi_{s(t)}(x_{n(t)})$

But as $\lambda_{n(t)} = \tau\left(x_{n(t)}\right) < 2\lambda$ and as $x_{n(t)} \in B_\delta \subseteq U \cap N$, it follows from the

definition of N that

$\phi_{s(t)}(x_{n(t)}) \in U_1$ i.e. $y \in U_1$

Then $U \subseteq U_1$

As U is positively invariant, we have

$\phi_t(U) \subseteq U \subseteq U_1$

We have $\phi_t(U) \subseteq U_1$.

We now show that $\forall y \in U, d(\phi_t(y), r) \to 0$ as $t \to \infty$

By the definition $y \in U \Rightarrow y = \phi_s(x)$, for some $x \in B_\delta(o)$

$\phi_t(y) = \phi_{t+s}(x)$, for some $x \in B_\delta(o)$

We can find $n(t)$ such that $\phi_t(y) = \phi_{t+s}(y)$ lies between $x_{n(t)}$ and $x_{n(t)+1}$ on

the trajectory through x.

Further $n(t) \to \infty$ as $t \to \infty$.

$\phi_t(y) = \phi_{s(t)}(x_{n(t)})$ for some $s(t), 0 \leq s(t) < \tau(x_{n(t+1)})$

If $g^n(x)$ is denoted by x_n then as for $x \in B_\delta(o)$

$|g(x)| \leq \mu|x|, |g^n(x)| \leq \mu^n |x|$

$$\therefore |x_n| \le \mu |x|$$

As u <1, we have $g''(x) \to 0$ as $n \to \infty$

i.e. $x_n \to \infty$ as $n \to \infty$

$$\therefore 0 \le d\big(\phi_t(y), r\big) \le d\big(\phi_{s(t)}(x_{n(t)}), 0\big)$$

i.e. $0 \le d\big(\phi_t(y), r\big) \le |\phi_{s(t)}(x_{n(t)}) - \phi_{s(t)}(o)| \to 0$ as $t \to \infty$

$(\therefore t \to \infty \Rightarrow n(t) \to \infty \ \& \ x_{n(t)} \to 0)$

Theorem 17 : Let r be a closed orbit of period λ of the dynamical system $x' = f(x)$. Let $p \in r$. Suppose that (n-1) of the eigen values of the Linear map $D\phi_\lambda(p): \in \to E$ are less than 1 in absolute value then r is asymptotically stable orbit.

Proof: Without loss of generality, we can assume that the origin i.e. o lies on the closed orbit r of the dynamical system.

So we can consider the linear operator $D\phi_\lambda(o)$, the linear part of the map ϕ_λ.

We note that ϕ_λ is an identity on r.

In case any of the eigen values of $D\phi_\lambda(o)$ is complex, we consider the complexified space namely E_c of E and complexification $\big(D\phi_\lambda(o)\big)_c$ of the linear operator $D\phi_\lambda(o)$.

Let B denote the direct sum of generalized eigen spaces corresponding to those eigen values whose absolute value is less than one.

It is obvious that the space B is invariant under the operator $\big(D\phi_\lambda(o)\big)_c$.

Consider the space $B \cap E$ and denote it by say H.

H is a real subspace of E of dimension n-1 and further it will be invariant under the operator $D\phi_\lambda(o)$.

Thus $D\phi_\lambda(o)|_H$ is an operator on H.

But we know that $(D\phi_\lambda(o))|_H = D_g(o)$

Thus the operator $D_g(o)$ has all its eigen values (the n-1 eigen values of $D\phi_\lambda(o)|_H$) with an absolute value less than 1.

Thus $D_g(o)$ is a linear contraction

As such o is a sink for the Poincare map $g:S_o \to S$.

r is an asymptotically stable orbit.

Definition: [PERIODIC ATTRACTOR]

If r is a closed orbit with period λ for a dynamical system $x' = f(x)$ such that at a point say o of r, the operator $D\phi_\lambda(o)$ has n-1 eigen values with absolute value less than one, then r is said to be a Periodic attractor.

Thus every periodic attractor is an asymptotically stable.

Theorem 18: Let r be a periodic attractor. If $\lim_{n\to\infty} d(\phi_t(x),r)=0$ then there is a unique point $z \in r$ such that $\lim_{n\to\infty}|\phi_t(x)-\phi_t(z)|=0$

Proof: Since we are given that $\lim_{n\to\infty} d(\phi_t(x),r)=0$,

the trajectory $\phi_t(x)$ is approaching the closed orbit r and as such it is approaching a point on r.

Thus we can assume the point x itself to be sufficiently closed to r and then show that $\phi_t(x)$ approaches a unique point on r as $t\to\infty$.

If the point x is sufficiently close to a point y on r and if $\phi_s(y)=0$ (we assume o to be a pt. on r), it follows that $\phi_{s1}(x)$ will be in a very small nbd of o for some value S_1 close to s.

Thus without loss of generality, we can choose x itself to be sufficiently close to o and then show that $\phi_t(x)$ tends to a unique point on r.

Thus we start with the assumption that the point x is very close to o.

In order to prove the thm, we shall show that the seq. $\phi_{n\lambda}(x), n = 0,1,2,\cdots$ tends to a unique point. on r.

For this we shall show that $\phi_{n\lambda}(x)$ is defined for all a and stays close to o.

This will be shown by showing that if $\phi_{o\lambda}, \phi_{2\lambda}, \cdots \phi_{n\lambda}$ are close to o for some positive integer n then $\phi_{(n+1)\lambda}, \phi_{(n+2)\lambda}, \cdots$ also remain close to o.

We shall show that $\phi_{n\lambda}(x) = \phi_{tn}\left(g^n(x)\right)$.

Where we define $t_o = o, t_1 = t_o - \tau\left(g^o(x)\right) + \lambda$ and inductively

$$t_n = t_{n-1} - \tau\left(g^{n-1}(x)\right) + \lambda$$

Obviously, the formula holds for n =0

We assume that the formula holds for n-1 and then show that if holds for n.

Thus consider $\phi_{n\lambda}(x)$.

$$\phi_{n\lambda}(x) = \phi_{\lambda+(n-1)\lambda}(x) = \phi_\lambda\left(\phi_{(n-1)\lambda}(x)\right)$$

$$= \phi_\lambda\left(\phi_{t_{n-1}}\left(g^{n-1}(x)\right)\right) = \phi_{\lambda+t_{n-1}}\left(g^{n-1}(x)\right)$$

But $t_n = t_{n-1} - \tau\left(g^{n-1}(x)\right) + \lambda$

$$\Rightarrow \lambda + t_{n-1} = t_n + \tau\left(g^{n-1}(x)\right)$$

$$\therefore \phi_{n\lambda}(x) = \phi_{\lambda+t_{n-1}}\left(g^{n-1}(x)\right)$$

$$= \phi_{t_{n+\tau\left(g^{n-1}(x)\right)}}\left(g^{n-1}(x)\right)$$

$$= \phi_{tn}\left(\phi_{\tau\left(g^{n-1}(x)\right)}\right)(g^{n-1}(x)) = \phi_{tn}\left(g(g^{n-1}(x)\right)$$

$$\phi_{n\lambda}(x) = \phi_{tn}\left(g^n(x)\right)$$

Thus we can write $\phi_{n\lambda}(x) = \phi_{tn}\left(g^n(x)\right)$

We have $t_n = t_{n-1} - \tau\left(g^{n-1}(x)\right) + \lambda$

$\therefore |t_n = t_{n-1}| = |\tau\left(g^{n-1}(x)\right) - \lambda|$

We note that $\tau(o) = \lambda$ and $D\tau(o) = 0$ and therefore if $|g^{n-1}(x)|$ is sufficiently small then for a given $\in > 0$, we have

$$|\tau\left(g^{n-1}(x)\right) - \tau(o) - D\tau(o)\left(g^{n-1}(x)\right)| < \in |g^{n-1}(x)|$$

i.e. $|t_n - t_{n-1}| < \in |g^{n-1}(x)|$, for an appropriate norm on E.

since r is a periodic attractor, $Dg(o)$ is a linear contraction (0 is a sink) and so

$|g(x)| \leq \mu |x|$, for some $\mu, 0 < \mu < 1$

$\therefore |t_n - t_{n-1}| < \in \mu^{n-1} |x|$, for $\in$ arbitrary small and $0 < \mu < 1$.

As such since $|g^n(x)| \leq \mu^n |x|$ and $|x|$ is sufficiently small, we have

$\phi_{tn}\left(g^n(x)\right)$ sufficiently close to o for n as large as we want.

For all n, we can use the formula $\phi_{n\lambda}(x) = t_n\left(g^n(x)\right)$

We can indeed show that the seq. $\{t_n\}$ is a convergent seq. it follows that the seq. $\{t_n\}$ is Cauchy seq. and therefore convergent.

Let $t_n \rightarrow s$. Here taking limit as $n \rightarrow \infty$, we have

$\phi_{n\lambda}(x) \rightarrow \phi_s(o)$ a pt. on r.

EXISTENCE, UNIQUENESS AND CONTINUITY FOR NON-AUTONOMOUS DIFFERENTIAL EQUATIONS

Let E be a normed vector space, $W \subseteq R \times E$ an open set and $F : W \rightarrow E$ a continuous map.

Let $(t_o, u_o) \in W$. A solution to the initial value problem

$$X'(t) = f(t, x(t))$$

$$X(t_o) = u_o$$

Is a differential function in E defined for t in some interval I having the following properties namely

$t_o \in J$ and $X(t_o) = u_o$

$(t, x(t)) \in W$, $X'(t) = f(t, x(t))$, $\forall t \in J$

We call the function $f(t, x)$ Lipschitz in x if there is a constant $k \geq 0$ such that

$$|f(t, x) - f(t, y)| \leq k |x - y|, \ \forall (t, x), (t, y) \in W$$

Suppose $x(t), t \in J$ is a solution to the initial value problem

$$x'(t) = f(t, x(t)), \ x(t_o) = u_o$$

Integrate $x'(t) = f(t, x(t))$ from t_o to t, for some $t \in J$

i.e. $\int\limits_{to}^{t} x'(s) ds = \int\limits_{to}^{t} f(s, x(s)) ds$

i.e. $x(t) - x(t_o) = \int\limits_{to}^{t} f(s, x(s)) ds$

i.e. $x(t) = u_o + \int\limits_{to}^{t} f(s, x(s)) ds$

Conversely, if for some J with $t_o \in J$, we have

$$x(t) = u_o + \int\limits_{to}^{t} f(s, x(s)) ds, t \in J$$

$x(t)$ is differentiable and we have

LIMIT SETS

$$x'(t) = f(t, x(t)), x(t_o) = u_o$$

Thus the integral equation namely

$$x(t) = u_o + \int_{to}^{t} f(s, x(s)) ds$$

Is equivalent to the given non-autonomous initial value problem

$$x'(t) = f(t, x(t)), x(t_o) = u_o$$

If $|.|$ is the norm on E then we define a norm $|.|$ on $R \times E$ as

$|(t, u)| = max \{ |t|, |u| \}$

Where, $|t|$ is the absolute value of t and

$|u|$ is the norm of u in E.

Thus the induced metric on RxE will be

$$d((t_1, u_1), (t_2, u_2)) = |(t_1 - t_2), (u_1 - u_2)|$$

$$= max \{|t_1 - t_2, u_1 - u_2|\}$$

Theorem 17: Let $W \subseteq R \times E$ be open and $F : W \to E$ a continuous map that is Lipschitz in X. if $(t_o, u_o) \in W$, there is an open interval J containing to and a unique solution to the non-autonomous initial value problem

$$x'(t) = f(t, x(t)), x(t_o) = u_o, t \in J$$

Proof: Let $k \geq 0$ be the Lipschitz constant for f in x.

Since $(t_o, u_o) \in W$ and W is open, we can find $b > 0$ such that

$$\{(t, u) \ s.t \ d((t, w, (t_o, u_o)) \leq b\} \subseteq W$$

Let $W_o \equiv \{(t, u) \text{ such that } d((t, w, (t_o, u_o)) \leq b\}$.

Since F is continuous on W and W_o is a compact subset of W. F is bounded on W_o.

Let $M \geq 1$ be chosen in such a way that

443

$|f(t,u)| \leq M, \forall (t,u) \in W_o$

We now define a new real number a such that

$$0 < a < \min\left\{\frac{b}{M}, \frac{1}{k}\right\}$$

We now define $J = [t_o - a, t_o + a]$

We now construct seq. of function $x_o, x_1, x_2, \cdots$ in the following way for $t \in J$

We define $x_o(t) = u_o, t \in J$

We then define $x(t) = u_o + \int\limits_{to}^{t} f(s, x(s))ds, t \in J$

We note that for all $t \in J, s \in J$ in the above expression and

$$d((s, x_o(s)), (t_o, u_o)) = \max\left\{|s - t_o|, |x_o(s) - u_o|\right\}$$

$$= \max\left\{|s - t_o|, 0\right\} \leq a \leq b$$

Since $M \geq 1$ we have a $a \leq b$

Thus the R.H.S. makes sense and $x_1(t)$ is well defined for all $t \in J$,.

We now define $x_2(t)$ as

$$x_2(t) = u_o + \int\limits_{to}^{t} f(s, x(s))ds \text{ , for } t \in J$$

We note that $d((s, x_1(s)), (t_o, u_o)) \leq b$

For this use the fact that

$$x_1(s) = u_o + \int\limits_{to}^{s} f(t, x_o(t))dt$$

$$\therefore |x_1(s) - u_o| \leq + \int\limits_{to}^{s} |f(t, x_o(t)| dt$$

444

LIMIT SETS

$$\leq |s-t_o|\cdot M \leq aM \leq b$$

$$d\big((s,x_1(s)),(t_o,u_o)\big)=\max\big\{|s-t_o|,|x_1(s)-u_o|\big\}\leq b$$

$$(s,x_1(s))\in W_o \text{ , for } s\in J$$

Thus $x_2(t)$ is well defined.

We note that $|x_2(t)-u_o|\leq aM\leq b$

We can thus proceed and define $x_3(t),x_4(t),\cdots$ and so on.

In general, $x_{n+1}(t)=u_o+\int_{to}^{t}f(s,x_n(s))ds, t\in J$

We next show that the seq. of functions $\{x_o(t),x_1(t),\cdots\}$ converges uniformly on J.

Consider $x_2(t)=u_o+\int_{to}^{t}f(s,x_1(s))ds$

$$x_1(t)=u_o+\int_{to}^{t}f(s,x_o(s))ds$$

$$\therefore x_2(t)-x_1(t)=\int_{to}^{t}\big(f(s,x_1(s))-f(s,x_o(s))\big)ds, \forall t\in J$$

$$\therefore x_2(t)-x_1(t)\leq\int_{to}^{t}|f(s,x_1(s))-f(s,x_o(s))|\,ds$$

$$\leq\int_{to}^{t}k\,|x_1(s)-x_o(s)|\,ds$$

Let $L=\max\{x_1(s)-x_o(s)|,s\in J\}$

$$\therefore x_2(t)-x_1(t)\leq\int_{to}^{t}k.L\ ds\leq akL\ \forall t\in J$$

Consider $x_3(t)-x_2(t)$

$$| x_3(t) - x_2(t) | \leq \int_{to}^{t} | f(s, x_1(s)) - f(s, x_o(s)) | \, ds$$

$$\leq \int_{to}^{t} k. \, akL \, ds \leq$$

$$\leq (ak)^2 L$$

We can generalize this to get

$$| x_{n+1}(t) - x_n(t) | \leq (ak)^n . L, n \geq 2$$

However, from the definition of 'a', we have

$$a < \frac{1}{k} \quad \text{i.e.} \quad ak < 1$$

Let ak be denoted by say α. Then we have $0 < \alpha < 1$.

Consider the series $\displaystyle\sum_{n=0}^{\infty} \alpha^n L = L \sum_{n=0}^{\infty} \alpha^n$

It is a convergent geometric series.

Hence for a given $\in > 0$, we can find N such that

$$\sum_{n=0}^{\infty} \alpha^n . L < \in$$

Thus for $n, m \geq N$,

$$| x_n(t) - x_m(t) | = | x_n(t) - x_{n-1}(t) + x_{n-1}(t) - x_{n-2}(t) + \cdots + x_{m+1}(t) - x_m(t)$$

$$\leq | x_n(t) - x_{n-1}(t) | + | x_{n-1}(t) - x_{n-2}(t) | + \cdots + | x_{m+1}(t) - x_m(t) |$$

$$\leq L\alpha^{n-1} + L\alpha^{n-2} + \cdots + L\alpha^m = L \left(\sum_{i=m}^{n-1} \alpha^i \right)$$

$$| x_n(t) - x_m(t) | \leq L \left(\sum_{i=N}^{\infty} \alpha^i \right) < \in, \forall t \in J$$

Thus for a given $\in > 0, \exists \, N \text{ s.t } \forall n, m \geq N$

LIMIT SETS

$|x_n(t) - x_m(t)| < \epsilon, \forall t \in J$

Thus the seq. $\{x_o(t), x_1(t), \cdots\}$ is uniformly Cauchy in J.

Since the function $x_n(t)$ are having values in E, a finite dimensional real space, the above seq. is uniformly convergent.

Let $x(t) = \lim_{n \to \infty} x_n(t), t \in J$

Since $x_{n+1}(t) = u_o + \int_{to}^{t} f(s, x_n(s)) ds$

We take limit as $n \to \infty$ and use the fact that $x_n \to x$ uniformly, f is Lipschitz continuous in X to get

$x(t) = u_o + \int_{to}^{t} f(s, x_1(s)) ds, \forall t \in J$

It follows that $x(t)$ is a solution to the given initial value problem.

Uniqueness: in order to prove the uniqueness.

Let us assume if possible $x(t)$ and $y(t)$ to be two solutions for $t \in J$.

Consider the function $|x(t) - y(t)|, t \in J$

As J is compact and $|x(t) - y(t)|$ is a continuous real valued function, $|x(t) - y(t)|$ will attain its maximum say Q at some pt. $t_1 \in J$.

Thus we have

$$0 \le Q = |x(t_1) - Y(t_2)| = \left| \int_{t_o}^{t_1} x'(s) ds - \int_{t_o}^{t} y'(s) ds - \right|$$

$$= \left| \int_{t_o}^{t} (x'(s) - y'(s)) ds \right|$$

$$= \left| \int_{t_o}^{t} (f(s,x(s)) - f(s,y(s))) \right| ds$$

As $(x'(t) = f(t,x(t)), x(t_o) = u_o$

$\qquad y'(t) = f(t,y(t)), y(t_o) = u_o).$

$$\therefore 0 \le Q \le \int_{t_o}^{t_1} |(f(s,x(s)) - f(s,y(s))| \, ds$$

$$\le \int_{t_o}^{t_1} k \, | \, x(s) - y(s) \, | \, ds$$

$$\le \int_{t_o}^{t_1} k \, Q \, ds \le ak \, Q$$

But $a \, k < 1$

Thus we have $0 \le Q \le akQ$

$\therefore Q = 0$ i.e. $| x(t) - y(t) | = 0, \forall t \in J$

i.e. $x(t) - y(t), \forall t \in J$

Theorem 18: Let $A : J \to L(E)$ be a continuous map from an open interval J to the space of Linear operators on E. Let $(t_o, u_o) \in J \times E$. Then the initial value problem $x'(t) = A(t)x, x(t_o) = u_o$

has unique solution an all of J.

Proof: Let J_o be any compact subset of J. consider the map

$t \to A(t) \to \| A(t) \|$

As J_o is compact and the map $t \to \| A(t) \|$ is continuous, $\| A(t) \|$ will be bounded for $t \in J_o$

Let k be s.t $\| A(t) \| \le k, \forall t \in J_o$

LIMIT SETS

Consider $A(t)x - A(t)y$.

$|A(t)x - A(t)y| = |A(t)(x-y)|$

$$\leq \|A(t)\| \cdot |x-y| \leq k|x-y|$$

Thus if we denote $A(t)x$ by $f(t, x)$, $t \in J_o, x \in E$

Then we have $|f(t,x) - f(t,y)| \leq k, \forall t \in J_o$

Thus f satisfies Lipschitz condition w.r.t. x. where

$$f(t,x) \equiv A(t)x|_{J_o xE}$$

Thus now if J_o is a compact subset of j containing the point t_0. We can consider the differential equation $x'(t) \equiv A(t)x$ with initial value $x(t) = u_o$, where $f(t,x) \equiv A(t)x$, restricted to $J_o \times E$.

By the thm, we can find a unique solution to this initial value problem in an interval say $[t_o - a, t_o + a]$ for some $a > 0$.

We now consider the point $t_o + a$ and consider the initial value problem

$$x'(t) = f(t,x), x(t_o + a) = x_1$$

Where x_1 is known to us.

By the above argument there will exist a unique solution in a nbd of point $t_o + a$ to the above initial value problem.

By the uniqueness of the solution, the previous solution can be extended to cover the above nbd of $t_o + a$.

This process can be repeated to get a solution defined over the interval J.

GRONWALL'S LEMMA

Let $u : [o, \alpha] \to R$ be continuous and non-negative. Suppose $c \geq 0, k \geq 0$ are such that

$$u(t) \leq c + \int_0^t k\, u(s)\, ds, \forall t \in [o, \alpha]$$

Then $u(t) \le ce^{kt}, \forall t \in [o, \alpha]$

Proof: Let us denote the expression $c + \int_0^t k\, u(s)\, ds$ by v(t)

Then we have $u(t) \ge 0, t \in [o, \alpha]$ and $u(t) \le v(t)$

We first assume that c > 0.

Hence for $t \in [o, \alpha]$, v(t) > 0.

Since u is continuous, it follows that v is differential and

$v'(t) = ku(t), \forall t \in [o, \alpha]$

As $v(t) > 0, \forall t \in [o, \alpha]$, we get

$\dfrac{v'(t)}{v(t)} = k\dfrac{u(t)}{v(t)} \le k$ as $u(t) \le v(t)$

Hence $\forall t \in [o, \alpha]$, we have $\dfrac{v'(t)}{v(t)} \le k$

$\therefore \int_0^t \dfrac{v'(t)}{v(t)}\, dt \le \int_0^t k\, dt$

i.e. $\int_0^t \left(\dfrac{d}{dt} \log v(t) \right) dt \le kt$

i.e. $\log v(t) - \log v(o) \le kt$

but v(o) = c

$\log v(t) \le \log c + kt$

$\qquad \le \log c + \log(\exp kt))$

$\qquad \le \log(ce^{kt})$

$v(t) \le ce^{kt}, \forall t \in [o, \alpha]$

$u(t) \le v(t) \le ce^{kt}$

$\Rightarrow u(t) \le ce^{kt}, \forall t \in [o, \alpha]$

LIMIT SETS

We now consider the case $c = 0$ i,e. we are given that

$$0 \le u(t) \le \int_0^t k\, u(s)ds \le \frac{1}{n} + \int_0^t k\, u(s)ds, n = 1,2,\cdots$$

$$\therefore 0 \le u(t) \le \frac{1}{n} + \int_0^t k\, u(s)ds$$

$$\Rightarrow 0 \le u(t) \le \frac{1}{n} e^{kt} \text{ (use } \frac{1}{n} \text{ as C in the previous part)}$$

$$\le \frac{1}{n} e^{k\alpha} \text{ as } t \in [o, \alpha]$$

Thus we have $\forall t \in [o, \alpha]$

$$0 \le u(t) \le \frac{1}{n} e^{k\alpha}$$

As we can take $n = 1,2,3,\cdots$, we must have

$$0 \le u(t) \le 0$$

i.e.. $u(t) \equiv 0$.

Theorem 19: Let $W \subseteq R \times E$ be open and $f, g : W \to E$ continuous. Suppose that for all $(t, x) \in W$

$$|f(t,x) - g(t,x)| < \in$$

Let k be Lipschitz constant in x for $f(t,x)$. If $x(t), y(t)$ are the solutions to $x'(t) = f(t,x)$ and $y'(t) = g(t,y)$ respectively on some interval J and $x(t_o) = y(t_o)$ then

$$|x(t) - y(t)| \le \frac{\in}{k} \left(\exp(k|t - t_o|) - 1 \right), \forall t \in J.$$

Proof: For $t \in J$, we consider x(t) and y(t)

We have

$$x(t) = \int_{t_o}^{t} x'(s)ds + x(t_o)$$

$$y(t) = \int_{t_o}^{t} y'(s)ds + y(t_o)$$

As $x(t_o) = y(t_o)$, we have

$$x(t) - y(t) = \int_{t_o}^{t} \left(x'(s) - y'(s) \right) ds$$

$$= \int_{t_o}^{t} \left(f(s, x(s)) - g(s, y(s)) \right) ds$$

$$= \int_{t_o}^{t} \left(f(s, x(s)) - f(s, y(s) + f(s, y(s)) - g(s, y(s)) \right) ds$$

$$\therefore |x(t) - y(t)| \leq \int_{t_o}^{t} | f(s, x(s)) - f(s, x(s)) | \, ds + \int_{t_o}^{t} | f(s, y(s)) - g(s, y(s)) | \, ds$$

$$\leq \int_{t_o}^{t} k | x(s) - y(s) | \, ds + \int_{t_o}^{t} \in ds .$$

For $\forall t \in J$, let $u(t) = | x(t) - y(t) |$, thus we have

$$u(t) \leq \int_{t_o}^{t} k \, u(s) ds + k \int_{t_o}^{t} \frac{\in}{k} ds = \int_{t_o}^{t} k \left(u(s) + \frac{\in}{k} \right) ds .$$

i.e. $\forall t \in J$, $u(t) \leq \int_{t_o}^{t} k \left(u(s) + \frac{\in}{k} \right) ds$

let $v(t) = u(t) + \frac{\in}{k}$

adding $\frac{\in}{k}$ to the above, we get

$$u(t) + \frac{\in}{k} \le \frac{\in}{k} + \int_{t_o}^{t} k \left(u(s) + \frac{\in}{k} \right) ds$$

i.e. $u(t) \le \frac{\in}{k} + \int_{t_o}^{t} k\, v(s)\, ds$

Applying Gronwall's Lemma with $c = \frac{\in}{k}$ to get

$$v(t) \le \frac{\in}{k} \cdot \exp\left((t - t_o)k\right) \le \frac{\in}{k} \cdot \exp\left(k \,|\, t - t_o \,|\right)$$

i.e. $u(t) + \frac{\in}{k} \le \frac{\in}{k} \cdot \exp\left(k \,|\, t - t_o \,|\right)$

i.e. $u(t) \le \frac{\in}{k} \exp\left(k \,|\, t - t_o \,|\right) - \frac{\in}{k}$

i.e. $u(t) \le \frac{\in}{k}\left[\exp\left(k \,|\, t - t_o \,|\right) - 1\right]$.

DIFFERENTIABILITY OF THE FLOW OF AUTONOMOUS EQUATIONS:

Consider an autonomous differential equation

$x' = f(x), f : W \to E$, W is open in E,   (8.1)

Where F is assume to be C^1.

If $\phi(\cdot)$ is the flow of this diff. equation then we can show that ϕ is C^1 in both the arguments.

Let y(t) be a particular solution of the diff. equation (8.1) namely

$x' = f(x)$, for t in some open interval J.

Fix to $t_o \in J$ and put $y(t_o)$

For each $t \in J$ put $A(t) = Df(y(t))$

Thus, $A : J \to L(E)$ is continuous. We define a non-autonomous Linear

equation $u'A(t)u$ $\qquad$ (8.2)

This equation (8.2) is known as variational equation of (8.1) along the solution $y(t)$.

We know that equation (8.2) has a solution on all of J for every initial condition $u(t_o) = u_o$.

If u_o is small, we can show that $t \to y(t) + u(t)$ is a good approximation to the solution $x(t)$ of diff. equation (8.1).

Theorem 20: Consider the non-autonomous system $x'(t) = f(t, x(t))$ defined over an open set $W \subseteq R \times E$ such that f satisfies Lipschitz condition w.r.t. the 2nd variable.

If $y(t)$ and $z(t)$ are two solution of the above diff. equation in a common domain J containing to, then for all

$$t \in J, |\, y(t) - z(t) \le |\, y(t_o) - z(t_o)| \cdot (\exp(k\,|\,t - t_o\,|))$$

i.e. the solution depends continuously on initial value (at $t - t_o$).

Proof: For $t \in J$, we have

$$y(t) - y(t_o) = \int_{t_o}^{t} y'(s)\, ds = \int_{t_o}^{t} f'(s, y(s)\,)ds$$

$$z(t) - z(t_o) = \int_{t_o}^{t} z'(s)\, ds = \int_{t_o}^{t} f(s, z(s)\,)ds$$

$$\therefore y(t) - z(t) = y(t_o) - z(t_o) + \int_{t_o}^{t} (f(s, z(s)\,) - f(s, z(s)\,))\, ds$$

$$\therefore |\, y(t) - z(t)\,| \le |\, y(t_o) - z(t_o)\,| + \int_{t_o}^{t} |\, f(s, z(s)\,) - f(s, z(s)\,)\, ds$$

$$\le |\, y(t_o) - z(t_o)\,| + \int_{t_o}^{t} k\,|\, y(s) - z(s)\,|\, ds$$

Where, k is the Lipschitz constant.

LIMIT SETS

Let us denote $|y(t)-z(t)|$ by $v(t), t \in J$

$$v(t) \leq v(t_o) + \int_{t_o}^t k\, v(s)\, ds$$

By Gronwall's Lemma, we have

$$v(t) \leq v(t_o) \cdot \exp\left(k\,(t-t_o)\right)$$

$$\leq v(t_o) \cdot \exp\left(k\,|(t-t_o)|\right)$$

i.e. $|y(t)-z(t)| \leq |y(t_o)-z\,(t_o)| \cdot \exp\left(k\,|(t-t_o)|\right)$

in a compact set J_o containing t_o, it follows that

$$|y(t)-z(t)| \rightarrow 0 \text{ as } |y(t_o)-z\,(t_o)| \rightarrow 0$$

We now consider the autonomous system.

$$x'(t) = f(x)$$

Define on an open set $W \subseteq E$.

We assume that $F \in C'$. Let $\phi_t(x)$ denote the flow of the diff. equation.

We show that $\phi_t(x)$ is also C^1.

Let $y_o \in W$. Let $y(t)$ denote a solution of the system

$$x' = f(x)$$

Defined in an interval J containing t_o s.t. $y(t_o) = y_o$.

Consider the variational differential equation along the solution y namely

$$u'(t) = Df(y)(u)$$

Let ξ be a point in E s.t. $y_o, y_o + \xi, \xi$ all lie in W.

We denote by $u(t,\xi)$, the solution at t of the variational equation such that $u(t_o) = \xi$.

We also denote by $u(t,\xi)$, the solution of the diff. equation $x'(t) = f(x(t))$ such that this solution takes value $y_o + \xi$ at $t = t_o$.

Theorem 21: Let $J_o \subseteq J$ be a compact interval containing to. Then

$$\lim_{|\xi|} \frac{|y(t,\xi) - y(t) - u(t,\xi)|}{|\xi|} = 0$$

uniformly in $t \in J_o$.

Proof: let $t \in J_o$. By the definition of $y(t,\xi), y(t)$ and $u(t,\xi)$, we have

$$y(t) = y(t_o) + \int_{t_o}^t f\left(y(s)\right) ds$$

$$y(t) = y_o + \int_{t_o}^t f\left(y(s)\right) ds$$

$$y(t,\xi) = y_o + \xi + \int_{t_o}^t f\left(y(s,\xi)\right) ds$$

$$u(t,\xi) = \xi + \int_{t_o}^t \left(Df\left(y(s) \cdot (u(s,\xi))\right)\right) ds$$

$$\therefore y(t,\xi) - y(t) - u(t,\xi)$$

$$= \int_{t_o}^t f\left(y(s,\xi)\right) ds - \int_{t_o}^t f\left(y(s,\xi)\right) ds - \int_{t_o}^t Df\left(y(s)\right) : u(s,\xi)\, ds$$

$$= \int_{t_o}^t \left(f\left(y(s,\xi)\right) - f\left(y(s)\right) - \int_{t_o}^t Df\left(y(s)\right) u(s,\xi) ds \right)$$

If $a, b \in W$ then $f(b) = f(a) + Df(a)(b - a) + R(a, b - a)$ such that

$$\lim_{|b-a| \to 0} \frac{|R(a, b-a)|}{|b-a|} = 0$$

$$\therefore f\left(y(s,\xi)\right) - f\left(y(s)\right) = Df(y(s))(y(s,\xi) - y(s)) + R(y(s), y(s,\xi) - y(s))$$

and $\quad \lim_{|y(s,\xi) - y(s)| \to 0} \frac{|R(y(s), y(s,\xi) - y(s))|}{|y(s), y(s,\xi) - y(s)|} = 0$

Thus for $t \in J_o$,

$$y(t,\xi) - y(t) - u(t,\xi) \int_{t_o}^t Df\left(y(s)\right)(y(s,\xi) - y(s)) +$$

$$R(y(s), y(s,\xi) - y(s)) - Df(y(s), y(s,\xi)) ds.$$

456

LIMIT SETS

$$= \int_{t_o}^{t} \left(Df\left(y(s)\right)\left[\left(y(s,\xi)-y(s)-u(s,\xi)\right)\right] + R\left(y(s),-y(s,\xi)-y(s)\right)\right) ds$$

$$\therefore |y(s,\xi)-y(t)-u(s,\xi)|$$

$$\leq \int_{t_o}^{t} \left| Df\left(y(s)\right)[y(s,\xi)-y(s)-u(s,\xi)]ds + \int_{t_o}^{t} |R\left(y(s),-y(s,\xi)-y(s)\right)| ds$$

$$\leq \int_{t_o}^{t} \| Df\left(y(s)\right)\| |y(s,\xi)-y(s)-u(s,\xi)| ds + \int_{t_o}^{t} |R\left(y(s),-y(s,\xi)-y(s)\right)| ds$$

Since $F \in C', J_o$ is compact, we have

$$\| Df\left(y(s)\right)\| \leq N, \forall t \in J_o \; S \in [t_o,t]$$

Further Let $|y(t,\xi)-y(t)-u(t,\xi)|$ be denoted by sat v(t).

Thus for all $t \in J_o$, we have

$$v(t) \leq \int_{t_o}^{t} Nv(s)\, ds + \int_{t_o}^{t} |R\left(y(s),-y(s,\xi)-y(s)\right)| ds$$

Since $\displaystyle \lim_{|y(s,\xi)-y(s)|\to 0} \frac{|R\left(y(s),y(s,\xi)-y(s)\right)|}{|y(s,\xi)-y(s)|} = 0$

For a given $\in > 0, \exists \, \delta_i > 0$ such that

$$|y(s,\xi)-y(s)| < \delta_1 \Rightarrow |R\left(y(s),y(s,\xi)-y(s)\right)| \leq \in \cdot |y(s,\xi)-y(s)|$$

Since $y(s,\xi)$ and $y(s)$ are solution of the some diff. equation $x' = f(x)$,

with value $y_o + \xi$ and y_o at $t = t_o$ we have

$$|y(s,\xi)-y(s)| \leq |(y_o + \xi)-y_o| \cdot \exp\left(k\,(t-t_o)\right)$$

$$\therefore |y(s,\xi)-y(s)| \leq |\xi| \cdot \exp\left(k\,|s-t_o|\right)$$

Since $s,t,t_o, \in J_o$, a compact interval,

$\exp\left(k\,|s-t_o|\right)$ is bounded, $\forall s \in J_o$ and

Thus $|\xi| \cdot \exp\left(k\,|s-t_o|\right) \to 0$ uniformly in J_o as $|\xi| \to 0$.

Hence for the above δ_1 we can find $\delta > 0$ such that

457

$|\xi| < \xi \Rightarrow |\xi| \cdot \exp\left(k\,|\,s - t_o\,|\right) < \xi_1,\, \forall s \in J_o.$

Thus for the given $\in > 0$, we can find $\delta > 0$ s.t.

$|\xi| < \xi \Rightarrow |\,R\left(y(s), y(s, \xi) - y(s)\right)| \le \in \cdot |\,y(s, \xi) - y(s)\,|$

Thus we have $\forall s \in J_o.$

$$v(t) \le N \int_{t_o}^{t} v(s)\,ds + \int_{t_o}^{t} |\,R\left(y(s), -y(s, \xi) - y(s)\right)|\,ds$$

$$\le N \int_{t_o}^{t} v(s)\,ds + \int_{t_o}^{t} \in |\,y(s, \xi) - y(s)\,|\,ds$$

$$\le N \int_{t_o}^{t} v(s)\,ds + \int_{t_o}^{t} \in \cdot |\xi| \cdot \exp\left(k\,|\,s - t_o\,|\right)ds$$

$$\therefore\ v(t) \le N \int_{t_o}^{t} v(s)\,ds + \in \cdot |\xi|\ \int_{t_o}^{t} e^{k\,|\,s-t_o\,|}\,ds$$

Since the expression $\int_{t_o}^{t} e^{k\,|\,s-t_o\,|}\,ds$ is bounded, $\forall s, t \in J_o$

We can find a constant c >0 (c will depend only on J_o and k).

Thus we have $\forall t \in J_o.$

$$v(t) \le N \int_{t_o}^{t} v(s)\,ds + C \cdot \in |\xi|$$

Thus by Gronwall's Lemma, we have $\forall t \in J_o.$

$$v(t) \le C \cdot \in |\xi| \cdot \exp\left(N\,|\,t - t_o\,|\right)$$

$$\frac{v(t)}{|\xi|} \le c \cdot \in \exp\left(N\,|\,t - t_o\,|\right)$$

But as J_o is compact (it is of finite length), the expression $\exp\left(N\,|\,t - t_o\,|\right)$ is bounded.

Thus $c \cdot \in e^{N|t - t_o|}$ can be made arbitrarily small by choosing $\in$ orbitrary small.

Thus $\lim\limits_{|\xi| \to 0} \dfrac{v(t)}{|\xi|} = 0$ uniformly in J_o

i.e. $\lim_{|\xi|\to 0} \dfrac{|y(t,\xi)-y(t)-u(t,\xi)|}{|\xi|} = 0$

Theorem 22: The flow $\phi(t,x)$ or $\phi_t(x)$ of the diff. equation $x'=f(x)$, where f is C^1, that is $\dfrac{\partial \phi}{\partial t}$ and $\dfrac{\partial \phi}{\partial x}$ exists and are continuous in (t, x).

Proof: We note that $\phi_t(x)$ is a solution of the diff. equation $x'=f(x)$

Which takes value x at $t = 0$.

By the definition, then $\dfrac{\partial}{\partial t}\phi_t(x) = f\big(\phi_t(x)\big)$

Thus $\phi_t(x)$ is differentiable w.r.t. t and its derivative w.r.t. t is given by $f\big(\phi_t(x)\big)$.

But f as a function of its argument is C^1 and thus obviously continuous.

Thus $\dfrac{\partial \phi}{\partial t}$ exists and is continuous.

In order to see if $\phi_t(x)$ is differentiable w.r.t. the variable x, we consider

$\phi_t(y_o + \xi) - \phi_t(y_o)$.

We must show that $\lim_{|\xi|\to 0} \dfrac{|\phi_t(y_o + \xi) - \phi_t(y_o)|}{|\xi|}$ exists and is continuous.

But $\phi_t(y_o + \xi) - \phi_t(y_o) = y(t+\xi) - y(t)$.

(We note that according to our notation $y(t+\xi)$ is a solution that takes the value $y_o + \xi$

At $t = t_o$. So we may just consider $\phi_{t-t_o}(y_o + \xi)$ and $\phi_{t-t_o}(y_o)$).

From the proposition, it is clear that

$\lim_{|\xi|\to 0} \dfrac{|y(t+\xi) - y(t)|}{|\xi|}$ is a linear map $\xi \to u(t+\xi)$

Further as the solution depend continuously on initial values, the map

459

$\xi \to u(t+\xi)$ is continuous.

Thus $\phi_t(x)$ is differentiable w.r.t. both the arguments and the derivatives are continuous.

i.e. $\phi_t(x) \in C^1$

Let $\phi_t(x)$ denote the flow of the diff. equation

$x' = f(x)$, where f is C^1.

We know that $\phi_t(x)$ is itself C^1 and $\dfrac{d}{dt}\phi_t(x) = f(\phi_t(x))$.

$$D\left(\frac{d}{dt}\phi_t(x)\right) = D(f(\phi_t(x)))$$

i.e. $\dfrac{d}{dt}(D\phi_t(x)) = D(f(\phi_t(x)))$

$$= Df(\phi_t(x)) \cdot D\phi_t(x).$$

Definition: A function $f: W \to E$ is called $C^r, 1 \leq k < \infty$ if f has r continuous derivatives. For $r \geq 2$, it is equivalent to f is C^1 and $Df: W \to L(E)$ is C^{r-1}. If f is C^r for all $r \geq 1$, we say that f is C^∞.

Theorem 23: Let C^1 be open and Let $f: W \to E$ be $C^r, 1 \leq r < \infty$. Then the flow $\phi: \Omega \to E$ of the diff. equation $x' = f(x)$ is also C^r.

Here $\Omega \equiv J \times E$, where J is the interval of time over which the flow field is defined.

Proof : The theorem has already been proved for the case r =1.

So we assume $r \geq 2$.

We assume that the thm is true for r-1 and then show that it holds for r.

i.e. we assume that whenever the R.H.S. of a diff. equation is r-1 times continuously differentiable then its flow field is also r-1 times continuously

LIMIT SETS

differential equation $\xi' = F(\xi)$

where $\xi = (x,u)$ and $F = (F_1, F_2)$, where $F_1(x,u) = f(x)$ & $F_2(x,u) = Df(x)u$

Thus the equation is equivalent to the two equations

$x' = f(x)$, $u' = Df(x)u$

Since $f \in C^r$, it follow that f as well as $Df(x)$ both are in C^{r-1}.

i.e. $\phi_t(x,u)$ is in C^{r-1}.

But $\phi_t(x,u) = (\phi_t(x), D\phi_t(x)(u))$

Thus both $\phi_t(x)$ and $u \to D\phi_t(x)u$ are both in C^{r-1}.

It follows that $\dfrac{\partial \phi}{\partial x}$ is in C^{r-1}.

Further, as $\dfrac{\partial \phi}{\partial t} = f(\phi_t(x))$ and f is in C^r.

We have $\dfrac{\partial \phi}{\partial t}$ in C^r.

Thus both $\dfrac{\partial \phi}{\partial t}$ and $\dfrac{\partial \phi}{\partial x}$ are in C^{r-1}.

$\therefore \phi \in C^r$.

PERSISTENCE OF EQUILIBRIA

Let W be an open set in a vector space E and $f : W \to E$ a C^1 vector field.

By a perturbation of F , we simply mean another C^1 vector field on W which can be considered as 'C^1 -close ' to f.

In define C^1-closeness. We may define a distance on all C^1-vector fields defined on W, using the fact that they are of class C^1.

Let v(w) be the set of all C^1 vector fields on w. if E has a norm, we define C^1 norm $\| h \|_1$ of a vector field $h \in v(w)$ to be the least upper bound of all the numbers.

DYNAMICAL SYSTEM – A SHORT COURSE

$|h(x)|, \|Dh(x)\|, x \in W$.

We allow the possibility $\|h\|_1 = \infty$ if these numbers are unbounded.

Let $f : W \to E$ be a C^1 vector field and $x_o \in W$ be such that the linear operator $Df(x_o) : E \to E$ is an invertible.

Theorem 24 : Assume E is normed. Let $v > \|Df(x_o)^{-1}\|$. Let $v \subseteq W$ be an open ball around x_o such that

$$\|Df(y)^{-1}\| < v \text{ and } \|Df(y) - Df(z)\| < \frac{1}{v}, \forall y, z \in v$$

Then $f|_v$ is one-to-one.

Proof: Let $Y \in v$ and $u \in E$, since $Df(y)$ is invertible, we have

$$u = \left(Df(y)^{-1}\right)\left(Df(y)u\right)$$

$$|u| = |Df(y)^{-1}\left(Df(y)\right)u|$$

$$\leq \|Df(y)^{-1}\| \, |Df(y)u|$$

$$|Df(y)u| \geq \frac{|u|}{\|Df(y)^{-1}\|}$$

But we are given that $\forall y \in v$,

$$\|Df(y)^{-1}\| < v \text{ i.e. } \frac{1}{\|Df(y)^{-1}\|} > \frac{1}{v}$$

Thus for any $u \in E$ and $Y \in V$, we have

$$|Df(y)u| > \frac{|u|}{v}$$

Let y and z be any two points of v s.t. $y \neq z$.

As v is an open ball centered around x_o, it is a convex set.

As $y, z \in v$, the line segment joining y and z is in v.

462

LIMIT SETS

As $y, z \in v \subseteq E$, we let $u = z - y$ i.e. $z = y + u$

We define a new function $\phi : [0,1] \to E$ as

$\phi(t) = f(y + tu), 0 \le t \le 1$

Obiviously, $\phi(o) = f(y)$ and $\phi(1) = f(z)$

As $f \in c^1$, ϕ is differentiable and for $t \in [o,1]$

$\phi'(t) = Df(y + tu) \cdot u$

$$\therefore f(z) - f(y) = \phi(1) - \phi(o) = \int_0^1 \phi'(t) \, dt = \int_0^1 Df(y + tu) u \, dt$$

$$\therefore f(z) - f(y) = \int_0^1 \left(Df(y + tu)u - Df(y)u + Df(y)u \right) dt$$

$$= \int_0^1 Df(y)u \, dt + \int_0^1 \left(Df(y + tu)u - Df(y)u \right) dt$$

$$= \int_0^1 Df(y)u \, dt - \int_0^1 -\left(Df(y + tu)u - Df(y)u \right) dt$$

As
$$\left| \int_0^1 -\left(Df(y + tu)u - Df(y)u \right) dt \right| \le \int_0^1 | Df(y + tu)u - Df(y)u | \, dt$$

We have

$$-\left| \int_0^1 -\left(Df(y + tu)u - Df(y)u \right) dt \right| \ge -\int_0^1 | Df(y + tu)u - Df(y)u | \, dt$$

Thus using the fact that $|a - b| \ge ||a| - |b||$, we have

$$| f(z) - f(y) | \ge \left| \int_0^1 Df(y)u \, dt \right| - \int_0^1 | Df(y + tu)u - Df(y)u | \, dt$$

$$= | Df(y)u | - \int_0^1 | Df(y + tu)u - Df(y)u | \, dt$$

$$> \frac{|u|}{v} - \frac{|u|}{v} = 0$$

Thus, $|f(z) - f(y)| > 0$

$\therefore f(y) \neq f(z)$

$\therefore f|_v$ is one-one.

Theorem 25: Suppose E is a normed vector space with norm denoted by an inner product. Let $B \subseteq W$ be a closed ball around x_o with boundary ∂B and $f : W \to E$ be a c^1 –map. Suppose Df(y) is invertible $\forall y \in B$.

Let $\min \{ |f(y) - f(x_o)| / y \in \partial B \} > 2\delta > 0$.

then $w \in f(B)$ if $|w - f(x_o)| < \delta$.

Proof: Let us assume that w is such that

$|w - f(x_o)| < \delta$.

For $y \in B$, we define a function

$$H(y) = \frac{1}{2} |f(y) - w|^2.$$

Since the function H is real valued continuous function defined on the compact set B. H will be bounded and will attain its minimum & maximum.

Let $y_o \in B$ be such that $H(y_o)$ is maximum.

We now show that $y_o \notin \partial B$

If possible, let us assume that $y_o \in \partial B$

Consider

$$|f(y) - w| = |f(y_o) - f(x_o) + f(x_o) - w|$$
$$= |(f(y_o) - f(x_o)) - (w - f(x_o))|$$

$$\geq |f(y_o) - f(x_o)| - |w - f(x_o)|$$

But we are given for every $y_o \in \partial B$,

$$|f(y) - f(x_o)| > 2\delta$$

In particular, $|f(y_o) - f(x_o)| > 2\delta$

Also, $|w - f(x_o)| < \delta$

$$|f(y_o) - w| > 2\delta - \delta = \delta$$

Thus if $y_o \in \partial B$, then

$$|f(y_o) - w| > \delta > |w - f(x_o)|$$

i.e. $|f(y_o) - w| > |f(x_o) - w|$

i.e. $H(y_o) > H(x_o)$, a contradiction as $x_o \in B$

and by our choice y_o is a pt. of B, where H attains its minimum.

Thus $y_o \in B$ but $y_o \notin \partial B$

We shall show that $f(y_o) = w$

We note that for $y \in B$,

$$H(y) = \frac{1}{2}|F(y) - w|^2 = \frac{1}{2}\langle F(y) - w, F(y) - w \rangle$$

Thus the derivative of $H(y)$ i.e. $DH(y)$ will be a linear operator on E such that for every $z \in E$,

$$DH(y)z = \langle F(y) - w, DF(y)z \rangle$$

Since y_o is a point of minimum of $H(y)$ and y_o is in the interior of B, we must have

$DH(y_o) = 0$ as a Linear operator.

Further we are given that $\forall y \in B, Df(y)$ is invertible

So, let $v \in E$ be such that $Df(y_o)v = F(y_o) - w = 0$.

Since y_o is the point of minima, we have $DH(y_o) = 0$

For the above v, $DH(y)v = 0$.

But $DH(y_o)z = \langle F(y_o) - w, DF(y_o)z \rangle$

Thus we have

$$0 = DH(y_o)v = \langle F(y_o) - w, DF(y_o)v \rangle$$
$$= \langle F(y_o) - w, F(y_o) - w \rangle$$
$$= | F(y_o) - w |^2$$

$F(y_o) = w$

Thus, $w \in f(B - \partial B)$.

Theorem 26: Let $f : W \to E$ be c^1 – and suppose $x_o \in W$ is such that the linear operator $Df(x_o) : E \to E$ is invertible. Then there is a nbd v of c^1 – vector fields with c^1 – norm of f and an open set $u \subseteq W$ containing x_o such that if $g \in v$ then

(a) $g |_u$ is one-one and (b) $f(x_o) \in g(u)$

Proof: (a) We are given that $Df(x_o)$ is invertible, as the collection of all invertible operators on E is an open set, it follows that $Df(x_o)$ is an interior point of the collection of all invertible operators.

We can thus find an $\alpha > 0$ s.t. whenever A is an operator on E s.t. $\| A - Df(x_o) \| < \alpha$ then A is invertible.

Since f is a c^1 – vector field, the map $x \to Df(x)$ is continuous for all $x \in W$. In particular, it will be continuous at x_o.

Thus for the above α, we can find a nbd say U_1 of $x_o \in W$ such that

LIMIT SETS

$$x \in U_1 \Rightarrow \| Df(x) - DF(x_o) \| < \frac{\alpha}{2}$$

We consider v_1, $a \dfrac{\alpha}{2}$ nbd of f in the $c^1 -$ norm i.e. $g \in v_1$ means g is a $c^1 -$

vector field defined on w and for $x \in W$, $\| g - f \|_1 < \dfrac{\alpha}{2}$

i.e. $\forall x \in W, | g(x) - f(x) | < \dfrac{\alpha}{2}$

$$| Dg(x) - f(x) | < \frac{\alpha}{2}$$

Thus for all $x \in U_1$ and $g \in V_1$, we have

$$\| Dg(x) - Df(x) \| < \frac{\alpha}{2}$$

Thus for all $x \in U_1$ and $g \in V_1$, we have

$$\| Dg(x) - Df(x_o) \| = \| Dg(x) + Df(x) + Df(x) - Df(x_o) \|$$

$$\leq \| Dg(x) + Df(x) \| + \| Df(x) - Df(x_o) \|$$

$$< \frac{\alpha}{2} + \frac{\alpha}{2} = \alpha \quad \text{(By using the definition of } U_1 \text{ and } V_1 \text{)}$$

Thus for all $x \in U_1$ & $g \in V_1$, $Dg(x)$ is invertible.

Since $Df(x_o)$ is invertible, $\| Df(x_o)^{-1} \|$ is finite and thus let $v>0$ be chose s.t.

$$\| Df(x_o)^{-1} \| < \upsilon$$

We not that the map $x \rightarrow Df(x)$ is continuous.

We can choose x to be close to x_o to have $Df(x)$ sufficiently close to $Df(x_o)$

For the invertible operator the map $A \rightarrow A^{-1}$ is continuous and hence $\| A \| \rightarrow \| A^{-1} \|$ is also continuous.

Further $\| Df(x) \|$ sufficiently close to $\| Df(x_o) \|$ will further imply $\| Df(x)^{-1} \|$

to be sufficiently close to $\| Df(x_o)^{-1} \|$.

Since $\| Df(x_o)^{-1} \| < \upsilon$, we can find a small nbd of x_o in U_1 such that for all x in this nbd of x_o,

$\| Df(x)^{-1} \| < \upsilon$

Thus, if we choose a small nbd of $f \subseteq U_1$ then $\forall x$ in the above nbd say $U_2 \subseteq U_1$ of x_o and for all g in the nbd $V_2 \subseteq V_1$ of f such that

$\forall x \in U_2, g \in V_2, \| Df(x)^{-1} \| < \upsilon$.

Again using the continuity of the map $x \to Df(x)$, we can find a small nbd of $x_o \in U_2$ s.t. for all x in this nbd, we have

$\| Df(x) - Df(x_o) \| < \dfrac{1}{2\upsilon}$

Thus, if y and z are two pts in this nbd then

$\| Df(y) - Df(z) \| \le \| Df(y) - Df(x_o) \| + Df(x_o) - Df(z) \|$

$$< \frac{1}{2\upsilon} + \frac{1}{2\upsilon} = \frac{1}{\upsilon}.$$

If for all y and z in this nbd of x_o contained in U_2, we have

$\| Df(y) - Df(z) \| < \dfrac{1}{\upsilon}$

We consider a closed ball centered at x_o lying within this nbd.

Since this ball is compact, $\| Df(y) - Df(z) \|$ will attain its maximum for some y, z in this closed ball.

This maximum will also be less than $\dfrac{1}{\upsilon}$

Let the max. be attained at pts. y_o and z_o of the closed bal.

We now take an $\%_2, c^1$ nbd of f contained in V_2.

We denote this nbd by V_3.

Thus $\forall g \in V_3$, $\| Dg(x) - Df(x) \| < \frac{\epsilon}{2}, \forall x \in W$.

Thus for y, z in the closed ball, we have

$$\| Dg(y) - Dg(z) \| \leq \| Dg(y) - Df(y) \| + Df(y) - Df(z) \| + \| Df(z) - Dg(z) \|$$

$$\leq \max . + \frac{\epsilon}{2} + \frac{\epsilon}{2}$$

$$\leq \max . + \epsilon < \frac{1}{\upsilon}$$

We denote any open ball centered at x_o and lying entirely within the above closed ball by U.

Thus, $\forall y, z \in U$ and $g \in V_3$ all the conditions of Lemma are satisfied.

So, $g\big|_U$ is one-one.

(b): We next show that for a small nbd of F,

$f(x_o) \in g(U)$.

$\forall g \in V_3$, $g\big|_U$ is one-one, it follows that $f\big|_U$ is also one-one.

Let B be a closed ball with centered at x_o s.t. B lies entirely within U.

Obviously, $\forall y \in \partial B, f(y) \neq f(x_o)$

Thus if we consider $\{ f(y) - f(x_o) \mid, y \in \partial B \}$ it will be a collection of positive real numbers.

Since the set ∂B is compact, minimum of these positive numbers will be attained at some $y \in \partial B$.

It thus follows that min $\{ f(y) - f(x_o) \mid, y \in \partial B \} > 0$

We choose $\delta > 0$ in such a way that V_3 such that $\forall g \in V$, we have min $\{|f(y)-f(x_o)|, y \in \partial B\} > 2\delta > 0$

we can choose a very small nbd say v of f contained in

min $\{|f(y)-f(x_o)|, y \in \partial B\} > 2\delta > 0$

so, if f and g are such that

$|f(x_o)-g(x_o)| < \delta$.

Then by Lemma , we have

$f(x_o) \in y(B) \subseteq g(U)$

Definition : Consider the differential equation $x' = f(x)$.

where, f is c^1 on $W \subseteq E$.

Let $\bar{x}$ be a point of equilibrium. If $\bar{x}$ is a hyperbolic point then none of the eigen values of $Df(\bar{x})$ have zero real part.

If all the n-eigen values have -ve real part then $\bar{x}$ is a sink.

By index of $\bar{x}$ we mean the number of eigen value (counted according to multiplicity) having –ve real part.

Theorem 27 : Let $f:W \to E$ be a c^1 vector field and $\bar{x} \in W$ an equilibrium of $x' = f(x)$ s.t. $Df(\bar{x})L \in (E)$ is invertible. Then there exist a nbd $U \subseteq W$ of $\bar{x}$ and a nbd V of F (in c^1) s.t. for any $g \in V$, there is a unique equilibrium $\bar{y} \in U$ of $\bar{y} = g(y)$.

Moreover, if E is normed, then for any $\epsilon > 0$, we can choose V so that $|\bar{y}-\bar{x}| < \epsilon$.

Proof: We are given that for $\bar{x} \in W$, $Df(\bar{x})$ is invertible.

So, by the proposition, with $\bar{x}$ instead of x_o, we can find a nbd U of $\bar{x}$ and a nbd V of F (in c^1) s.t.

$\forall g \in V, g|_U$ is one-one and $f(\bar{x}) \in g(U)$.

LIMIT SETS

But $\bar{x}$ being an equilibrium point of the diff. equation $x' = f(x)$

We have $f(\bar{x}) = 0$

Thus, $\exists\, \bar{y} \in U$ such that $g(\bar{y}) = 0$.

Since $g\big|_U$ is one-one, such a $\bar{y} \in U$ is unique.

Obviously, $\bar{y}$ is a pt. of equilibrium of $\bar{y} = g(y)$.

2nd part:

Since W is any open subset of E on which f is c^1.

We can replace W by an $\in$-nbd of $\bar{x}$ so that U will lie within this $\in$-nbd of $\bar{x}$.

$\therefore |\bar{x} - \bar{y}| < \in$.

Theorem 28 : Suppose that $\bar{x}$ is a hyperbolic equilibrium. In thm , the nbds V and U can be chosen so that if $g \in V$, then the unique equilibrium $\bar{y} \in U$ of $\bar{y} = g(y)$ is hyperbolic and has the same index as $\bar{x}$.

Proof: By theorem , we can choose nbd V of F in such a way that in a nbd of $\bar{x}$, there is a unique equilibrium point $\bar{y}$ of the differential equation $\bar{y} = g(y)$.

Further we can choose $\bar{x}$ and $\bar{y}$ sufficiently close to each other. As such $Df(\bar{x})$ and $Dg(\bar{y})$ will also be sufficiently close to each other.

As such their eigen values will also be sufficiently will remain close to each other.

Thus the number of eigen value of $Df(\bar{x})$ having –ve real part will be same as the no. of eigen values of $Dg(\bar{y})$ with –ve real part.

Thus the index of $\bar{x}$ is same as that of $\bar{y}$.

DYNAMICAL SYSTEM – A SHORT COURSE

PERSISTENCE WITH CLOSED ORBITS

We consider a dynamical system $x' = f(x)$

Where $f : W \subseteq E \to E$ is a c^1-vector field.

Let r be a closed orbit with period λ. Let 0 be a point on r. Let H be a hyper plane and $S \subseteq H$. Be a local at O.

Theorem 29: Let $u : S_o \to S$ be a Poincare map for a Local section S at O. Let $U \subseteq W$ be nbd of r. suppose that 1 is not an eigen value of Du(o). then there exists a nbd v of f (with c^1 norm) such that every vector field $g \in V$ has a closed orbit $B \subseteq U$.

Proof: We have the diff. equation $x' = f(x)$, where F is c^1 on $W \subseteq E$

r is closed orbit of this diff. equation and $o \in r$.

S is a local section at O w.r.t. the diff. equation $x' = f(x)$ i.e. S is an open subset of H (a hyper plane of E) containing o s.t. S is transverse to F.

i.e. $f(x) \notin H, \forall x \in S$

by the definition S_o is the intersection of a small nbd (in E) of o with S.

If we take a very small closed ball in E around o and denote by S_o its intersection with S, then we can consider S_o to be a compact subset of H containing o.

For any $x \in S_o$. let $\tau(x)$ be the time at which the trajectory of the diff. equation $x' = f(x)$ starting at x meets S for the first time.

If $\phi_t(x)$ is the flow of the diff. equation $x' = f(x)$ and as u is the Poincare map for $x' = f(x)$, we have

$u(x) = \phi_{\tau(x)}(x), x \in S_o$

LIMIT SETS

We define a map say ξ on S_o as

$\xi(x) = u(x) - x$

As o lies on a closed orbit, we have $u(o) = 0$.

i.e. $\xi(o) = 0$

Further for any $x \in S_o$, $D\xi(x) = Du(x) - I$

Where I denotes the identity linear operator.

So, $D\xi(o) = Du(o) - I$.

As none of the eigen values of $Du(o)$ are 1, it follows that none of the eigen values of $Df(o)$ are o.

Thus, $D\xi(o)$ is invertible.

Let g be a c^1-vector field defined on w such that g is sufficiently close to f in c^1-norm.

From the definition of c^1-norm, it follows that $|f(x) - g(x)|$ as well as $|Df(x) - Dg(x)|$ are sufficiently small for all $x \in w$.

We now show that S can be considered as a local section for the diff. equation $y' = g(y)$.

Since for every $x \in S$, $f(x) \notin H$, where H is a hyper plane, choosing g in such a way that g(x) is very close to $f(x)$, $\forall x \in S$, will ensure that $g(x) \notin H$.

(we can presume that H is that subspace of E whose first component of f(x) is non-zero.

Thus if $|f(x) - g(x)|$ is very small will imply first component of g(x) also non-zero.)

Thus we can consider S as a local section at O for the diff. equation $y' = g(y)$.

Again, because $\| f - g \|_1$ is very small, the trajectories $x(t)$ and $y(t)$ of the diff. equation $x' = f(x)$ and $y' = g(y)$ will remain close to each other over a finite interval of time.

Let $\psi_t(x)$ denotes the flow of the diff. equation $y' = g(y)$ then we can say that for a finite interval of time

$|\psi_t(x) - \phi_t(x)|$ will be very small.

Hence, if $\sigma(x)$ denotes the time at which the trajectory of the diff. equation $y' = g(y)$ starting at $x \in S_o$, meets S, then $|\sigma(x) - \tau(x)|$ will also be very small.

Let us define $u : S_o \to S$ as $v(x) = \psi_{\sigma(x)}(x) v$ is then the Poincare map for the diff. equation $y' = g(y)$.

Also, we have $\forall x \in S_o | u(x) - vx) |$ sufficiently small.

We next consider the variational equations namely

$$\frac{dA}{dt} = Df(x)A \text{ and } \frac{dA}{dt} = Df(x)A$$

We note that $\frac{\partial \phi}{\partial x}(t, x)$ and $\frac{\partial \psi}{\partial x}(t, x)$ are the solutions of there variational equations respectively.

Thus $\| f - g \|_1$ sufficiently small will ensure that over a finite interval of time $\frac{\partial \phi}{\partial x}$ and $\frac{\partial \psi}{\partial x}$ will remain close to each other.

Since $\frac{\partial \phi}{\partial x}(t, x)$ and $\frac{\partial \psi}{\partial x}(t, x)$ are given by $f(\phi(t, x))$ and $g(\psi(t, x))$ respectively, $\| f - g \|_1$ sufficiently small will imply $\frac{\partial \phi}{\partial x}(t, x)$ and $\frac{\partial \psi}{\partial x}(t, x)$ to be sufficiently close to each other a finite period of time.

LIMIT SETS

Thus, $\|f-g\|_1$ sufficiently small implies $\|u-v\|$ sufficiently small as well as $\|D\phi - D\psi\|$ sufficiently small.

But as $u(x)=\phi_{\tau(x)}(x)$ and $v(x)=\psi_{\sigma(x)}(x)$, it follow that $\|D\phi(t,x)-D\psi(t,x)\|$ sufficiently small implies $\|Du-Dv\|$ sufficiently small.

Thus, $\|f-g\|_1$ sufficiently small implies $\|u-v\|_1$ is also sufficiently small.

Now, we define $\eta(x)=v(x)-x$.

If we are able to show that $\exists$ some $\bar{y}$ on S_o such that

$$v(\bar{y})=\bar{y}$$

Then the trajectory through $\bar{y}$ is a closed orbit for the diff. equation $y'=g(y)$.

Thus in order to show existence of a closed orbit for the diff. equation $y'=g(y)$, we need to show the existence of some $\bar{y}$ such that $\eta(\bar{y})=0$

If we choose $\|f-g\|_1$ sufficiently small then $\|u-v\|_1$ is also sufficiently small

As $\xi(x)-\eta(x)=u(x)-v(x)$ and $D\xi(x)-D\eta(x)=Du(x)-Dv(x)$

$$\|\xi-\eta\|_1 = \|u-v\|_1$$

Thus choosing g sufficiently close to f in c^1 – norm ensures η sufficiently close to ξ in c^1 – norm.

Using the proposition, it follows that if η is very close to in ξ in c^1 – norm then in a nbd of o, $\xi(o)$ will lie in the image of this nbd under η.

As $\xi(o)$, it follows that $\exists$ $\bar{y}$ in a nbd of o.

s.t. $\eta(\bar{y})=0$ i.e. $v(\bar{y})=\bar{y}$

i.e. the trajectory through $\bar{y}$ is a closed orbit for the diff. equation $y' = g(y)$.

Further, since η can be chose so close to ξ s.t.

$|o - \bar{y}| < \in$, for any $\in > 0$,

It follows that the closed orbit through $\bar{y}$ can also be taken sufficiently close to the orbit r through o.

STRUCTURAL STABILITY

Consider a diff. equation $x' = f(x)$,

Where f is a c^1 – vector field. Suppose $\phi_t(x)$ denotes the flow of this diff. equation.

Let $y' = g(y)$ be a perturbed diff. equation i.e. g is c^1 – vector field and g is close to f in an appropriate norm.

Let $\psi(t, x)$ denote the flow of this perturbed equation. If the flow of the diff. equation $y' = g(y)$ is topologically equivalent to the flow of the diff. equation $x' = f(x)$ then we say that $x' = f(x)$ is structurally stable.

Thus for a diff. equation $x' = f(x)$ to be structurally stable.

whenever $y' = g(y)$ is its 'small' perturbation, there should exist a homeomorphism h such that $h(\phi_t(x)) = \psi_t(h(x))$

under this homeomorphism, equilibrium pts, closed orbits are all preserved.

Definition : Let $D^n = \{x \in R^n \ s.t. \, |x| \leq 1\}$ and $\partial D^n = \{x \in R^n \ s.t. \, |x| = 1\}$

Consider a c^1 – vector field $f : W \to R^n$ defined on some open subset W of R^n containing D^n s.t. $\langle f(x), x \rangle x > < 0$, for $x \in \partial D^n$

LIMIT SETS

Such an f is called structurally stable on D^n if there exists a nbd v of f (in c^1-norm) if $g : W \to R^n$ is in V then the flow of f and g are topologically equivalent on D^n.

This means there exists a homeomorphism $h : D^n \to D^n$ such that for each $x \in D^n$,

$$h(\{\phi_t(x), t \geq 0\}) = \{\psi_t(h(x))/t \geq 0\}$$

Where, ψ_t is the flow of g and if x is not an equilibrium, h preserves the orientation of the trajectory.

The orientation of a trajectory is simply the direction of that points move along the curve as t increases.

* Measure of an open set is not zero.

* A set is dense if its closure is entire space.

* Boundary of Rational numbers Q is Real numbers R.

i.e. $b(Q) = R$

* $F : [a, b] \to R$, sometimes we take $F : (a, b) \to R$, in continuous mapping we take interval open because we can take pt. from left and Right. But in closed interval if take end pts i.e. a & b then we can't take pts. from Left & right respectively.

In other words, closed interval avoid approaching from the boundary.

* Plane Autonomous system, $x' = AX$

*Non Autonomous system, $x' = f(t, x)$

*Gronwall's Lemma $\Rightarrow$ uniqueness of solution of ODE.

*1) Real V.S., Real operator $\Rightarrow$ Jordan form

2) Real V.S., complex eigen values $\Rightarrow$ Real form

3) Complex V.S., complex eigen values $\Rightarrow$ Real canonical form

LEVEL CURVES

At different pts, any function is not constant but if we take that function along a curve (e.g. $x^2 + y^2$) then it will be constant throughout.

* $f(x, y) = 0$ then we can express one variable (x or y) in terms of other

iff $\dfrac{\partial f}{\partial x} \neq 0$ or $\dfrac{\partial f}{\partial y} \neq 0$.

BIBLIOGRAPHY

[1] **Abraham R.:** Foundation of Mechanics (New York: Benjamin, 1967).

[2] **Bartle R.:** The Elements of Real Analysis (New York: Wiley, 1964).

[3] **Chern S.S. and Smale S.:** Proceedings of the Symposium in Pure Mathematics XIV, Global Analysis (Providence, Rhode Island: Amer. Math. Soc., 1970).

[4] **D'Ancona U.:** The Struggle for Existence (Leiden, The Netherlands: Brill, 1954).

[5] **Desoer C. and Kuh E.:** Basic Circuit Theory (New York: McGraw-Hill, 1969).

[6] **Feynman R., Leighton R., and Sands M.:** The Feynman Lectures on Physics, Vol. 1 (Reading, Massachusetts: Addison-Wesley, 1963).

[7] **Goel N. S., Maitra S.C., and Montroll E.W.:** Nonlinear Models of Interacting Populations (New York: Academic Press, 1972).

[8] **Halmos P.:** Finite Dimensional Vector Spaces (Princeton, New Jersey: Van Nostrand,1958).

[9] **Hartman P.:** Ordinary Differential Equations (New York: Wiley, 1964).

[10] **Hrsch Morris W. and Smale Stephen:** Differential Equations, Dynamical Systems, and Linear Algebra, (Academic Press: New York and London, 1970).

[11] **Lang S.:** Calculus of Several Variables (Reading, Massachusetts: Addition-Wesley, 1973).

[12] **Lang S.:** Analysis I (Reading, Massachusetts: Addition-Wesley, 1968).

[13] **Lang S.:** Second Course in Calculus, 2nd ed. (Reading, Massachusetts: Addition-Wesley, 1964).

[14] **La Salle J. and Lefschetz S.:** Stability by Liaunov's Direct method with Applications (New York : Academic Press,1961).

[15] **Lefschetz S.:** Differential Equations, Geometric Theory (New York : Wiley(Interscience), 1957).

[16] **Loomis L. and Sternberg S.:** Advanced Calculus (Reading, Massachusetts: Addition-Wesley, 1968).

[17] **Montroll E. W.:** On the Volterra and other nonlinear models, Rev. Mod. Phys. 43(1971).

[18] **Newman M.H.A.:** Topology of Plane Sets (London and New York: Cambridge Univ. Press, 1954).

[19] **Nitecki Z.:** Differential Dynamics (Cambridge, Massachusetts: MIT Press, 1971).

[20] **Peixoto M.M.:** Dynamical Systems (New York: Academic Press, 1973).

[21] **Pontryagin L.:** Ordinary differential equations (Reading, Massachusetts: Addition-Wesley, 1962).

[22] **Rescigno A. and Richardson I.:** The struggle for life; I, Two species, Bull. Math. Biophysics 29 (1967), 377-388.

[23] **Smale S.:** On the Mathematical Foundations of Electrical circuit theory, J. Differential Geom. 7 (1973), 193-210.

[24] **Synge J. and Griffiths B.:** Principles of Mechanics (New York: McGraw-Hill, 1949).

[25] **Thom R.:** Stabilite Structurelle et Morphoenese: Essai d'une theorie generale des modeles (Reading, Massachusetts: Addition-Wesley, 1973).

[26] **Wintner A.:** The Analytical Foundations of Celestial Mechanics (Princeton, New Jersey: Princeton Univ. Press, 1941).

BIBLIOGRAPHY

[27] **Zeeman E.:** Differential equations for heartbeat and nerve impulses, in Dynamical systems (M. M. Peixoto, ed.), p. 683 (New York: Academic Press, 1973).

⊙ 编辑手记

本书是一部英文版的数学教程，中文书名或可译为《动力系统——短期课程》.本书的作者为南德奥·柯布拉加德（Namdeo Khobragade），R.T.M 那格浦尔大学数学系教授，在他的指导下有 17 名学生获得了博士学位.他已经发表了 220 多篇研究性文章，出版了 25 部著作.

动力系统按其相空间维数的多少，分为有限维动力系统和无穷维动力系统.此外，动力系统又有离散与连续两种形式之分.本书侧重于连续形式的动力系统.

对于有限维动力系统，其相空间为有限维，由常微分方程（组）来描述.因为线性的常微分方程（组）已有完整的理论，所以人们没有太大的兴趣.因为其复杂性不够，以研究非线性居多.

编辑手记

对于无穷维动力系统,其相空间为无穷维,它可以是泛函常微分方程(组)(如时滞常微分方程(组)等),但应用上最常见的是非线性发展型数学物理偏微分方程(简称非线性发展方程).

确定性的动力系统是指系统的行为遵从确定性的规律.三百多年前建立的牛顿力学所描写的力学系统就是典型的确定性动力系统.

作者在本书前言中介绍到:

本著作包含了八章内容,是为了本科生和研究生而写.

第一章包含矩阵和算子,子空间、基和维数,行列式,迹和秩,直和分解,实特征值,具有不同实特征值的微分方程,复特征值的相关内容.

第二章包含复向量空间,具有复特征值的实算子,复线性代数对微分方程的应用,对 $\mathbf{R}^n$ 中拓扑的回顾,新范数,算子的指数.

第三章包含齐次线性方程组,非齐次方程,高阶方程组,准素分解,S+N 分解,幂零典范形式.

第四章包含若尔当典范形式和实典范形式,典范形式和微分方程,函数空间中的高阶线性方程,汇点与源,双曲流,算子的一般属性,一般性的意义.

第五章的内容包括动力系统和矢量场,基本定理,存在性与唯一性,初始条件下解的连续性,扩张解,整体解,微分方程的流.

第六章包括非线性汇点,稳定性,李雅普诺夫函数,梯度系,梯度与内积.

编辑手记

第七章包括极限集,局部截面和流盒,平面动力系统中的单调序列,庞加莱—本迪克森定理,庞加莱—本迪克森定理的应用,一个种类,捕食者和猎物,竞争种类.

第八章包括闭合轨道的渐近稳定性,离散动力系统,稳定性和闭合轨道,非自治方程与流的可微性,平衡的持久性,闭合轨道的持久性,结构稳定性.

从前言中可见本书中的一个最重要的定理就是庞加莱—本迪克森定理.粗略地说,这个定理表明,如果当 $t \to +\infty$(或 $-\infty$)时,平面动力系统的某条轨线始终在某个平衡点的有界区域内,则它是一条闭轨或者盘旋地趋近某闭轨.这个定理的严格证明要用到若尔当曲线定理.它的证明可以参见陈建功先生所著的《实函数论》(第 262 页).

本书是作者为本科生和研究生而写.现在的研究生教育问题有很多,其中之一是:研究生提不出问题.孙新波教授指出:

> 研究生提不出问题的重要原因是缺乏想象力,而想象力往往来源于仰望星空的历练和脚踩大地的实践.就目前而言,前者比后者更重要.
> 卡尔·波普有言:宇宙学的问题是任何一个有思考的人都感兴趣的问题.接着卡尔·波普的话来讲,研究生提不出问题、不会提问题、批判思维的缺失等,都可以归因于过度关注狭义的研究本身而忽略了广义的为什么研究,这显然是舍本逐末的结果.师生为什么不具备无界的想象力?是什么控制了师生的想象力?可

编辑手记

以从哪些方面帮助师生破解这道难题？众所周知，西方博士生的培养，学业开始阶段通常要学习较多的哲学类课程，而毕业的学位往往是 PhD(哲学博士)，由此可见本体论、方法论和认识论对研究生培养的重要性．在此仅谈谈方法论和方法对师生提问题及批判思维的价值和意义．

借用西蒙·斯涅克的黄金圈思维，将研究生学会提问的方法论界定为"Why(内)—How(中)—What(外)"的同心圆法则，将研究生学会提问的方法界定为"看—听—说—写"四步法．方法论与方法有什么区别？从学术角度来看有很多定义，不一一列举．一般认为方法论是生成方法的方法，二者相辅相成、互为一体．

什么是同心圆法则？具体而言，What 代表现象和成果的外在呈现，对研究生而言主要是科研成果数量的多少和质量的高低，一般应该是并重的．但是，现实中重视数量远大于重视质量，这就是结果和现象的外显．How 代表方法和措施的中间探索，对研究生而言主要是创新研究方法和研究措施，进一步创新研究内容．就研究方法而言，目前往往采用追随西方研究方法的策略，而研究举措则多用模仿的方式，这也是产生问题的重要原因．Why 代表目的和理念的内在追求，是对研究意义和研究价值的长期坚守．这是研究生研究的灵魂所在，也是最重要的，但现实中却成为最容易被忽视和被抛弃的，因为不能从灵魂本源出发做研究工作．很多研究是为了研究而研究，不会提问题、思维缺乏批判性等根本问题自然成为研究生研究生涯的

编辑手记

常态.

希望阅读本书会对本科生和研究生有所裨益!

刘培杰

2021 年 6 月 23 日

于哈工大

刘培杰数学工作室
已出版(即将出版)图书目录——原版影印

书　　名	出版时间	定价	编号
数学物理大百科全书.第1卷	2016—01	418.00	508
数学物理大百科全书.第2卷	2016—01	408.00	509
数学物理大百科全书.第3卷	2016—01	396.00	510
数学物理大百科全书.第4卷	2016—01	408.00	511
数学物理大百科全书.第5卷	2016—01	368.00	512
zeta函数,q-zeta函数,相伴级数与积分	2015—08	88.00	513
微分形式:理论与练习	2015—08	58.00	514
离散与微分包含的逼近和优化	2015—08	58.00	515
艾伦·图灵:他的工作与影响	2016—01	98.00	560
测度理论概率导论,第2版	2016—01	88.00	561
带有潜在故障恢复系统的半马尔柯夫模型控制	2016—01	98.00	562
数学分析原理	2016—01	88.00	563
随机偏微分方程的有效动力学	2016—01	88.00	564
图的谱半径	2016—01	58.00	565
量子机器学习中数据挖掘的量子计算方法	2016—01	98.00	566
量子物理的非常规方法	2016—01	118.00	567
运输过程的统一非局部理论:广义波尔兹曼物理动力学,第2版	2016—01	198.00	568
量子力学与经典力学之间的联系在原子、分子及电动力学系统建模中的应用	2016—01	58.00	569
算术域	2018—01	158.00	821
高等数学竞赛:1962—1991年的米洛克斯·史怀哲竞赛	2018—01	128.00	822
用数学奥林匹克精神解决数论问题	2018—01	108.00	823
代数几何(德文)	2018—04	68.00	824
丢番图逼近论	2018—01	78.00	825
代数几何学基础教程	2018—01	98.00	826
解析数论入门课程	2018—01	78.00	827
数论中的丢番图问题	2018—01	78.00	829
数论(梦幻之旅):第五届中日数论研讨会演讲集	2018—01	68.00	830
数论新应用	2018—01	68.00	831
数论	2018—01	78.00	832

刘培杰数学工作室
已出版(即将出版)图书目录——原版影印

书　名	出版时间	定　价	编号
湍流十讲	2018—04	108.00	886
无穷维李代数:第 3 版	2018—04	98.00	887
等值、不变量和对称性:英文	2018—04	78.00	888
解析数论	2018—09	78.00	889
《数学原理》的演化:伯特兰·罗素撰写第二版时的手稿与笔记	2018—04	108.00	890
哈密尔顿数学论文集(第 4 卷):几何学、分析学、天文学、概率和有限差分等	2019—05	108.00	891
偏微分方程全局吸引子的特性:英文	2018—09	108.00	979
整函数与下调和函数:英文	2018—09	118.00	980
幂等分析:英文	2018—09	118.00	981
李群,离散子群与不变量理论:英文	2018—09	108.00	982
动力系统与统计力学:英文	2018—09	118.00	983
表示论与动力系统:英文	2018—09	118.00	984
分析学练习.第 1 部分	2021—01	88.00	1247
分析学练习.第 2 部分,非线性分析	2021—01	88.00	1248
初级统计学:循序渐进的方法:第 10 版	2019—05	68.00	1067
工程师与科学家微分方程用书:第 4 版	2019—07	58.00	1068
大学代数与三角学	2019—06	78.00	1069
培养数学能力的途径	2019—07	38.00	1070
工程师与科学家统计学:第 4 版	2019—06	58.00	1071
贸易与经济中的应用统计学:第 6 版	2019—06	58.00	1072
傅立叶级数和边值问题:第 8 版	2019—05	48.00	1073
通往天文学的途径:第 5 版	2019—05	58.00	1074
拉马努金笔记.第 1 卷	2019—06	165.00	1078
拉马努金笔记.第 2 卷	2019—06	165.00	1079
拉马努金笔记.第 3 卷	2019—06	165.00	1080
拉马努金笔记.第 4 卷	2019—06	165.00	1081
拉马努金笔记.第 5 卷	2019—06	165.00	1082
拉马努金遗失笔记.第 1 卷	2019—06	109.00	1083
拉马努金遗失笔记.第 2 卷	2019—06	109.00	1084
拉马努金遗失笔记.第 3 卷	2019—06	109.00	1085
拉马努金遗失笔记.第 4 卷	2019—06	109.00	1086
数论:1976 年纽约洛克菲勒大学数论会议记录	2020—06	68.00	1145
数论:卡本代尔 1979:1979 年在南伊利诺伊卡本代尔大学举行的数论会议记录	2020—06	78.00	1146
数论:诺德韦克豪特 1983:1983 年在诺德韦克豪特举行的 Journees Arithmetiques 数论大会会议记录	2020—06	68.00	1147
数论:1985—1988 年在纽约城市大学研究生院和大学中心举办的研讨会	2020—06	68.00	1148

刘培杰数学工作室
已出版（即将出版）图书目录——原版影印

书　名	出版时间	定　价	编号
数论:1987 年在乌尔姆举行的 Journees Arithmetiques 数论大会会议记录	2020—06	68.00	1149
数论:马德拉斯 1987:1987 年在马德拉斯安娜大学举行的国际拉马努金百年纪念大会会议记录	2020—06	68.00	1150
解析数论:1988 年在东京举行的日法研讨会会议记录	2020—06	68.00	1151
解析数论:2002 年在意大利切特拉罗举行的 C. I. M. E. 暑期班演讲集	2020—06	68.00	1152
量子世界中的蝴蝶:最迷人的量子分形故事	2020—06	118.00	1157
走进量子力学	2020—06	118.00	1158
计算物理学概论	2020—06	48.00	1159
物质,空间和时间的理论:量子理论	2020—10	48.00	1160
物质,空间和时间的理论:经典理论	2020—10	48.00	1161
量子场理论:解释世界的神秘背景	2020—07	38.00	1162
计算物理学概论	2020—06	48.00	1163
行星状星云	2020—10	38.00	1164
基本宇宙学:从亚里士多德的宇宙到大爆炸	2020—08	58.00	1165
数学磁流体力学	2020—07	58.00	1166
计算科学:第 1 卷,计算的科学(日文)	2020—07	88.00	1167
计算科学:第 2 卷,计算与宇宙(日文)	2020—07	88.00	1168
计算科学:第 3 卷,计算与物质(日文)	2020—07	88.00	1169
计算科学:第 4 卷,计算与生命(日文)	2020—07	88.00	1170
计算科学:第 5 卷,计算与地球环境(日文)	2020—07	88.00	1171
计算科学:第 6 卷,计算与社会(日文)	2020—07	88.00	1172
计算科学.别卷,超级计算机(日文)	2020—07	88.00	1173
代数与数论:综合方法	2020—10	78.00	1185
复分析:现代函数理论第一课	2020—07	58.00	1186
斐波那契数列和卡特兰数:导论	2020—10	68.00	1187
组合推理:计数艺术介绍	2020—07	88.00	1188
二次互反律的傅里叶分析证明	2020—07	48.00	1189
旋瓦兹分布的希尔伯特变换与应用	2020—07	58.00	1190
泛函分析:巴拿赫空间理论入门	2020—07	48.00	1191
卡塔兰数入门	2019—05	68.00	1060
测度与积分	2019—04	68.00	1059
组合学手册.第一卷	2020—06	128.00	1153
—代数、局部紧群和巴拿赫—代数丛的表示.第一卷,群和代数的基本表示理论	2020—05	148.00	1154
电磁理论	2020—08	48.00	1193
连续介质力学中的非线性问题	2020—09	78.00	1195
多变量数学入门(英文)	2021—05	68.00	1317
偏微分方程入门(英文)	2021—05	88.00	1318
若尔当典范性:理论与实践(英文)	2021—07	68.00	1366
伽罗瓦理论.第 4 版(英文)	2021—08	98.00	1408

刘培杰数学工作室
已出版(即将出版)图书目录——原版影印

书　名	出版时间	定　价	编号
典型群,错排与素数	2020—11	58.00	1204
李代数的表示:通过 gln 进行介绍	2020—10	38.00	1205
实分析演讲集	2020—10	38.00	1206
现代分析及其应用的课程	2020—10	58.00	1207
运动中的抛射物数学	2020—10	38.00	1208
2—纽结与它们的群	2020—10	38.00	1209
概率,策略和选择:博弈与选举中的数学	2020—11	58.00	1210
分析学引论	2020—11	58.00	1211
量子群:通往流代数的路径	2020—11	38.00	1212
集合论入门	2020—10	48.00	1213
酉反射群	2020—11	58.00	1214
探索数学:吸引人的证明方式	2020—11	58.00	1215
微分拓扑短期课程	2020—10	48.00	1216
抽象凸分析	2020—11	68.00	1222
费马大定理笔记	2021—03	48.00	1223
高斯与雅可比和	2021—03	78.00	1224
π 与算术几何平均:关于解析数论和计算复杂性的研究	2021—01	58.00	1225
复分析入门	2021—03	48.00	1226
爱德华·卢卡斯与素性测定	2021—03	78.00	1227
通往凸分析及其应用的简单路径	2021—01	68.00	1229
微分几何的各个方面.第一卷	2021—01	58.00	1230
微分几何的各个方面.第二卷	2020—12	58.00	1231
微分几何的各个方面.第三卷	2020—12	58.00	1232
沃克流形几何学	2020—11	58.00	1233
彷射和韦尔几何应用	2020—12	58.00	1234
双曲几何学的旋转向量空间方法	2021—02	58.00	1235
积分:分析学的关键	2020—12	48.00	1236
为有天分的新生准备的分析学基础教材	2020—11	48.00	1237
数学不等式.第一卷.对称多项式不等式	2021—03	108.00	1273
数学不等式.第二卷.对称有理不等式与对称无理不等式	2021—03	108.00	1274
数学不等式.第三卷.循环不等式与非循环不等式	2021—03	108.00	1275
数学不等式.第四卷.Jensen 不等式的扩展与加细	2021—03	108.00	1276
数学不等式.第五卷.创建不等式与解不等式的其他方法	2021—04	108.00	1277

刘培杰数学工作室
已出版(即将出版)图书目录——原版影印

书 名	出版时间	定 价	编号
冯·诺依曼代数中的谱位移函数:半有限冯·诺依曼代数中的谱位移函数与谱流(英文)	2021—06	98.00	1308
链接结构:关于嵌入完全图的直线中链接单形的组合结构(英文)	2021—05	58.00	1309
代数几何方法.第1卷(英文)	2021—06	68.00	1310
代数几何方法.第2卷(英文)	2021—06	68.00	1311
代数几何方法.第3卷(英文)	2021—06	58.00	1312
代数、生物信息和机器人技术的算法问题.第四卷,独立恒等式系统(俄文)	2020—08	118.00	1199
代数、生物信息和机器人技术的算法问题.第五卷,相对覆盖性和独立可拆分恒等式系统(俄文)	2020—08	118.00	1200
代数、生物信息和机器人技术的算法问题.第六卷,恒等式和准恒等式的相等 问题、可推导性和可实现性(俄文)	2020—08	128.00	1201
分数阶微积分的应用:非局部动态过程,分数阶导热系数(俄文)	2021—01	68.00	1241
泛函分析问题与练习:第2版(俄文)	2021—01	98.00	1242
集合论、数学逻辑和算法论问题:第5版(俄文)	2021—01	98.00	1243
微分几何和拓扑短期课程(俄文)	2021—01	98.00	1244
素数规律(俄文)	2021—01	88.00	1245
无穷边值问题解的递减:无界域中的拟线性椭圆和抛物方程(俄文)	2021—01	48.00	1246
微分几何讲义(俄文)	2020—12	98.00	1253
二次型和矩阵(俄文)	2021—01	98.00	1255
积分和级数.第2卷,特殊函数(俄文)	2021—01	168.00	1258
积分和级数.第3卷,特殊函数补充:第2版(俄文)	2021—01	178.00	1264
几何图上的微分方程(俄文)	2021—01	138.00	1259
数论教程:第2版(俄文)	2021—01	98.00	1260
非阿基米德分析及其应用(俄文)	2021—03	98.00	1261
古典群和量子群的压缩(俄文)	2021—03	98.00	1263
数学分析习题集.第3卷,多元函数:第3版(俄文)	2021—03	98.00	1266
数学习题:乌拉尔国立大学数学力学系大学生奥林匹克(俄文)	2021—03	98.00	1267
柯西定理和微分方程的特解(俄文)	2021—03	98.00	1268
组合极值问题及其应用:第3版(俄文)	2021—03	98.00	1269
数学词典(俄文)	2021—01	98.00	1271
确定性混沌分析模型(俄文)	2021—06	168.00	1307
精选初等数学习题和定理.立体几何.第3版(俄文)	2021—03	68.00	1316
微分几何习题:第3版(俄文)	2021—05	98.00	1336
精选初等数学习题和定理.平面几何.第4版(俄文)	2021—05	68.00	1335

刘培杰数学工作室
已出版(即将出版)图书目录——原版影印

书　　名	出 版 时 间	定　价	编号
狭义相对论与广义相对论:时空与引力导论(英文)	2021－07	88.00	1319
束流物理学和粒子加速器的实践介绍:第2版(英文)	2021－07	88.00	1320
凝聚态物理中的拓扑和微分几何简介(英文)	2021－05	88.00	1321
混沌映射:动力学、分形学和快速涨落(英文)	2021－05	128.00	1322
广义相对论:黑洞、引力波和宇宙学介绍(英文)	2021－06	68.00	1323
现代分析电磁均质化(英文)	2021－06	68.00	1324
为科学家提供的基本流体动力学(英文)	2021－06	88.00	1325
视觉天文学:理解夜空的指南(英文)	2021－06	68.00	1326
物理学中的计算方法(英文)	2021－06	68.00	1327
单星的结构与演化:导论(英文)	2021－06	108.00	1328
超越居里:1903年至1963年物理界四位女性及其著名发现(英文)	2021－06	68.00	1329
范德瓦尔斯流体热力学的进展(英文)	2021－06	68.00	1330
先进的托卡马克稳定性理论(英文)	2021－06	88.00	1331
经典场论导论:基本相互作用的过程(英文)	2021－07	88.00	1332
光致电离量子动力学方法原理(英文)	2021－07	108.00	1333
经典域论和应力:能量张量(英文)	2021－05	88.00	1334
非线性太赫兹光谱的概念与应用(英文)	2021－06	68.00	1337
电磁学中的无穷空间并矢格林函数(英文)	2021－06	88.00	1338
物理科学基础数学.第1卷,齐次边值问题、傅里叶方法和特殊函数(英文)	2021－07	108.00	1339
离散量子力学(英文)	2021－07	68.00	1340
核磁共振的物理学和数学(英文)	2021－07	108.00	1341
分子水平的静电学(英文)	2021－08	68.00	1342
非线性波:理论、计算机模拟、实验(英文)	2021－06	108.00	1343
石墨烯光学:经典问题的电解解决方案(英文)	2021－06	68.00	1344
超材料多元宇宙(英文)	2021－07	68.00	1345
银河系外的天体物理学(英文)	2021－07	68.00	1346
原子物理学(英文)	2021－07	68.00	1347
将光打结:将拓扑学应用于光学(英文)	2021－07	68.00	1348
电磁学:问题与解法(英文)	2021－07	88.00	1364
海浪的原理:介绍量子力学的技巧与应用(英文)	2021－07	108.00	1365
多孔介质中的流体:输运与相变(英文)	2021－07	68.00	1372
洛伦兹群的物理学(英文)	2021－08	68.00	1373
物理导论的数学方法和解决方法手册(英文)	2021－08	68.00	1374
非线性波数学物理学入门(英文)	2021－08	88.00	1376
波:基本原理和动力学(英文)	2021－07	68.00	1377
光电子量子计量学.第1卷,基础(英文)	2021－07	88.00	1383
光电子量子计量学.第2卷,应用与进展(英文)	2021－07	68.00	1384
复杂流的格子玻尔兹曼建模的工程应用(英文)	2021－08	68.00	1393
电偶极矩挑战(英文)	2021－08	108.00	1394
电动力学:问题与解法(英文)	2021－09	68.00	1395
自由电子激光的经典理论(英文)	2021－08	68.00	1397

刘培杰数学工作室
已出版(即将出版)图书目录——原版影印

书 名	出版时间	定 价	编号
曼哈顿计划——核武器物理学简介(英文)	2021—09	68.00	1401
粒子物理学(英文)	2021—09	68.00	1402
引力场中的量子信息(英文)	2021—09	128.00	1403
器件物理学的基本经典力学(英文)	2021—09	68.00	1404
等离子体物理及其空间应用导论.第1卷,基本原理和初步过程(英文)	2021—09	68.00	1405
拓扑与超弦理论焦点问题(英文)	2021—07	58.00	1349
应用数学:理论、方法与实践(英文)	2021—07	78.00	1350
非线性特征值问题:牛顿型方法与非线性瑞利函数(英文)	2021—07	58.00	1351
广义膨胀和齐性:利用齐性构造齐次系统的李雅普诺夫函数和控制律(英文)	2021—06	48.00	1352
解析数论焦点问题(英文)	2021—07	58.00	1353
随机微分方程:动态系统方法(英文)	2021—07	58.00	1354
经典力学与微分几何(英文)	2021—07	58.00	1355
负定相交形式流形上的瞬子模空间几何(英文)	2021—07	68.00	1356
广义卡塔兰轨道分析:广义卡塔兰轨道计算数字的方法(英文)	2021—07	48.00	1367
洛伦兹方法的变分:二维与三维洛伦兹方法(英文)	2021—08	38.00	1378
几何、分析和数论精编(英文)	2021—08	68.00	1380
从一个新角度看数论:通过遗传方法引入现实的概念(英文)	2021—07	58.00	1387
动力系统:短期课程(英文)	2021—08	68.00	1382
几何路径:理论与实践(英文)	2021—08	48.00	1385
论天体力学中某些问题的不可积性(英文)	2021—07	88.00	1396
广义斐波那契数列及其性质(英文)	2021—08	38.00	1386
对称函数和麦克唐纳多项式:余代数结构与 Kawanaka 恒等式	2021—09	38.00	1400
杰弗里·英格拉姆·泰勒科学论文集:第1卷.固体力学(英文)	2021—05	78.00	1360
杰弗里·英格拉姆·泰勒科学论文集:第2卷.气象学、海洋学和湍流(英文)	2021—05	68.00	1361
杰弗里·英格拉姆·泰勒科学论文集:第3卷.空气动力学以及落弹数和爆炸的力学(英文)	2021—05	68.00	1362
杰弗里·英格拉姆·泰勒科学论文集:第4卷.有关流体力学(英文)	2021—05	58.00	1363

刘培杰数学工作室
已出版(即将出版)图书目录——原版影印

书　名	出版时间	定　价	编号
非局域泛函演化方程:积分与分数阶(英文)	2021—08	48.00	1390
理论工作者的高等微分几何:纤维丛、射流流形和拉格朗日理论(英文)	2021—08	68.00	1391
半线性退化椭圆微分方程:局部定理与整体定理(英文)	2021—07	48.00	1392
非交换几何、规范理论和重整化:一般简介与非交换量子场论的重整化(英文)	2021—09	78.00	1406
数论论文集:拉普拉斯变换和带有数论系数的幂级数(俄文)	2021—09	48.00	1407

联系地址:哈尔滨市南岗区复华四道街 10 号　哈尔滨工业大学出版社刘培杰数学工作室

网　　址:http://lpj.hit.edu.cn/

邮　　编:150006

联系电话:0451—86281378　　13904613167

E-mail:lpj1378@163.com